Forum for Interdisciplinary Mathematics

The **Forum for Interdisciplinary Mathematics** is a Scopus-indexed book series. It publishes high-quality textbooks, monographs, contributed volumes and lecture notes in mathematics and interdisciplinary areas where mathematics plays a fundamental role, such as statistics, operations research, computer science, financial mathematics, industrial mathematics, and bio-mathematics. It reflects the increasing demand of researchers working at the interface between mathematics and other scientific disciplines.

More information about this series at https://link.springer.com/bookseries/13386

Suhrit Dey · Charlie Dey

Mathematical and Computational Studies on Progress, Prognosis, Prevention and Panacea of Breast Cancer

 Springer

Suhrit Dey
Department of Mathematics
and Computer Science
Eastern Illinois University
Charleston, IL, USA

Charlie Dey
Texas Advanced Computing Center
The University of Texas at Austin
Texas, TX, USA

ISSN 2364-6748   ISSN 2364-6756 (electronic)
Forum for Interdisciplinary Mathematics
ISBN 978-981-16-6079-5   ISBN 978-981-16-6077-1   (eBook)
https://doi.org/10.1007/978-981-16-6077-1

This Springer imprint is published by the registered company Springer Nature Singapore Pte Ltd.
The registered company address is: 152 Beach Road, #21-01/04 Gateway East, Singapore 189721, Singapore

Dedication Describing the Dictum that Directed Me

At the outset, I heartily congratulate those who are cancer survivors. I met only a few of them. Information in this book may give them some ideas on what battle they might have fought. And what lifestyle they may try to adopt. They are gallant fighters. Their strength of mind, their earnest encouragement enthused me and became a constant reminder that there must be a way out to keep cancer under control. I am very grateful to them. I passionately pray for their best health.

This book, however, relates to a very sad chapter of my own life, happened almost thirty years ago. This is no ordinary book in my eyes. This is the book of my life. Telling my tales of tears.

Day after day, month after month, for almost two long years, I witnessed as a solemnly sad, silent, sobbing, spectator, many grieving reels of real-life scenarios relating to untold pain and unending sufferings of many cancer patients in hospitals after hospitals. Often as those prodigiously painful events unfold one after another in front of my doleful, fearful, tearful eyes, calmly I cry inside being overwhelmed with unspeakable agony and unthinkable anguish that still hurt every cell of my body. I try to forget. Try not to think. But I can't. Gloomy, gruesome, grim faces of many departed souls flock inside my fleeting thoughts, darkening my daydreams, and becoming my nightmares. I think in vain that all they prayed for is providential mercy, to have a life to live. To hug. To be hugged. To love. To be loved. To kiss. To be kissed. That mercy was not granted. Benign lord was brutally unmerciful.

I also thought that the glamor and glory of enlightening, ostentatious scientific achievements failed them completely. I could see in my vision hundreds of the Nobel Laureates standing around these doleful, dying young ladies, who were at most in their mid-forties. Standing with pale faces, facing down. Speechless. Sleepless. Motionless. Helpless observers observing their painful untimely demise.

I remember a famous oncologist gravely saying to me, "It may take twenty-five years or more before cancer could be traceable after it began. By that time, it could be untreatable." I understood that I understood nothing. Wondered ignorantly, "Why so long? What is cancer?" Utter stupidity was my only identity. Utter ignorance was my only knowledge.

I remember a radiologist's technician telling me, "We try to kill the tumor, but we cannot kill cancer." I was dumbfounded. Stupefied. Trembling. Innocently, I thought that killing a tumor means killing cancer. I had no idea what cancer is.

I remember a prominent physician telling me, "Angiograms, blood tests, X-rays, CT-scans, even MRIs, all negative do not mean cancer isn't there." I was mystified. I knew that the accuracy of these tests is strong enough to identify cancer. I was blanketed by blank thoughts of despair. I wondered with woe, "What is cancer?" "How can it confuse such an experienced doctor?"

I remember a reputable radiologist looking at a CT scan in color and saying softly, "This shows massive bone metastases, an indication that the patient will die within three months." I couldn't control my tears. I was scared. Very scared. I lost all my strengths. I couldn't even walk. I sobbed silently. Sobbed with profound sighs. The light of my mathematical logic was out. Totally out. Darkened in seconds. I wondered with desperation, "What is cancer?" "How can a doctor make such a grim prediction?"

Painstakingly, I remember a senior nurse once telling me when I was with a patient who gave me a sweet smile, "Every time she smiles is a time to celebrate. Every time she hugs is an unforgettable moment." I looked at her with a lifeless look. Her statements moved my heart and at the same time scared me. I stood motionless and emotionless. All my thoughts were frozen. Couldn't think that I can think. Couldn't make out what the senior nurse said and why.

I remember a dying patient who held my helpless hands and gently spoke with a feeble voice, "You are so smart, Suhrit; can you do something and save me? I do not want to die. I do not want to die, Suhrit." And just after a few hours, I saw her face, pale and painless. She breathed her last in deep sleep.

I remember a good friend of mine. She loved to talk. Loved to make jokes. I saw her in the hospital bed. Her deathbed. She was in excruciating pain even with intravenous pain relieving medications. With a moaning voice, she said, "Doctors said that my smoking is the cause of my suffering. My grandma smoked all her life. She lived for 93 long years." She paused. Took a shallow breath. And said. Her voice was trembling, "I am only 43. My cancer is all over my body. I won't live, Suhrit." She paused again. And began, "Can somebody save me? I know you can figure out." I had nothing to say. My eyes were soaked with silent, sententious, tears. Tears that I was not aware of. Her last few words were, "I want to live, Suhrit. Please save me, Suhrit." But soon after all that stopped. Stopped forever. Nurses stepped in. I stepped out. A long lugubrious event was about to happen. This stark tragedy blurred my vision. Soaked away all my tears. Choked my dim voice. Paralyzed my wishful thoughts. Silenced my breath.

I remember the wobbling words of many dying patients who just desperately wanted to live. Wanted to live to be loved. Wanted to breathe fresh air. To see a starry sky shining above on the vast dark floor of a silent holy night. Wanted to watch a flock of flying beautiful butterflies. To walk barefoot on dew dipped green grass on a sunlit morning. Listen to birds chirping in pairs. Watch tiny silvery ripples playing merrily near river banks. Watch the soothing orange rays of a sunset. I heard them saying meekly that they wanted to ring a church bell. Wanted to sing hymns from the Gospel to glorify the blessings of God. Wanted company of friends for a sumptuous Thanksgiving dinner. Wanted to hug their kids. Place their heads softly on their beloveds' shoulders.

These are the day-to-day living legends of longings of a life. Talking tales of deep-seated dreams and desires. All remained unfulfilled. Unheard. Unnoticed. Like dry leaves of a brutal snowy winter. Deadly demonic cancer forcibly extinguished their lamps of life. Forcibly took them away from their loved ones. To a dark dreary, doomed destination, forever unknown.

I remember pale, sleepless, speechless, flaccid faces of children, spouses and parents. Unaware of tears silently rolling down drop by drop on their dry cheeks. Hunger did not bother them. Insomnia did not bother them. They often got nights and days all mixed up. I remember.

I saw them praying. Praying for mother. Praying for sister. Praying for grandma. Praying for aunt. Praying for daughter. Praying for the dearer than the dearest spouse. Prayers that didn't work. All was a silent witness. All noticed that justice was unjust. Kindness was unkind. Wisdom of the wise was wisdom of ignorance. Prayers were futile. God was absent.

I remember, one morning I entered the room of a patient. Gently went near her bed. Silence all around was so deep that I could hear her feeble heartbeats as she slowly went into a deep sleep on her hospital bed. The very last everlasting sleep. The very last bed for the very last breath. A profound silence silenced even the natural flow of air in the room. Lights in the room became darkness of night. I noticed it all without noticing. The most pathetic scene of life leaving life is beyond any description.

I witnessed with awe and horror how slowly sweet somber smiles of these passing away patients faded away. Last somber words of love faded away. Last somber sight of passion of compassion faded away. Last hope of life faded away. Last words of love faded away. I heard a lone, long last lingering breath. Soon breathing stopped. Eyes blinked no more. Heart beat no more. Head turned aside. Life became lifeless. Became absurd. Body became colder. Prayers of friends and family members stopped.

At this moment, all words of scriptures suddenly became words of nothing. All words of consolation, all words of condolence suddenly became meaningfully meaningless. They are just a few sonorously significant words in a dictionary, now signifying nothing. Now, the universe is solemnly silenced. Silenced forever. And remained peacefully peaceless. This is a reality. Lost paradise of pantheism.

Nurses came in. I heard their sounding shoes. No words were spoken. They disconnected one by one all the life-supporting tubes. The patient did not turn. Did not talk. Did not show any pain or discomfort. Doctors entered the room with pieces of papers to sign.

I saw with awe from a distance. This is the ultimate certificate of life on its doomsday. All will get this certificate. No prerequisites are needed. No preparations are needed. No tests will be given. The person receiving this certificate will never read what is written on this last certificate of life. Certifying lifelessness as an integral part, a legal document of lifelessness of a living life.

This is death. It has a beginning but no end. One cannot describe it. Here precision is forever imprecise. Conclusion is forever inconclusive. Definition is forever undefined. Yet, from times immemorial, it is the integral, inherent and innate nature of Nature. It is her merciless mandate. Her truculent tenet. Her defiant decree. Her unkindest ultimatum.

Death is the destiny of all. Neither beauty nor bounty, neither prayer nor pity can save anyone from this cacophonous, incongruous, irreversible law of Nature.

I forgot to cry. Forgot to say a word. Forgot to feel life of my own life. The penultimate truth of life is so heartrending. A story of terror and trepidation.

The dazzling day became a gloomy night right in front of me. This is the deepest darkness of the fathomless feeling of helplessness. The lingering legacy of loneliness. Deepest darkness of my state of confusion. I was bewildered. Only one question was ravaging my mind: Where did she go? Felt deep emptiness in my heart. As if I am living a life with no life. I am feeling a feeling with no feeling. I am breathing a breath with no breath.

These stark tragedies happened again and again, and I was a solitary witness, stilled, static, like a stone statue. My grief stricken heart spoke with wordless words: Cancer, thou art the cruelest criminal ever created by the Creator.

These tantalizing tales of torture and torment of a cold-blooded murderer will never ever leave my memory. With a painful heart and a pensive mind, I watched all these and moaned inside. Still now the agonizing anamnesis of these distressing thoughts on untold, unimaginable, unspeakable sufferings of these affectionate friends of mine kept on haunting me and hurting my mind. Again and again. So healing never began.

As I lonely lament and languish for the loss of my loving friends, they enkindled in me an exigent eagerness to look for an answer, an answer to my enigmatic, earnest enquiries that could be hidden in mathematics, my first love. I asked myself, "What I could have done to help them? Am I totally incapable? Can I not analyze medical facts and figures on cancer by applying statistics and mathematical modeling to play a relevant role to cure or control cancer, regardless how insignificant my present knowledge is on medical science?" I was confident that mathematics can overcome most of my all hurdles, all hindrances hurting my heavyhearted mind. Mathematics is the foundation of science. I gained more and more confidence.

My studies started slowly but steadily and merged with my professional life. Soon it was financed, fanned and fueled by funding from NASA. My research moved forward in full swing.

The final truth is: Those departed, doting, dignified souls gave me an insatiable inspiration to set up my mind and put the language of mathematics on my pen to write in this monograph, the story of their strenuous struggle to survive and their painful, pitiful, pathetic passing away at the end. They installed in me a vigorous voice, an unyielding will to write this book applying stringent logic of mathematics to tell all health professionals that mathematicians are determined to join their team declaring a war against dreadful cancer, especially breast cancer, and eventually win a coveted victory turning the victims into victors. Loss of even one life is indeed too much, an irremediable and irreparable loss.

This book is respectfully dedicated to the loving memories of all my very affectionate friends who were innocent victims of vicious cancer. I often think of them.

I miss them. Miss them very much. My mood changes abruptly from pleasant to painful.

This monograph is my heart's communion with these departed souls. A narration of my concern, my respect and reverence for them. It is also a sublime solace for my melancholic mind. And finally, a gentle reminder to the world that these noble souls did not die in vain.

Georgetown, Texas, USA   Suhrit Dey
November 2020

Foreword

I was honored to receive the invitation to write this foreword to Prof. S. K. Dey's monograph on Mathematical and Computational Studies on Progress, Prognosis, Prevention and Panacea of Breast Cancer. I have the unique privilege of knowing Prof. Dey for over 50 years: first as an undergraduate student at Eastern Illinois University, completing my first differential equations course under his instruction. He was my mentor and advisor through my B.S. and M.A. studies. I was his student when his son and coauthor S. Charlie Dey was born. I have worked with Prof. Dey at Argonne National Laboratory and NASA Ames Research Center. Professor Dey is the finest applied mathematician I have had the privilege to learn from and work with.

One enjoyable encounter was in 1978 at Argonne National Laboratory. I was working with Dr. Marvin Breig, a visiting physics professor also from Eastern Illinois University. Professor Dey walked in very excitingly and asked Prof. Breig "what is a plasma?" We asked why do you want to know? Professor Dey said, "Because I have just told the Physics Division that I can solve a system of equations that model a plasma problem that they have been stuck on for many months. Please tell me what plasma is?" He soon provided the Physics Division with a solution to their problem.

His career has documented the successful application of mathematics toward solutions of some of the most challenging physics, engineering, and medical challenges of the past few decades. His work in bioscience and cancer was set into motion with his publication on the mathematical modeling of chemical kinetics at the University of California, Berkeley, in 1982. Over the past four decades, he has been invited to lecture on mathematical models to fight the war on breast cancer at numerous universities, research institutions, and hospitals internationally. Oncologists and medical researchers have consulted with Prof. Dey on how to improve his models, and in particular, he has received enthusiastic comments and suggestions from the oncologists Dr. Charles Wiseman of St. Vincent Hospital in Los Angeles, California, and Dr. Characiejus of Vilnius Hospital in Vilnius, Lithuania.

I have heard experts in cell biology saying Prof. Dey has the science correct in his journal publication, but I can't understand all the mathematics, and experts in mathematics have said the mathematics is correct but I don't understand all of the biology. This monograph hopefully will help readers bridge their knowledge across both disciplines.

Moffett Field, California   Johnny P. Ziebarth, Ph.D.
April 2020

Preface

The first author started conducting studies on cancer when as a recipient of awards from the National Academy of Science of the U.S.A he joined NASA Ames Research Center, Moffett Field, California, in 1980, and numerically solved a mathematical model on chemical kinetics on irradiation of water, developed by Dr. Magee and Dr. Chatterjee, at the Biomedical Division of University of California, Berkeley, in 1982. The second author at the age of ten started working on his dad's computer codes. The gist of what they observed is:

> Cancer strikes the body at a subquantum biological state. One or more genes are altered due to a number of reasons causing the cell to mutate and grow beyond control violating all principles of homeostasis. It takes several mutations before a cell becomes a cancerous cell. To protect itself from the clutches of the immune system, it can hide inside an adipose tissue, or under the skin for years. It never dies so long it can get glucose and oxygen.

Later for almost two years (1988 and 1989), the first author worked round the clock almost like an intern with oncologists in several hospitals in St. Louis, Missouri; Champaign, Illinois; and Charleston, Illinois, and observed the mode of cancer treatments for several patients conducted by several oncologists in as much details as possible. Two of these doctors are Dr. Charles Roper (Thoracic Surgeon, Barnes Hospital, St. Louis) and Dr. Latinville (Neurosurgeon, DePaul Hospital, St. Louis). He asked all doctors several questions about their modes of treatment. In general, he noticed that the procedures of treatments are: cut (surgery), burn (radiation), and poison (chemotherapy). He saw mostly tragedies and hardly any triumph and wondered as a mathematician what can be done to better such treatments.

Cancer appeared deadlier every moment at that time, because standard treatment procedures, which sometimes appeared to be promising, failed within just a few years. To a patient and her family, life appeared to be hopelessly meaningless. With the sad demise of a mother, a family fell inside the dungeon of utter distress and disintegration. Horror of losing a mother seems to hang over forever.

When the project on Breast Cancer Cure from NASA, Ames, came in 2001, he poured his heart into it. What he has learned is that instead of expecting the practicing doctors to follow any mathematical model for treatment of cancer as set up by mathematicians, mathematicians should follow the mode of treatments conducted by the practicing physicians and recommend changes which the physicians could recognize and might be able to adopt in their practice. In fact, many oncologists are quite willing to listen to recent mathematical models which could be applied in their profession. In this regard, most physicians understand statistical modeling more easily. So, we have introduced some statistical studies in a simplistic way.

Mathematics is not just made for mathematicians, just like science is not made just for scientists. In cancer research, many mathematicians made valuable contributions. Yet, when I spoke directly with oncologists, they hardly read or even saw these articles. In most cases, however, the level of mathematics is often too high for most medical practitioners to read and understand, and also, they hardly have any extra time to do so. So, the group conducting mathematical studies and the group practicing medicine in the practical field seem to be two wildly disjoint sets. Doctors often take into notice the available statistical data published in reputable medical journals in their practice.

However, primarily because this work has been oriented toward practical applications, when the authors lectured on this topic at several hospitals in the oncology department (especially St. Vincent Hospital, Los Angeles, and Vilnius Hospital, Lithuania), several doctors attended, participated in the discussions, and gave us excellent inputs. Also, the researchers from NASA, Ames; NASA, Johnson Space Center, Division of Biological Sciences, Houston, Texas; Los Alamos Scientific Laboratory, New Mexico; Indian Institute of Technology Delhi; Indian Institute of Technology Roorkee; Indian Statistical Institute, Kolkata; Bose Institute, Kolkata; Indian Association of Cultivation of Science, Kolkata; Indian Institute for Chemical Biology; Lithuanian Academy of Sciences; University of Rome, Applied Mathematics Department; CNR Naples, Italy; Harvey Mudd College, California; University of Dundee, Scotland; Eastern Illinois University; Mississippi State University; etc., gave him many constructive suggestions when he visited these institutions as a visiting scientist and/or an invited lecturer. Every attempt was taken to accommodate all the helpful suggestions that we received from the experts in this monograph.

In 2014, he was invited to lecture on mathematical modeling of breast cancer at the International Conference on Applied Mathematics, Kolkata, by the organizers from the Applied Mathematics Department of the University of Calcutta. Thereafter the conference he came to know Mr. Shamim Ahmad, Editor of Springer Nature, who requested him to write a monograph on this topic. That was inspiring. Authors are very thankful to him, but at the same time, this task is truly humongous—an uphill project.

The aim of this monograph is to study mathematically appropriate treatments to cure and/or keep breast cancer under control, from a very practical standpoint. The word "cancer" is still one of the most scariest and, in a way, most mysterious words in our dictionary. When researchers like Dr. Floris Barthel—a recipient of a prestigious award from the National Institute of Health, USA—from The Jackson Laboratory predicted that in 2018, more than 1.7 millions of Americans will be diagnosed with cancer, everyone gets scared, regardless of the advancement of treatments for this disease. Mathematical modeling and computational studies on these models could be helpful with regard to most efficient treatments of breast cancer. That is the aim of this monograph.

The book is practically meant for all who are majoring in medical biology or physiology and has a good background of statistics, calculus, differential equations, and numerical analysis. In this age, most graduate students in applied computational science are required to take these courses for graduation. Computational biology is a major topic of applied computational science. So, we expect that both graduate students and researchers in this area will get benefit by using this book. Attempts were made to keep mathematical notations to a minimum.

Our study reveals that looking into all different aspects of this deadly disease in practically all forms and examining its assay and administration from a mathematical standpoint, there are remedies at every level. A person should be able to live a normal life taking regular medications like those who are living with diabetes and asthma.

Several times, we had to mention the same cancer scenarios to make them easier for the readers to understand our discussions. There are several repetitions of many essential aspects of cancer in several chapters so that readers will find it easy to read and understand.

Cancer will be cured only when scientists will find a way to reverse the process of mutation of genes and transform them back to their original forms, respectively. Then, mutated cells will become normal cells and the disease will be cured. Cancer is a genetic disease, not an infectious or contagious disease.

We hope this book will be valuable to the readers. Any suggestion for a betterment will be welcomed by the authors.

Organization of Chapters

We have designed our monograph of 11 chapters and 3 appendixes:

Chapter 1 deals with the motivation and methodology that we have planned to adopt and follow in the monograph. Modeling in any aspect of science and engineering requires a thorough study about what is happening in the real world that will motivate a mathematician or an engineer to mathematize those scenarios. The deeper one can observe and analyze the better will be for modeling. The topic here is to analyze scientifically how body fights the war against cancer. So, we need to look into the nature and tactics of war of the attacker, the foreign antigen carried by cancer cells and the body's built-in defensive forces. We must keep in mind that in the true sense of the term, cancer is not really foreign. Due to genetic mutations again and again, the domestic cells (cells of the body itself) become malignant and attack the body. So, very likely they know all the whereabouts of the body, namely where to hide, where to metastasize, how to metastasize, who could help them, how to befool the defense, etc. These aspects of cancer cells have been looked into before any modeling. And we did that in this chapter.

Cancer cells do not always follow the same pattern of attack. They often display very heterogeneous nature confusing the body, confusing even expert doctors. Uncertainty is the very nature of cancer cells. So, principles of statistics need to be studied. In Chap. 2, we have discussed the basis of this research which is statistics. At the beginning, some very elementary concepts of set theory have been mentioned as prerequisites. Fundamentals of statistics which we have used are briefly discussed. We analyzed how a practical treatment is undertaken to bring panacea after prognosis is done and treatments proceed. After solving mathematical models computationally, medical professionals may apply statistical principles to test all the data statistically to estimate their validity and applicability. So, a short note on data analysis has been given.

In Chap. 3, preliminaries regarding the war on breast cancer are studied. Like in any regular war, there are attackers and there are defenders. Cancer is the attacker. Body's immune system is our first line of defense. In this regard, we studied the performance of inhibitors. Simple time-dependent models have been introduced.

Body is three-dimensional and changes with time. In Chap. 4, three-dimensional time-dependent models of the fight between cancer cells and other inhibiting biochemicals have been derived mathematically forming a set of reaction–dispersion equations. This is mathematization of the war in the entire computational field representing a human body. Cancer cells fight, metastasize, start new battlefields, and also hide to escape attacks by the defenders. Immune systems together with medications fight within their limited capabilities. This is indeed a regular war apparently unseen. The battlefield is the entire body. Most intense battle zones are near the tumor sites.

In Chap. 5, mathematically, we wanted to observe how and when the fight gets worse, or it gets better meaning individual treatments are effective. In this respect, we have modeled advanced treatments with stronger immunotherapy applying the use of monoclonal drugs.

Chapter 6 is the central part of this monograph. Mathematically, we have tried to model the present treatment methodologies and the consequences of various forms of breast cancer treatments. Here, we got excellent qualitative agreements with existing data. Computed data were analyzed and presented both in tabular forms and in graphs and through computer simulations.

In Chap. 7, mathematical modeling on gene therapy has been briefly discussed. Our emphasis is how poisonous proteins released by mRNAs due to mutations in genes could be corrected.

In Chap. 8, we have introduced some smartest fighters. When fighting gets very intense and standard drugs are failing, then some smart combat drugs SCDs and skilled killer drugs SKDs have been applied. These are theoretical. The basic differences between them are how they fight. While SCDs fight along with the other fighting agents, SKDs fight independent of other fighters. But they have one objective "victory must be achieved and patients must survive." It is true that at the present time this may sound to be just a theoretical approach, but in the future this could be achievable.

In Chap. 9, we have discussed nutritional therapy, mathematically. From times immemorial, all over the world men and women used nutritions to fight against all kinds of ailments, including cancer, and certainly it did work. Why? Because it kept them alive and we are their lineage.

In Chap. 10, the main theme is to introduce a broader and more universal concept regarding growth, dispersion, and movements of cancer cells as they swim through blood, lymph, and interstitial fluid. We have described the model using three partial differential equations. They are linear, and as such, we have solved them analytically. We found that all apparently very different equations share the same format of solutions revealing the validity of all the models described earlier.

Chapter 11 is our conclusion. We have briefly discussed how in some cases cancer could pose a threat to life which is irreversible. We also stated why necrotic tumors often pose a severe metastasis. On a positive side, we discussed how meditation and healthy lifestyles could be effective to control cancer. A lifestyle change is required so that to a great extent cancer can be kept under control and patients should be able to live a normal life.

There are three appendices in the book. In Appendix A, we have answered questions of reviewers why the CDey-Simpson method was applied to solve stiff models in numerical analysis. We discussed the algorithm in detail, analyzed its stability properties in detail, and drew graphs of stability contours. In Appendix B, we have discussed why parabolic partial differential equations which represent our mathematical models are stiff, especially when the computational field is large. In Appendix C, we have shown how in the future, the CDey-Simpson algorithm, which is fully vectorized, may be implemented as massively parallel solvers and the models could be solved by using clusters of supercomputers as synchronous computing in hospitals and doctors' offices. This was a question asked by a senior reviewer.

Austin, Texas   Suhrit Dey
March 2020   Charlie Dey

Acknowledgements

This project has been funded by RIACS of NASA Ames Research Center, Moffett Field, California, USA, and NASA Cooperative Agreement NCC2-1006, and by the Faculty Research Council of Eastern Illinois University, USA. We are very thankful to Dr. John Ziebarth, Former Director of the NAS Division of NASA, and Mr. Bud May, Director of Research of Eastern Illinois University, for all the help we received from them.

We are thankful to Dr. Saunak Roychoudhary and Mr. Giedrius Lingis, for their editorial works. We thank Mr. Chanchal Pramanik helping us during eproof reading. We thank Mr. Shamim Ahmad, Senior Editor of Springer Nature, for his invitation and encouragement to write this monograph. We are thankful to Ms. Rupali Gudji, Mr. Vishal Pande, and Ms. Austeja Ilginyte who drew some figures for this book. The Baltic Tours of Vilnius, Lithuania, was very helpful in getting some figures on physiology from the Internet for us.

We are thankful to the personnel of SouthStar Bank, here in Round Rock, USA, who help us print our manuscripts for corrections.

We are very thankful to many medical professionals who wanted to remain anonymous for many valuable suggestions. We are especially thankful to the teams of oncologists from St. Vincent Hospital, Los Angeles; Vilnius Hospital, Lithuania; and Chittaranjan Cancer Hospital, Calcutta, India, which gave us very strong encouragement and answered some of our questions regarding cancer treatments.

Finally, we are very thankful to Prof. Roma P. Dey for doing most strenuous statistical data analyses and correcting our manuscripts continuously.

January 2021
Suhrit Dey
Charlie Dey

A Congratulatory Note from Mohit De[1]

My very best wishes and congratulations on writing the book *Mathematical and Computational Studies on Progress, Prognosis, Prevention and Panacea of Breast Cancer* by my brother Dr. Suhrit Dey and his accomplished son Mr. S. Charlie Dey.

My brother Dr. Suhrit Dey is a versatile genius. He is at once a poet, a philosopher, a musician, and above all a mathematician. I have read quite a few of his articles which speak volumes of his mathematical talent. This new book, which is an application of mathematics in medical science, has moved me. He has been assiduous of learning mathematics since his childhood. He finds mathematics in every walk of life. Whenever he comes to pay us a visit, he discusses his research works touching on mathematical modeling and statistical analysis of huge data collections and how these have helped him in his present works.

I am also very much research minded. I taught mathematics here in London for years. So, I enjoyed his discussions thoroughly. During his stay with me, he has visited several universities in Europe and in the UK as an invited lecturer. He delivered lectures on varieties of topics on mathematical models, including cancer treatments. His power of delivery is very fascinating. Mathematics seems to be his religion. His book Vedanta: The Science of Consciousness & Divinity published by the Roerich Society of Lithuania, is a masterpiece.

My recent research on music and mathematics reveals that mathematicians are not going to be fun or cuddy folks. But, I can tell from my experience that studying mathematics can also offer many opportunities to analyze many diverse events of the nature. Looking at my brother, I feel mathematicians enjoy both romance and challenge studying both physiology and psychology in light of mathematics. It can lead a mathematician to a romantic life, which is on par with being a poet or a musician. My brother is an embodiment of all these.

I wish him all success and a very healthy long life.

M.A. (Kolkata), M.Phil. (London)   Mohit Dey
May 2020

---

[1] Suhrit Dey's Elder Brother

Words of Gratitude from a Patient's Spouse on the Success of our Modeling

The following are the two emails from Mrs. Sandra Reeds, Secretary of our Mathematics Department, who typed all my articles on cancer treatment where I emphasized immunotherapy as our first line of defense.

On September 16, 2016

Hi, Dr. Dey,

How are you? I just wanted you to know that Jack is on a new immunotherapy drug right now, and he is doing great. His liver cancer is shrinking. He is on REGN2810, a Phase-1 immunotherapy study. We are excited about that, and he has very few side effects. Basically, he is just tired. That is the only side effect we notice. But actually, he was always tired anyway. He gets 1 infusion every Tuesday.

The way it came about is that they were going to put him in a new viral therapy Phase-1 trial. Anyway, I remembered your research papers on immunotherapy, so I asked them, did they not have any immunotherapy trials. And they really did, Jack was actually the last one to be able to get into this study at the Indiana University hospital. After being on this drug for 2 months, they did an MRI on him and they could not find any cancer to biopsy. Isn't that wonderful? We are so excited. He started it about 2 months ago, thanks to you, as I typed your T-cell papers! I remembered! So, thank you, Dr. Dey!

Sandy

On January 14, 2019

Hi, Dr. Dey,

Jack is still cancer free, and now has had a successful liver transplant. This was a miracle because he initially had a 5-centimeter, a 3-centimeter plus several 1-centimeter cancerous tumors across his liver, which made him ineligible for a liver transplant. After the immunotherapy, the cancer was gone. There were no noticeable side effects to the medication. Jack had one infusion that lasted 20 minutes every other week of RGN2810; it was a phase-1 study. The immunotherapy, which was a cure for Jack's cancer, was the only reason Jack was able to get a liver transplant.

Before this study, he had been given chemoembolization, Y90 beads, and Nexavar in pill form. None of these worked. In fact, Barnes Hospital in St. Louis gave Jack only 6–9 months to live. We did not listen to them though. Instead, we went to the Indiana University Hospital in Indianapolis for help and they started Jack in this immunotherapy study. He is now cancer free. It is only because of immunotherapy.

I always remembered all of the papers I typed for you about immunotherapy and when Jack was offered the opportunity to get in this study, I knew there was hope for him. Thank you so much, Dr. Dey, and may God bless you!

Sandy

Introduction and Summary of the Research

The entire content in the book can be categorized into four simple models describing the growth of malignancy and its antidotes. It is a war between the attacker of the body (that is, cancer) and the defenders of the body (that is, the leukocytes, medicines, radiation, etc.). It is indeed an unseen dreadful war. At the very beginning it was undetected. We will study how to win the coveted victory and stop the ruthless aggression of this ferocious attacker.

Our modelings are quite different from those available on the Internet and references. Many such references have been mentioned in the main text. We tried to mathematize the procedures that we have observed doctors conducting in various hospitals and cancer treatment centers. We presented our works at various universities and hospitals around the world and received words of encouragement.

Model 1: The Uncontrolled Growth

In one line, cancer means uncontrolled growth of malignant cells. Mathematically, uncontrolled growth of any species is $\frac{du}{dt} = au$, *where* $u = $ *mass* of the species and a = the rate of growth. If $a = 0$, u will not grow anymore. It will either die or remain the same. Let at $t = 0$, $u = u_0$. The solution is $u = u_0\, exp(at)$. If $a > 0$, growth will be uncontrolled. If $a = 0$, growth will stop. If $a < 0$, u will die as $t \to \infty$. So, we need to look into how that could happen when it is related to cancer. When cancer starts, the body does resist applying both genetic methods and the immune system. Sometimes, that does not work.

Model 2: A Noticeable Scenario of Uncontrolled Growth

We consider another model like $du/dt = (\xi - \lambda)u$, where $\xi = $ a constant, the rate of growth of u and $\lambda = $ *a constant*, an inhibitor of that growth. Let at $t = 0$, $u = u_0$. Then, the solution is: $u(t) = u_0 \, exp\,[(\xi - \lambda)t]$. Thus, so long $\xi > \lambda$, $u(t)$ will keep on growing. But, if $\xi \leq \lambda$, this growth will be stopped.

The biggest hurdle is, if we have a model like this:

At $t = t_0, u_0 = U \, exp\,(-\lambda)$, implying that the larger the value of λ, the smaller the value of u_0 and the growth rate of u is given by $du/dt = \xi u$ (where λ and ξ are constants). Then, the solution is $u = U \, exp\,(\xi t - \lambda)$, and if λ is large in comparison with ξ, the growth of u could be significantly small at the beginning and practically unnoticeable until the time when $\xi t > \lambda$. Until that time, u will not grow (although it has a will to grow). In fact, the rate of growth is being thwarted continuously. So, u will grow at a slower rate. At $t = \lambda/\xi$, it will be U. And then, as t exceeds (λ/ξ), it will increase steadily. Still, depending on the value of ξ the growth could be unnoticed for a long time. So, that might give u enough time to metastasize. The presence of λ just delays the happening of that dangerous situation, and it did happen to many. The first author has witnessed that himself.

Model 3: The Most Realistic and Simplistic Model on the Growth of Cancer

Let us consider: $du/dt = uF(t)$, where $F(t) = \xi'(t) - \lambda'(t)$, subject to the condition, at $t = 0$, $u = U_0$. The solution is: $u(t) = U_0\{exp[\xi(t) - \lambda(t)]\}$. This means $\xi(0) = \lambda(0)$.

Let us briefly analyze this solution more. If at $t = t_1 > 0, \xi(t_1) = \varepsilon$, and $\lambda(t_1) >> \varepsilon$, and U_0 is very small at a time $t_1 = t_0 + \Delta t$, *where* $t_0 = 0$, *and* Δt *is very small*, that means $u(t)$ is displaying a tendency to grow, but $\lambda(t_1)$ is resisting that effort. So, $\lambda(t)$ is playing the role of the immune response.

As time progresses, and the body's resistance capacity declines, due to the lack of appropriate health care, passion for unhealthy food and drinks, pollutants and carcinogens in the environment, lack of exercise, sickness, side effects from taking certain medications, aging, stress, etc., $\lambda'(t) < 0$. So, body's defense mechanism is declining, cancer resumes its attack taking full advantage of that situation, and $\xi(t)$, the growth parameter of u, starts increasing so $\xi'(t) > 0$, *is an increasing function of time*.

This leads to cancer, which will appear and diagnosed by a doctor.

Model 4: A Method of Treatment of Uncontrolled Growth

To prevent such a scenario, we should look for a model where we could introduce two medications, v and w, such that $\xi = \xi(v, t)$ and $\lambda = \lambda(w, t)$, and choose v and w such that as t *increases* $\xi(v, t) \to 0$ and $\lambda(w, t)$ slowly increases till v and w going into steady states, making $u(t) \to 0$ as $t \to \infty$. To answer the question of a reviewer regarding how this could be explained, we have decided to do the following:

Let $du/dt = [\xi(v, t) - \lambda(w, t)]u$. The initial condition is given. Namely, at $t = 0$, $[\xi(v, 0) - \lambda(w, 0)] = 0$, giving $du/dt = 0$, $u = a\ constant$. It is clear from the description of this model that $\lambda(w, t)$ is positive, for all values of $\lambda(w, t)$, and it will reduce the rate of growth of $u(t)$. But in a three-dimensional setup, u could be all over the body and each u must be attacked. They have some individual characteristics too. So, one set of values of λ may not be sufficient to destroy all. Also, at each time step a unique λ may be badly needed to check the progress of cancer.

Then integrating from t_{n-1} *to* , t_n, we get

$$u(t_n) = u(t_{n-1})exp\left\{\int_{t_{n-1}}^{t_n} [\xi(v, t) - \lambda(w, t)]dt\right\}$$

To integrate, we may use a numerical method, like a trapezoidal rule. That gives,

$$\int_{t_{n-1}}^{t_n} \xi(v, t)dt = \frac{\Delta t}{2}\left[\xi(v_n, t_n) + \xi(v_{n-1}, t_{n-1})\right]$$

and the same is true for $\lambda(w, t)$. Both ξ and λ must be given or estimated by medical professionals. So, applying these integrals we will be able to compute values of u at each time step starting from $n = 1$. This could be accomplished applying immunotherapy with monoclonal drugs or some other appropriate drugs.

Necrotic Tumors

Cancer cells keep on growing so long they managed to get a supply of glucose and oxygen. They do not die. But, as the tumor grows, the cells at the core of the tumor often start dying. However, they may not really die. They may grow upto a certain size and stop their growth. This is just a remote possibility. This model is presented by a logistic equation,

$$\frac{du}{dt} = au - bu^2, a > 0\ and\ b > 0\ at\ u_0 = \frac{a}{1 + b}$$

The solution is

$$u = \frac{a}{b + e^{-at}}$$

So as $t \to \infty$, u attains its maximum, which is $u = a/b$. The most scary aspect of this growth is that if $a >> b$, although the tumor will increase linearly and decay quadratically, yet it will still keep on increasing to a maximum size. Most malignant tumors do not stop growing and when they do after being a necrotic tumor, they spread to other parts of the body. So, an assumption that it will stop growing is next to a myth.

Conclusion

It is not hard to comprehend that out of one hundred trillion cells in the body, at least a very few could mutate at any moment. Immune system destroys their activities, and life goes on. But, as the immune system gets weaker over time, the body gets weaker due to aging and other factors, mutations are not corrected and become a threat to life. Also, for various reasons like smoking, environmental pollutants, inheritance of oncogenes from the family, etc. cancer may be caused. The challenge is how to solve this crisis medically and how mathematical models can help the medical professionals in this regard.

This is the gist of the entire monograph where all these and more have been achieved in three-dimensional time-dependent mathematical models. These equations have been solved numerically. The models show that we may not be able to cure cancer, but in many cases we should be able to keep it contained. Our techniques are theoretically applicable to all forms of cancer.

In fact, for treatment of practically any viral or bacterial infection also the models described above could be applied just by modifying the growth rates of the pathogens and rate constants of medications. The models are very general in nature.

Since there are various kinds of breast cancers, there should be several different kinds of models for treatments that will mathematically simulate fighting between cancer and they are helpers on one side and the immune systems and their helpers on the other side. The first such mathematical model was developed in 2001 at NASA Ames Research Center by the first author. The model consists of a set of nonlinear reaction–dispersion parabolic partial differential equations. Dr. Peter Lee, from Stanford School of Medicine, Palo Alto, invited him several times for seminar presentations. He liked the models, gave many suggestions for betterment, and strongly encouraged him to continue with this very practical approach to treat breast cancer which strikes one out of every eight women in the USA.

Another most significant tool that we have used is statistics to analyze huge data obtained from the computer outputs regarding the success of the predictions of the models. These and more we will study in this monograph. This is a preliminary research report related to progress, prognosis, and panacea of this deadly disease.

References

1. Dey, S.K.: Computational modeling of the breast cancer treatment by immunotherapy, radiation and estrogen inhibition. In Scientiae Mathematicae Japonicae Online, Vol. 8, (2003)
2. Dey, S.K., Dey, S.C.: (Semantic Scholar). Mathematical modeling of breast cancer treatment, 2015

Contents

Notations

$\odot$	an operator, defined in the text
$\forall$	for all values
$\in$	An element of
$\notin$	Not an element of
$\subset$	Subspace of
$\subseteq$	subspace or the same space as
$F : D^m \subset R^n \rightarrow D^n \subset R^n$	F is a mapping which maps D^m to D^n where both D^m and D^n are subspaces of R^n, R^n is a real $n-$ dimensional space
$[a_{ij}]$	a matrix with the element a_{ij} on the *ith* row and the *jth* column
[ ]	A matrix notation where the matrix is arranged rowwise. The advantage is: Any row could be easily taken out as a Transpose of a vector which readers may find it easy to understand
$[\,]^T$	transpose of a matrix
n(A)	number of elements in the set A
$\|.\|$	Norm of a vector or matrix
U^n_{ijk}	$U(x_i, y_j, z_k; t_n)$
U^n_{max}	max value of U^n_{ijk} at t_n
$exp(x)$	e^x
a^n_1	is in some cases written like a_1 for simplification
#	stands for number of points, number of entries etc

Chapter 1
Introduction

Abstract This chapter deals with the motivation and methodology that we have planned to adopt and follow in the monograph. Modeling in any aspect of science and engineering requires a thorough study about what is happening in the real world that will motivate a mathematician or engineer to mathematicize the incident. The deeper one can observe and analyze the incident, the better will be for mathematical modeling. The topic here is to visualize scientifically how body fights the war against cancer. So, we need to look into the nature and tactics of war of the attacker, the foreign antigen carried by cancer cells, and the body's built-in defensive forces. We must keep in mind that in the true sense of the term, cancer is not really foreign. Due to repeated genetic mutations, the domestic cells (cells of the body itself) become malignant and attack the body. So very likely, they know all the whereabouts of the body, namely where to hide, where to metastasize, how to metastasize, who could help them, how to befool the defense, etc. These aspects of cancer cells have been looked into in some details before any modeling. We did that in this chapter.

1.1 Rationale

With her infinitely versatile charismatic skills, nature created all that we see all around us. She created an uncountable number of contradictory events. Positive and simultaneously negative events are abundant everywhere. So, we strongly believe that when she created genes, DNA, and living cells, she created cancer where some genes malfunction to destroy her own creations violating all laws that protect life. Cancer must be as old as the time when life began. It is an outcome of repeated genetic mutations. Scientists looked into many causes that lead to such genetic mutations.

Aging, stress, malnutrition, air pollutants, water pollutants are some of the causes for cancer, which existed in all ages. Ancient history of India, Greece, and Egypt recorded the existence of this deadly disease. The earliest known descriptions of cancer appeared in several papyri from Ancient Egypt [1, 2]. In Ancient Greece, cancer gets referenced in the Hippocratic Corpus texts, said to have been written by the "father of medicine" Hippocrates between 410 and 360 B.C. He used the

© Springer Nature Singapore Pte Ltd. 2021
S. Dey and C. Dey, *Mathematical and Computational Studies on Progress, Prognosis, Prevention & Panacea of Breast Cancer*, Forum for Interdisciplinary Mathematics, https://doi.org/10.1007/978-981-16-6077-1_1

words carcinos and carcinoma (these words refer to a crab), definitely by observing a real malignant tumor. The Roman physician, Celsus (28–50 BC), later translated the Greek term into "cancer." Again these scientists of antiquity definitely noticed the growth of cancer in some form. The Edwin Smith Papyrus was written around 1600 BC (possibly a fragmentary copy of a text from 2500 BC) and contains a description of cancer, as well as a procedure to remove breast tumors by cauterization. A natural question is why scientists working over a millennium have still not been able to find medications to cure this dangerous disease?

The shortest answer is: It is not just one disease. It is a generic name for hundreds of diseases with one common feature: rapid replications of cells, violating the dynamic equilibrium of the body. There are about 30,000 genes in a human genome. Over-expression/underexpression of just a few genes could lead to cancer. Cancer could appear anywhere in the body. They are very different in nature. That is what made cancer so difficult to treat medically, and that is what we will be dealing with through mathematical modeling in this monograph.

Regarding the number of genes, nobody can say for sure exactly how many there are. The estimates that we have found are between 20,000 and 35,000. These are just estimations. Some other estimations are different though.

Cancer may not be curable, but certainly in many cases it could be controllable. Validity of this statement is based upon the fact that the number of cancer survivors is on the rise. The rationale is how mathematicians could play a major role in that regard.

Science follows mathematics. Mathematical models have been in vogue for years to predict facts and figures related to practically any event in nature before, during and after it has taken place. Our aim is to reveal facts and figures related to cancer and discover avenues mathematically which could make cancer patients live like diabetics or asthmatics with regular intake of some medicine/ medicines in most cases.

1.2   The Nature of a Cancer Cell

There are some very huge differences between normal cells of the body and cancer cells. Whereas normal cells mature into very distinct cell types with specific functions, cancer cells do not. Normal cells always follow the law of dynamic equilibrium (homeostasis) of the body, and cancer cells do not. Possibly, the most conspicuous and most serious is that while a normal cell gets most of their energy in the form of a molecule known as adenosine triphosphate (ATP), the molecule that carries energy (obtained from breakdown from food molecules), cancer cells notoriously derive energy from multimitochondria, glucose, even in an hypoxic environment. Their efforts are endless. So, killing them is an outlandish undertaking. They could tolerate burning and poisoning. They can survive practically all tortures.

Essentially cancer is a genetic disease. It is in fact not just one disease. It is a collection of hundreds of diseases with one common property: Genes mutate and make cells grow uncontrolled, violating the fundamental principle of homeostasis (dynamic equilibrium) on which our lives depend. They do not die. They invade neighboring tissues and turn neighboring cells into cancerous cells. They even turn some soldiers of the immune system, namely the dendritic cells and macrophages, into their friends, and these cells help them to metastasize. They travel in groups. So, it will be difficult for the defenders to kill all of them. Some will escape and secretly invade other remote tissues. They even could pretend that they are not "foreign," and they are "self" having a gene of the "self." In case of danger, they hide in adipose tissues, under the skin and wherever defense is not very active. They release various proteins to befool the aggressive immune system. Even when they are killed, their remains create new cancer cells. According to [3], there are over 100 oncogenes discovered in the human body. Tumor suppressor genes (TSGs) are also present in the body. These are anti-oncogenes. Loss, suppression, or inactivation of anti-oncogenes leads to activation of oncogenes which help develop cancer. Although survival rates of cancer are increasing every year, yet it remains a very deadly disease.

Breast cancer is the second leading cause of cancer death in women (only lung cancer kills more women each year). The chance that a woman will die from breast cancer is about 1 in 38 (about 2.6%). Some types of benign breast lumps may increase a woman's risk of getting breast cancer. Absolutely, no lump in the breasts is tolerable.

The signs of danger that all women should be aware of are: any new lump in the breast or in the armpit, any swelling in the breast, any dimpling or irritation on the breast, redness or pain in these organs, any unexpected discharge from a nipple, or any discomfort.

About 85% of breast cancers occur in women who have no family history of breast cancer. These occur due to genetic mutations that happen as a result of the aging process and life style in general, rather than inherited mutations. The most significant risk factors [breastcancer.org] for breast cancer are sex (being a woman) and age (growing older).

1.3   A Need for a Practical 3D Mathematical Modeling

There are colossal differences between bioscience and physical science. Bioscience studies behaviors of animate objects, whereas physical science deals with inanimate objects. Biomolecules do not behave like inanimate chemical molecules. Also when chemicals are applied in a physiological environment, they are under full control of the laws of physiology. So, biochemicals and chemicals are two entirely different substances. Heterogeneity is the innate property of all chemicals in a human body. This necessitates the introduction of statistics to analyze huge biological data. Cancer cells are extremely intelligent biochemicals. Most carefully and vigorously, they set up a network of strategic attacks.

The entire bioscience, especially, studies on cancer treatments, depends mostly on observations, experimentations, and statistical analyses of huge amounts of data obtained from experimental studies on heterogeneous populations varying racially, internationally, gender-wise, age-wise, food intake, environmentally, and mental setups. Mathematics can handle all these variables and more in modeling. Certainly, this is not easy but doable. With the availability of supercomputers tied in super-clusters, these models could be solved in a real-time scenario and machine learning combined with artificial intelligence could give diagnostic analyses. We believe that will be of enormous help to the doctors.

At present, the Internet is literally flooded by mathematical models related to studies on cancer. But we have yet to find an oncologist or a medical practitioner who has used any one of these mathematical articles in treating a patient, because these are mostly theoretical. Theoretical descriptions on the growths, angiogenesis, metastasis on tumors are available in many texts like [11, 12, 17]. Regardless of their intrinsic theoretical appeal, health professionals hardly apply these works in practice. At present, technology is very powerful, very advanced to detect growths and spread of tumors even on a microscopic scale. Yet, there is a dire need for a practical mathematical description of tumors in their developmental stages in the language of mathematics. And more importantly, there is a dire need to develop tools which doctors could apply to study progress, prognosis, prevention, and panacea of this deadly disease. Just like X-rays, ultrasounds, MRIs, use of chemicals and radiation, mathematical models could be applied someday to treat cancer.

So, our objective is to develop and solve models which doctors could use without knowing any advanced mathematical analysis. Some texts were helpful in our studies. These are [3–7]. Modeling predator and prey could be found in every book on mathematical biology. This idea has been implemented in our study.

Although the number of survivors of cancer is on the rise, yet most people tend to think that cancers are death sentences. The main point of frustration is that: Most medical professionals do not seem to be very confident when they estimate the outcomes of their methods of treatments. These attitudes of uncertainty are the root cause of most frustrations. Practically, there is still no standardized treatment for cancer. One oncologist here in Austin, Texas, told the first author that "We change our strategies of treatment every now and then."

This medical challenge is also a mathematical challenge: How to model treatments of cancer which could be implemented in the medical practice? There are too many hurdles to resolve this mathematical issue. More than twenty years ago, the first author thought of such mathematical models observing the untold sufferings of cancer patients in several hospitals. This monograph is a practical attempt by which mathematicians could cooperate with the doctors in searching of an individualized treatment of cancer. Treatments of cancer require teamwork in which mathematicians should participate.

Mathematical modeling in biology is possibly one of the toughest challenges that applied mathematicians have been facing for years. Because unlike most other branches in science and engineering, parameters used in models could drastically differ between two statistically similar biological systems as time changes. Two patients, who could even be identical twins, may have very different internal structures. Because although their bodies may be alike, minds which control many physical aspects could be very different. The environments where they live could be very different. So, in practical situations, when it comes to cancer, one patient may require a host of software that could predict the mode of treatment for her, which could be quite different from her twin sister, even though they may have inherited the same mutated genes that led to having this disease. So, there hardly exists at the present time any proper mathematical tool for generalizations. Cancer modeling certainly falls in that category. Yet it is very much necessary these days. Because, when such models are computerized, with the advancement of fast parallel computers with multiple processors and coprocessors, different input parameters may be used again and again in the codes and modeling may be done for the treatment of every individual case which is equivalent to personalized treatment of cancer. Here, an attempt has been made in this regard. The major emphasis in this monograph is how to treat breast cancer for individual patients through individualized tailored treatments. This is a preliminary attempt.

1.4 The Mathematical Mapping Behind Our Computational Studies

The first challenge is that our body is three-dimensional. An adult person has about 10^{14} number of cells in the body. Realistically, cancer cells can move anywhere. So do the immune cells and every minute element of medications. Modeling treatments mathematically requires applications of supercomputers, possibly clusters of supercomputers for computational solutions. So, a huge amount of computer memory and a huge amount of computations are needed. For every strategy of treatment and for each patient, these huge codes will be run, possibly several times. Within the constraints of mathematical modeling, all these requirements may not be fulfilled. So, we must look into how it could be done. In this regard, it should be observed that examining a tiny droplet of blood, doctors can estimate the glucose level of the entire volume of blood. So, a smaller computational field could serve our purpose. Mathematics explains this through applications of surjective mapping.

1.4.1 Surjective Mapping

Let us consider a simple example:

$$f(x) = 1 \text{ if } 0 \le x \le 2 \quad \text{and} \quad f(x) = x + 1, \text{ if } x > 2$$

So, the mapping f takes all the values of $x \in [0, 2]$ to 1. Or in other words, the value $y = 1$ carries the information of the entire set of values of x so long $0 \le x \le 2$. That happens when we consider one droplet of blood carrying the information of glucose of the entire volume of blood in a human body. This is the fundamental concept behind surjective mapping. So when we are looking into the properties of computed values of certain variables in a small cube that really relate to values of those variables in a large part of the body, we are looking into a surjective mapping.

Let $F : D^3 \rightarrow \Omega^3$ be a surjective mapping (onto but not one-to-one) from D^3 (a real 3-dimensional Euclidean space) to Ω^3, where Ω^3 is a box consisting of $I + 1, J + 1, K + 1$ number of discrete points on the x - axis, y - axis, and z - axis, respectively. If these points are such that $I = J = K$, then we will call it a cubic box or simply a cube. Let us draw equispaced lines parallel to $x-, y-,$ and z - axes, respectively. So, there are many small cubes that make the big cube. Let each point of Ω^3 represent the location of a biochemical, for instance, a cancerous mass because cancer cells move in groups. If u^n_{ijk} is a *mass of cancer cells at a location* (i, j, k) *in the cube at time* $= t_n$, then $u^n_{ijk} = \alpha$, means the amount of mass of the malignant cell at that location and at that time is α. This notation is valid for all variables used in this monograph. So, a cancer cell is apparently cured if and only if $u_{ijk} \rightarrow 0$, at every point of the field of computation as time goes on. This is the aim of the mathematical studies on the treatment of cancer in this monograph. We did this by applying computational methods. This is assumed to hold true because of the properties of surjective mappings (more on this topic is discussed in Chap. 6).

1.4.2 Some Deadly Aspects of Cancer Cells

While discussing mathematical modeling and cancer, Mackenzie [8] mentioned the works of Fister and Panetta (SIAM, Vol. 63, No. 6). Their model is much simpler than ours. They considered three versions in their modeling:

(a) You kill the most cancer cells when the tumor is largest.
(b) When the tumor is small but has the highest growth rate (applicable to Hodgkin's disease).
(c) A saturation effect, which has the drug becoming less potent as fewer proteins are available for binding.

We practically disregarded all these. Working round the clock with oncologists for almost two years, the first author learned (i) cancer cells must be searched for and destroyed wherever and whenever they could be found because just one cancer cell has the ability to make many normal cells cancerous. (ii) Under no condition, a tumor shall be allowed to grow to become the largest. Truly, they grow and grow so long they could get glucose, oxygen, and nutrients from blood supply. So the moment they are detected they must be killed. That is how they could be kept under control. In fact, the moment cancer is found in one organ, oncologists order for a complete body scan. Even if it appears to be benign like an adenoma, still doctors want to see a complete body scan. (iii) Simply, if a group of cells do not seem to behave normally, they cannot be trusted. If a drug kills some cancer cells at some given locations, they spread fast and metastasize more. So, the aim of all drugs should be not just to destroy the tumors but to annihilate each and every cancer cell wherever they may be, the runaways too. That should be the aim of all treatments.

When cancer cells are well differentiated, they look almost like normal cells in the microscope. These cells grow relatively slowly. However, if they are poorly differentiated, looking very abnormal, they grow faster. But no growth rate can be trusted. Cancer cells could change their growth rates at any moment.

"For example, a slow growing tumor may 'takeoff' and begin to grow and spread rapidly" (Kenneth Blank, MD, OncoLink, Editorial Assistant).

In this monograph, all these and more have been studied through mathematical models.

Mathematicians should look into models and must look into how to destroy these relentless enemies of a human body under all possible conditions. Cancer knows no law. No mathematical law can ever predict how a malignant tumor will grow or the precise process of metastases. Adjustments and scientific inferences are needed possibly at every step of treatment. What pathologists or oncologists say is just an estimation based upon their experience.

Scientifically, cancer is a genetic disease and cannot be completely cured in general, because there could be multiple mutations in many genes, which may be impossible to correct, even if they were detected at a very early stage. However, they could be kept under control by proper medications. How much medication is needed? How to know that progress of cancer has been thwarted ? This monograph attempts to answer these fundamental questions. Working with several oncologists, the first author developed these models as some practical tools to treat cancer, in particular, breast cancer. The more the delay in treatment, the more difficult it is to overpower them and keep them under control because they could mutate more and spread. Our aim is: absolutely no waiting. Attackers must be attacked back as soon as possible and as meticulously as possible. Mathematically, we have found that practically at any stage of this disease, the attackers could be kept under control. We determined this through computational studies of our models. However, we also pointed out that in reality, these findings should be judged by expert oncologists.

This monograph analyzes these and more through mathematical findings. It is an attempt to work with doctors and other medical professionals, forming a team with one aim, how to keep cancer under control and save lives. Our models are very general. Regardless of how a tumor grows, our primary objective is how to destroy or at least keep them under control even when they are migrating toward tissues and organs outside the primary site. Our models attempt to follow the protocols of recent treatments and may further be generalized.

These models are made of nonlinear reaction–dispersion-type partial differential equations and were solved by numerical methods which we developed for the three-dimensional environment simulating a human body. These numerical methods follow a vectorized algorithm which may be implemented in any high-speed super-computer or clusters of supercomputers, through the message passing interface (MPI) where large-scale computations could be done. We hope someday, in every hospital or even in a doctor's office this could be used as a software.

Our methodology is very practical and application-oriented following recent advancements of cancer treatments. It is more general than the works in [9, 10]. Let us first study some facts about this disease.

We have noted that in an adult human body there are about one hundred trillion cells (10^{14} cells). Each cell is guided by a DNA comprising a large number of genes in a mannerly, orderly way. This is homeostasis. A state of dynamic equilibrium. Body changes constantly but cells should always behave in an orderly way. Cells in the lungs must stay in the lungs and do their respective jobs. The same is true for the cells in the stomach, in the kidneys, in the pancreas, in the liver, in the breasts, in the brain, etc. They replicate themselves whenever needed. These are our normal cells. In general, if there is a minor damage to a cell, this is repaired. If the damage is too much, or the cell is dysfunctional because of aging or any disease, the gene p53 causes apoptosis (programmed cell death) and the cell dies. Gene p53 is a tumor suppressor gene. This gene is inactive in cancer cells. Cancer cells totally ignore the law of homeostasis. As they grow, they invade the neighboring cells and change them into cancerous cells. They form tumors. They form groups to invade other tissues all over the body. They secrete various proteins to confuse the body's immune system and/or to put them to sleep. For example, breast cancer cells release the protein HSP27 (heat shock protein 27) which inhibits programmed cell death. This protein also makes T-cells, the relentless fighters of the immune system, unresponsive. Furthermore, it makes macrophages, the phagocytic fighters of the immune system, change their activities from being soldiers of the immune system into "extremely proangiogenic, inducing significant neovascularization" (building new blood vessels so that cancer may metastasize [11]). Every time a normal cell divides, the strips of DNA at the end of chromosomes, known as telomeres, get a bit shorter. This is the normal aging process of a cell. When they become too short, the gene p53 is activated and the cell dies. Cancer cells use the enzyme telomerase, repair the telomeres, and keep on replicating. p53 remains inactive or dysfunctional. This process could be eternal. "This process allows cancer cells to live forever. Cells from Henrietta Lacks, an American woman who was diagnosed with cervical cancer in 1951, are still growing," long after her death (http://whoami.sciencemuseum.org.uk/whoami).

When cancer starts, it generates its microenvironment which is undetectable. A cancerous cell is the output of one or more defective DNA. The cells not only form tumors; using blood vessels and lymph, they move to distant locations to form more tumors creating capillary tubes all around them through which they could swim, draw blood, get their nutrients, etc. These look like hands and legs. They also reach the nearest arteries to draw blood to get a lot of nutrients. They look like crabs. This is angiogenesis. Even when chemotherapy is conducted, breast cancer cells which nestle in the tiny blood vessels inside the bone marrow could survive [12].

There are many microstatic features of these cells. Many of them are very hard to detect. In our models, we have assumed that the cancer cells form tumors and are in motion and our defense team is able to detect them and fight back.

In order to model any event in nature, one has to understand and look into very carefully and critically different aspects of it, as many as possible, knowing fully well that future investigations will add more complexities in mathematical modeling. To look for an effective treatment of any disease, a primary search will be, to look into the role of the immune system, our first line of defense. And then look for the agents which help the perpetrator (cancer) and the medications which will help the immune mechanism. Fighting a disease is to take a comprehensive look into a scenario of such a battlefield where the defenders are fighting back the attackers. That is our primary initiative to look into the attacker/defender mathematical model.

Here, attackers are the cancer cells and their helpers and associate inside the body of a patient and around the environment, and defenders are the immune system, medications, food, etc. Unfortunately, whereas the attackers are uncontrolled in the body, the defenders have severe limitations. For example, the number of leukocytes (defensive forces) should be under control, else the body will not function normally. Thus, a proper treatment should be, satisfying all the limitations, the defenders must keep the uncontrolled enemies, controlled, and/or eliminate them from the body.

1.5 A Fundamental Drawback of the Use of Logistic Equations

Several authors [3, 7, 13] have applied the logistic equation in order to model growth of a cancerous tumor. That is not quite correct in general. Such a model may be applicable to benign tumors which grow generally up to certain size and stop growing. The cancerous tumors stop growing if and only if it fails to have enough supplies of food drawn from the blood vessels. They start dying at the core in general. The core becomes necrotic. Even then they remain dangerous. "The dead cells trigger a reaction that strengthens the cancer cells that have just escaped death by drugs. These become more like stem cells, which are resilient and robust, and eventually cause recurrence of the tumor" (Dr. Sui Huang, professor and cancer biologist at Institute for Systems Biology (https://isbscience.org/bio/sui-huang-md-phd/) along with former mentee and longtime collaborator Dr. Dipak Panigrahy at Beth Israel Deaconess Medical Center in Boston and colleagues at Harvard Medical School). So, the attack on cancer applying any therapy should not be local. It must be global. That implies that a realistic model should be time dependent and three-dimensional describing various stages of cancer as it grows and spreads. In most of our models, this approach has been adopted. Destroying the castles (tumors) of the enemies is not enough, and the runaways must be caught and killed. Because when the tumors are attacked, many malignant cells run away to hide and to form tumors. These runaways are more clever and more powerful. Using angiogenesis, and other means, they consume glucose and oxygen as much as possible depriving other organs of the body from getting any food and form tumors again. Even if a tumor is avascular, by the time it becomes necrotic at the core, it metastasizes at other parts of the body and grows. Uncontrolled growth never changes both at the primary site and at secondary sites. In the language of the expert neurosurgeon Dr. Latinville (St. Louis), how a cancerous tumor will behave is very hard to predict. But they must be nipped in the bud.

For nutritional therapy, logistic equations may be applied. Most nutritionists look into the growths of tumors and focus on how to keep those under control by strengthening the body, while other therapies are going on. We looked into that scenario in Chap. 9. However, an assumption that a malignant tumor has a steady-state value is a myth. These tumors grow and grow and grow so long they could get a supply of glucose and oxygen. Cancer cells can deprive other healthy cells of these nutrients and grow often killing them altogether.

White blood cells (leukocytes) grow when they find that the body is attacked by foreign pathogens. Sometimes they overgrow too, which could cause autoimmune disease (immune system fighting against itself). At that point, their growth is controlled by medications. So, mathematically it means that if and when growths of tumors are controlled, blood counts and the amounts of medications must go to a steady state, giving the doctors an idea regarding how much medications should be used to keep cancer under control and/or to destroy it. Most nutritionists take this into account. Our computational procedures always check upon these aspects of treatments.

1.6 The Dynamics of Violence Started by Cancer Cells

Before developing the models, it is essential to find out as much information as possible about the dynamics of the topics. We have made an attempt in this regard in this chapter. We will repeat some in other chapters as needed. In order to model any event in nature, one has to understand and look into very carefully and critically different aspects of it, as much as possible knowing fully well that such investigations could add more complexities in mathematical modeling. To look for a cure of any disease, a primary search will be to look into the role of the immune system, our first line of defense. And then look for the agents which help the perpetrator (cancer) and the medications which will help the immune mechanism as well as those which could hurt it. Fighting a disease is to take a comprehensive look into a scenario of such a battlefield where defenders are fighting back the attackers. Here, attackers are the cancer cells and their helpers and associate inside the body of a patient and around the environment, and defenders are the immune system, medications, food, etc. Unfortunately, whereas the attackers are uncontrolled in the body, defenders have severe limitations. White blood counts should not exceed 10,000 and must not be less than 5000. Medications cannot exceed certain specific amounts etc. Thus, a proper treatment should satisfy all the limitations, while the defenders must keep the uncontrolled enemies, controlled, or eliminate them from the body. We have mentioned before that a human body always tries to maintain a stable dynamic equilibrium, known as the homeostasis of the body. For example, the average body temperature is 98.6 °F (37 °C). There are some studies showing that throughout the day temperature may vary from 97 to 99 °F. An average fasting blood glucose varies from 70 to 130 mg/deciliter of blood. An average resting heart rate varies from 60 to 100. These are some of the examples of the principle of homeostasis that the body strictly maintains. As cells divide and get old and/or damaged, they are replaced by new cells which must follow the law of homeostasis performing the same job at the same location for proper maintenance of the body. The cell death is also preprogrammed. This is called apoptosis. A disease means that the state of homeostasis is violated. Body has a built-in mechanism to make attempts to restore homeostasis. If and when the body fails to achieve this state, we say the body has become sick and look for external help from doctors. As the body grows, genes inside cells replicate themselves. If there are any mistakes in replications, the body corrects it by destroying that cell. But sometimes the body fails. These mutated genes cause cell mutations. Mutated cells disobey the law of homeostasis and grow uncontrollably, eventually forming a tumor. A tumor with the size of a sphere with 2 mm in diameter contains approximately 10^6 number of malignant cells [www.ncbi. nim.nih.gov]. They attack neighboring cells. They travel through blood vessels and lymphatic vessels to distant organs to build colonies (another tumor). They grow tiny blood vessels all around the tumor to reach main arteries. From there, they draw oxygen and nutrients. This is angiogenesis. The picture of a tumor looks like a *crab* (cancer) (Fig. 1.1). They draw as many nutrients as possible, depriving other organs from getting food. So other organs soon start getting emaciated. Mutated cells carrying "foreign" antigens release abnormal proteins. Their presence in the blood activates a counter attack by the immune system in general. Some tumors could be benign. They stop growing at some point. They also do not metastasize. However, some of them could be life threatening. Cancer can happen at any part of the body. Detection at an early stage is crucial for survival. Yet such a detection is extremely subtle and often extremely hard. Recently a patient, who survived ovarian cancer, reported that her oncologist [14] detected a one-millimeter malignant cell in her blood test. In such cases, mathematical models for a cure could play a crucial role. A cancerous tumor is shown in the Fig. 1.1.

Cancer cells are very cruel and cunning. They may not display, at the very beginning, any uncontrolled growth at all and hide themselves from their enemies under adipose tissues or scar tissues [15], or somewhere in the body. Sometimes, they stay in plain sight [16] by hiding their antigens and appearing to be well differentiated like normal cells. Here, the author stated that "Cancer cells conceal themselves from the immune system not by barricading themselves in an impenetrable shell, but by the biological equivalent of hiding in plain sight. Their strategy is to make themselves inconspicuous by blending in, as far as possible, with their normal neighbors. They accomplish this by displaying proteins, known as antigens, on their surface, that are similar to those well-behaved normal cells nearby. This uniformity can mislead the body's T-cells, the patrol officers and duty inspectors of the immune system. T-cells conduct their interrogations of the body's cells with specialized protein hooks called receptors. The receptors interlock with different types of antigens. If the antigen is normal, the cell is left alone; if it signifies a cancerous or infected cell, the cell is destroyed. The trouble is that T-cells may not carry receptors for certain cancer-associated antigens and therefore don't recognize the cells as cancerous. Such T-cells

Fig. 1.1 A tumor

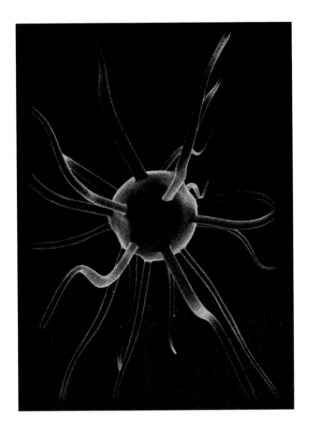

are in the position of a colorblind detective searching for a criminal known to be wearing a red scarf." Cancer cells hide in scar tissues [17]. So sometimes, scar tissue can appear suspicious containing a lesion with irregular, spiculated margins. There may be no central mass to this lesion. MRI and biopsy will hopefully validate what the lesion is.

1.7 Awakening of the Dynamics of the Immune System

Our immune system is our first line of defense. It conducts immunosurveillance inside the body round the clock. Whenever a foreign (non-self) element is detected, it meticulously launches an attack immediately. Bacteria (from outside), viruses, worn out cells, cancer cells, etc., exhibit, produce, or release non-self-substances, which are "non-self or foreign" antigens. Each antigen triggers a specific immune response against itself.

Immune system is triggered spontaneously for certain types of cancer when cancer cells appear to be highly abnormal, thus representing themselves as "non-self."

They release a host of proteins which appear on their surfaces through the major histocompatibility complex (MHC) molecule [60].

Then, the immune system recognizes them as foreign antigens and launches an attack. Sometimes, cancer cells hide this identity, especially, breast cancer cells and thereby they are able to suppress the immune response. Also, sometimes, due to a process known as immunosenescence, which happens naturally as the body ages, the immune system cannot recognize cancer cells as "non-self." Some researchers suggest that a special virus may be injected in the body which will reveal the true identity of the cancer cell [18]. In our model, it has been assumed that there exist drugs that can inhibit immunosenescence and reveal the MHC proteins identifying the cancer cells and enabling the immune system to recognize the foreign antigen (Fig. 1.2).

Let us first briefly discuss how our immune system works. Blood has about 40–45% erythrocytes or red blood cells. Only about 1% of blood consists of white cells or leukocytes and about 55% plasma. Leukocytes are our protectors. We will briefly study their activities so that we may simulate them in our models. Lymphocytes are kinds of leukocytes. They are born in the bone marrow and get settled in lymphoid organs. They travel through blood vessels and lymph vessels. They live in thymus, lymph nodes, spleen, etc. They are relentless fighters and can fight against all kinds of antigens (foreign attackers).

The body possesses two distinct kinds of defense mechanisms each coordinating with the other constantly. They are highly organized and strongly motivated to defend the body against all foreign invasions. The first kind is our innate immune response. It

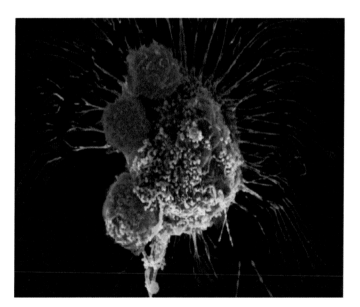

Fig. 1.2 The blue cells are T-cells attacking a mass of cancer cells

is non-specific and does both cellular and humoral responses to defend the body from all kinds of pathogens. These defenders are neutrophils, eosinophils, and natural killer lymphocytes (NK-cells). They attack every pathogen (non-self) and destroy it. The humoral response is very antigen specific. This is conducted by B and T lymphocytes. They are very aggressive fighters, well organized, and release specific antibodies and cytokines to destroy that specific non-self-antigen. The complete immune response consists of an intricate network of interactions with leukocytes and foreign pathogens. Dendritic cells found on lymph nodes, thymus, spleen, and other secondary lymphoid organs are antigen presenting cells. They are relatively large, adherent, motile cells. They present the antigen to helper T-cells. Helper T-cells do not kill the antigens themselves. They activate cytotoxic T-cells and B-cells, two most notorious fighters to destroy the enemies. Monocytes, a kind of leukocytes, get out of circulation of blood, taking shelter on tissues, and mature into macrophages and dendritic cells. Macrophages fight all antigens. They are another kind of leukocytes, very motile and phagocytic. They release lymphocyte-activating factor interleukin 1 (IL-1) and tumor necrosis factor (TNF). Dendritic cells also produce low level IL-1 which promotes lymphocyte differentiation and stimulation of functions of T-cells. IL-1 activates helper T-cells which releases the cytokine IL-2. This is a T-cell growth factor. Some of the helper T-cells also release IL-3 (which also supports growth factor of T-cells) and GM-CSF (granulocyte monocyte colony-stimulating factor), a kind of protein secreted by macrophages, NK-cells, and T-cells as a cytokine. Cytokines are large groups of proteins which regulate immune response. B-cells are activated when their cell receptor (BCR) binds to the antigen. After activation, B-cells differentiate into plasma cells which release only a specific kind of antibody to destroy that specific attacker and memory B-cells which remember the specific fighting techniques that the mother B-cells did to destroy the antigen (this is how vaccines work). Helper T-cells, which do not directly fight, stimulate B-cells to secrete specific antibodies to destroy the antigen. The immune system also destroys abnormal cells which could become cancerous later. Unfortunately, some hormones and enzymes help tumors to grow and sometimes the immune system fails to recognize the antigens because of their presence. These ideas and more have been taken into consideration in our modeling.

Recently, an epoch-making invention has been done by Dr. Daniel Chen, a Stanford oncologist, and Dr. Jim Allison, a researcher at MD Anderson, Cancer Center at Houston, Texas. CTLA-4 is a T-cell protein. It acts as a brake (inhibitor) on activities of cytotoxic T-cells, the tough cancer killers. The protein being present, cytotoxic T-cells do not become aggressive against malignant cells. Dr. Allison developed an antibody that blocks CTLA-4 activities. This blockage disengages the brakes and let cytotoxic T-cells attack cancer cells. He worked with Dr. Chen, and both got a Nobel Prize in 2018. These studies are online from MD Anderson Cancer Center, Houston, Texas.

1.8   The Role of Thymus [Ref. Rejuvenate Your Thymus Gland-Dr. J .E. Williams]

We talked about T-cells. Their origin is the lymphoid organ thymus. It is a bilobed organ located in the superior mediastinum posterior to the sternum and between the lungs. All blood cells originate from the bone marrow. Those which enter the thymus (T) become the T-cells. There are four different kinds of T-cells: (i) cytotoxic T-cells, which relentlessly destroy host cells which harbor any non-self (foreign) antigen. They are also called CD8 cells. They gain cytotoxicity from the cytokines which they produce; (ii) helper T-cells which modulate and regulate activities of a host of other leukocytes, including lymphocytes B-cells and CD8 cells; (iii) suppressor T-cells which turn off the immune response when the war is won, and (iv) memory T-cells which jump-start the precise immune response which they lodged previously against the same antigen in the past. A picture of thymus is given below (Fig. 1.3).

A strong thymus is very essential to fight any infection and cancer. Unfortunately, being relatively large at birth, about 32 g., thymus continues to grow until puberty, then gradually atrophies. Most medical professionals state that thymus becomes mostly a cluster of fatty tissues and practically ineffective by the age of 50. Some doctors believe that before thymus becomes inactive, enough T-cells have been produced to protect the body. However, recently, researchers believe that T-cells become less and less powerful with aging and that is why older people become sick more often. This could be prevented to some extent, if their thymus could be revived [19, 20]. Both studies are very comprehensive and clearly point out how thymus could be rebuilt and made most effective to fight all ailments including cancer. Thymosin is the hormone that thymus secretes to maintain lineage of T-cells. In this regard, a synthetic peptide thymosin alpha 1 [21, 22] has been manufactured to enhance the power of T-cells and dendritic cells and antibodies. According to Dr. Diamond [23], the importance of the thymus gland cannot be overestimated. It has many functions. Some of the most important ones are:

Fig. 1.3   Thymus

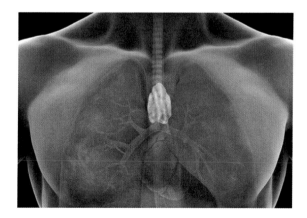

1. It plays a vital role in the body's immunological functions.
2. It performs endocrinological functions.
3. It is actively involved in the supervision of lymphatic drainage. It monitors and balances the body energy systems.
4. It serves as the seat of the life energy (the thymos).

"When the thymus monitoring system 'breaks down' under stress, then the meridian energy imbalances go uncorrected and the patterns for disease are set" [23]. Thymus atrophies when a person is in a stressfull situation.

Some works have also been reported [24] regarding application of this synthetic hormone to inhibit growth of breast cancer both in vivo (in a human body) and in vitro (in the laboratory) suppressing proliferations of malignant cells and inducing cell apoptosis.

1.9   Role of Entropy

Still there remains a question, and that is: How the immune system could be 100% sure that the cells they are attacking are "non-self" because sometimes cancer cells cleverly hide their antigen. In this regard, the following suggestion could be significant. Malignant cells, from the very beginning, start growing without any control violating the law of homeostasis. This growth is definitely very chaotic. So right at that location, entropy definitely starts increasing fast generating thereby an extra amount of heat causing inflammation. So, entropy detection technique is very essential to study the development of tumors. "Breast cancer is a serious problem and a common form of cancer diagnosed in women. Computer-aided diagnosis (CAD) is a tool which can assist the radiologists in the detection of abnormalities in medical images. In this study, a CAD system for breast cancer using X-ray mammography is presented with a high level of sensitivity by wavelet entropy features. The discrete wavelet transform (DWT) of a digital mammogram provides a multiresolution representation of it. The characteristics of a mammogram at different resolution levels are represented by computing wavelet entropy and used as features for the corresponding mammogram. Then, ensemble classification using K-nearest neighbor (KNN), Bayes, and support vector machine (SVM) is employed to classify the abnormalities as benign/malignant. The experiments show promising results with a high level of sensitivity, and hence, it is feasible for mammogram classification" [25].

So, a drug/electronic sensor looking for inflammation may examine the site and make an assessment whether the antigen is foreign and causing irregular growths of cells.

1.10 Few Special Proteins Released by Cancer Cells

Cancer cells release several different kinds of special proteins. "In over a hundred breast and brain cancer patients and tumor specimens, researchers found that the more hypoxic the tumor was, the more cadherin-22 (protein) it had" [26]. Dr. Nils Halberg [26], from University of Bergen, says "We discovered the aggressive cancer cells that are spreading in colon, breast, and skin cancer contained a much larger portion of the protein PITPNC1 than the non-aggressive cancer cells." Matrix metalloproteinases (MMP) remodel extracellular matrix necessary for many physiological conditions like mediation of immune response, growth of organs, wound repair, and a host of other constructive mechanisms of the body. They also participate from the initial phases of cancer onset to the settlement of a metastatic niche in a second organ. They play some critical role [27] in cancer metastasis. "MMPs facilitate tumor cell invasion and metastasis by at least three distinct mechanisms. First, proteinase action removes physical barriers to invasion through degradation of extracellular matrix (ECM) macromolecules such as collagens, laminins, and proteoglycans. This has been demonstrated in vitro through the use of chemo invasion assays and in vivo by the presence of active MMPs at the invasive front of tumors. Second, MMPs have the ability to modulate cell adhesion. For cells to move through the ECM, they must be able to form new cell–matrix and cell–cell attachments and break existing ones… Finally, MMPs may act on ECM components or other proteins to uncover hidden biologic activities. For example, the angiogenesis inhibitor angiostatin may be produced from plasminogen by MMP action and laminin-5 is specifically degraded by MMP-2 to produce a soluble chemotactic fragment" [27].

The cancer antigen CA27.29 is a valuable marker for breast cancer. The higher it is the more advanced the breast cancer is and bigger is the tumor [www.medscape.com. Jan 4, 2021]. Cancer antigen test reveals this information.

1.11 A Short Outline of Breast Cancer

This disease attacks both men and women. It is the second most common cause of death from cancer in women in the USA, next to lung cancer. Fortunately, because of early detection and treatment improvements, since 1989, the number of deaths has steadily decreased. Most breast tumors which are found to be cancerous are vascular. There are thin blood vessels inside the tumors (Fig. 1.4).

Rarely, there are some cancerous tumors in the breast with no blood vessels inside. They are avascular. As they grow, due to the lack of blood vessels inside, malignant cells start dying. So, the core of these tumors becomes necrotic. Avascular breast tumors are mostly benign. Let us first look into the anatomy of the breast and analyze the mechanism of the onset of cancer and why it could be deadly.

Breasts are called the mammary glands, made of milk-producing glandular structures made of lobules and several ducts which transport milk to babies. They are

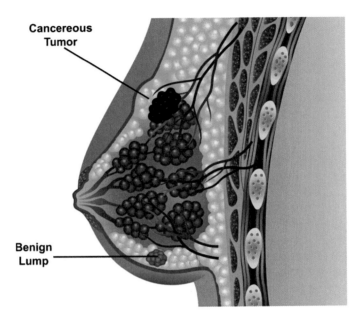

Fig. 1.4 Breast cancer [Ref. Indiamart.com]

surrounded by a sequence of lymph nodes, connected by lymphatic vessels which drain any excessive fluid. Lymph nodes are castles of the immune system. In between the glandular tissues, there are fatty tissues and connective tissues. There is no muscle inside the breasts only at the very bottom. There are lots of blood vessels supplying nutrients to these organs. Fatty tissues supply plenty of energy to the cancer cells so that they can proliferate at random. Lymphatic vessels and blood vessels help them to metastasize. That may be the reason why 80% breast cancers infiltrate into other organs and become invasive [28].

Broadly, there are two types of breast cancer, (1) ductal carcinoma in situ (DCIS) and (2) lobular carcinoma in situ (LCIS). In situ means "in place," or in other words cancer has not spread but could be invasive. DCIS refers to abnormal cells replacing the normal epithelial cells on the ducts. They could expand the ducts and the lobules. Since they could be invasive, in our model we treated them as if they had just started their invasion. Mathematically, this may be modeled as transportation through dispersion. However, most oncologists told me that almost from the beginning, malignant cells resume departing from the primary sight to metastasize at a distant location. The metastatic process, which is an extremely complex process, takes place in a microenvironment. Malignant cells face a complex network of hurdles to fulfill their wicked efforts. Dr.Roper, Barnes Hospital, St. Louis, told me that even when a patient seems to be cancer-free, after twenty years cancer may return and the reason is some malignant cells keep hiding inside adipose tissues for years. Unfortunately, in the breast there are plenty of such tissues. So we did our modeling accordingly.

Tumors in situ should be investigated very cautiously. Long-term follow-up of patients with such tumors, originally misdiagnosed as benign (non-cancerous), revealed that 20–53% were later, after about 10 years, diagnosed as having an invasive breast cancer [28]. Ductal carcinoma is the most invasive type of breast cancer. Doctors believe these invasive breast cancers are not curable, but can be treated well. Our mathematical models shall validate this statement.

Early detection of almost any kind of cancer, before it has metastasized to other organs, may be cured in general, at least in the case of breast cancer. So, women should watch out for any change, regardless how minor it may appear to be, in the appearance of their breasts. The following are some points to note in this regard: (i) a skin dimpling, (ii) a visible or an invisible lump, (iii) an orange patch on a breast, (iv) some patchy redness on top of a breast, (v) an ulceration, (vi) a nipple moving backward, (vii) any unexpected discharge from a nipple, (viii) any eczema around the nipple, and (ix) any swollen lymph node under the armpit.

1.12 Some Tests to Detect Cancer

Doctors perform several tests. Some of these are:

1. Blood tests to check for spread to the liver or bones.
2. Bone scans to check for spread to the bone.
3. X-rays and/or CT scans to check for spread to the chest, abdomen, and liver.
4. 3D mammograms.
5. MRI with contrasts.

Tumors are classified into three fundamental stages denoted by (IB) (LN) (ME).

Stage IB implies that it is within the breast and at most in an adjacent tissue. Stage LN implies that it has affected the adjacent lymph nodes. Stage ME implies that it has metastasized at a distant organ. The common sites where it spreads are: bones, liver, lungs, and brain. At stage ME, breast cancers often are serious threats to life.

1.13 Some Risk Factors: Carcinogens

For mathematical modeling, it is essential that we should know some of the causes for this malignancy. Then, the model is applied accordingly. Let us discuss some causes and risk factors for cancer and in particular for breast cancer.

With rampant industrial growth all around the globe, the environment is getting more and more polluted. Hundreds of life-threatening chemicals enter into our food chain continuously. Pollutants in air and water, even UV rays of the sunlight, are often threats to our health. Most of these are carcinogens. Ten most dangerous ones are:

Tobacco: There are at least 70 chemicals in tobacco, which damage our DNA.

Radon: It is a natural gas released by the earth. It causes lung cancer to non-smokers.

Asbestos: The tiny fibers that get trapped inside lungs and could cause cancer.

Crispy Brown Fried Foods: They contain acrylamide, a strong carcinogen.

Formaldehyde: Found in plywood and some fabrics. People working in these industries are exposed to formaldehyde, and many of them get cancer.

Ultraviolet Rays (UV): It causes skin cancer.

Alcohol: A major cause for breast and liver cancer.

Engine Exhaust: From trucks, busses, cars are strongly carcinogenic.

Processed Meat: Bacon is number one. Others are salami, pepperoni, and sausage often causing colon cancer.

Some of these can certainly be controlled. Unfortunately, some politicians bring up social issues, religious issues, bigotry against some special groups of people, selfish moneymaking business policies, etc., and intentionally ignore the principles of science making cancer flourish.

There are some chemicals built in our agricultural products. We should be aware of their presence and be careful when controlling our intakes. In general, cancer cells fuel their growth with glucose (sugar). Sugar is the source of energy. Interestingly, two scientists, Dr. Charles Perou and Dr. Lisa Carey, from UNC, School of Medicine, found in March 2018 that the spread of breast cancer could be halted by blocking amino acids found in the vegetable asparagus.

Amino acid itself is definitely not carcinogenic. We need them to be just alive. In fact, we need twenty essential amino acids for our survival. Nine of them, we must get from our diets.

However, it is a helper of cancer too. "In vitro and in vivo data suggest that selective amino acid deprivation may have anticancer activity. Controlled amino acid therapy (CAAT) is a protocol developed for patients with cancer that includes strict dietary guidelines and nutritional supplements that focus on controlling amino acid and carbohydrate intake" Andrea S. Blevins Primeau, Ph.D. [https://www.cancertherap yadvisor.com/home/tools/fact-sheets/controlled-amino-acid-therapy-and-cancer/].

1.14   Free Radicals

These are highly reactive chemicals which could be potentially dangerous to our cells. In an atom, electrons are stable when they are in pairs. But because of carcinogens in the environment and/or consumption of certain foods (foods overly grilled, red salty meats, processed foods like bacon, salami and sausages, deep fried foods, etc.) and alcohol, an electron (negatively charged particle) inside an atom or in a molecule (which has two or more atoms) becomes unpaired. Then the atom becomes unstable. Then it becomes a free radical. Then to stabilize them, often an oxygen molecule donates an electron to the atom and then the donor oxygen molecule itself becomes a free radical. This process is called oxidation. Antioxidants avert this process. In this regard, antioxidants, like vitamin C, which repair a damaged cell, work against cancer. Repairing a damaged cell means repairing the process of secreting proteins by malignant cells for abnormal growth and metastasis. So, the role of the antioxidant C as an anticancer drug cannot be underestimated. More will be discussed later. Free radicals could play a significant role as a cause for breast cancer because they damage DNAs.

1.15 Estrogen and Progesterone

Estrogen and progesterone are female hormones released by the ovaries. There are primarily three kinds of estrogen, namely, estrone (E1), estradiol (E2), and estriol (E3). First two of them estrone or E1 and estradiol or E2 are produced by the ovaries, and the third one estriol or E3 is produced by placenta during pregnancy. E1 production reduces drastically, during pregnancy. Thereby, it reduces the overall lifetime exposure to estrogen and the chance for breast cancer. However, E2 is the strongest steroid among the three and it contributes to a variety of health issues like breast cancer, endometriosis, fibroid, and endometrial cancer. Statistically, it has been found that E3 poses no threat of cancer [29]. They are instrumental for breast development, fat distribution in breasts and several other organs, and development of reproductive organs. Almost all breast cells both normal and malignant have receptors to receive these hormones of growth.

Progesterone prepares breasts for development and lactation. It also balances any negative effects of the growth hormone estrogen. Cancer cells in the breast, like any other cells of the breast, have receptors for both estrogen and progesterone. These hormones make them grow and stay healthy.

Estrogen turns on the gene BCL2. If and when it is expressed at a high level, it makes cancer cells in the breast grow fast aggressively. 80% of breast cancer cases fall in that category. They are denoted by ER + (estrogen positive). On the other hand, (natural) progesterone increases activities of the p53 gene, which is a tumor-suppressive gene that also turns on apoptosis or programmed cell death. But we have to keep in mind that for over 50% of cancer cells the gene p53 is inactive that is why they could grow widely. In general, if p53 is mutated "or altered in expression in human breast tumors" [30], tumors become very aggressive. However, exactly what role progesterone plays in human carcinogenesis is not well known. So, this is a controversial issue and as such we did not incorporate that in our model. What has been found is that if a tumor has receptors for both of these two hormones (ER+/PR+) prognosis is better than if it is ER+/PR−. Drugs, like tamoxifen and toremifene, have been used for over 30 years to inhibit functions of estrogen receptors for treatment of breast cancer for both ER+/PR+ and ER+/PR−. However, ER+ and/or PR+ are relatively slower growing and cancer could quite successfully be treated by hormone therapy where only estrogen is inhibited [10]. This indicates that progesterone must have a positive role to play on constraining growth and proliferation of breast cancer.

Another interesting finding was reported by Dr. Andrew Dannenberg, director of cancer prevention at the Sandra and Edward Meyer Cancer Center at Weill Cornell Medicine (CNN, December 10, 2018). He wrote: "We do find that excess body fat in those who are postmenopausal with a normal body mass is associated with about doubling in the risk of estrogen-dependent breast cancer." So fat intake should be minimized. That implies sugar intake should be minimized. So blood glucose should be strongly monitored.

1.16 HER2 Protein

HER2 is a gene expanded as human epidermal growth factor receptor 2. Genes carry recipes of various proteins a cell needs to function properly. Some of these proteins influence how cancer cells in the breast grow. HER2 genes release proteins which are receptors on the breast cells. They control breast cells regarding their growth and repairments. In about 20% of breast cancer patients, the HER2 gene makes too many copies of itself and causes cells to grow uncontrolled leading them to form malignant tumors. This is called the HER2 amplification process or overly expressed HER2 gene. These are very aggressive tumors but could be well controlled by medications. In general, the prognosis is very good.

Oncologists often classify breast cancers on the basis of hormone receptors and HER2 receptors to determine what kind of specific treatment a patient needs.

Hormone receptor-negative (ER−/PR−) means neither estrogen nor progesterone is playing any role for the cancer. So, hormone therapy drugs are not useful for this patient. Such kinds of breast cancers are more common in premenopausal women.

1.17 Mutations of Genes

According to a report of American Cancer Society [31], "About 5% to 10% of breast cancer cases are thought to be hereditary, meaning that they result directly from gene defects (called mutations) inherited from a parent."

p53 is a tumor suppressor gene. It inhibits formation of tumors. However, those who inherit mutations of this gene have an increased risk of breast cancer. Fortunately, according to the American Cancer Society such cases are rare. "The most common cause of hereditary breast cancer is an inherited mutation in the BRCA1 and BRCA2 genes. Both BRCA1 and BRCA2 are two tumor-suppressive genes. In normal cells, these genes help prevent cancer by making proteins that keep cells from growing abnormally" [32]. However, if a woman inherits a mutated copy of any of these genes, she has a high risk, as high as 80% to develop breast cancer during her lifetime. Frank and Teich [33] stated that "In 1994, the BRCA1 gene, which is responsible for 45% of hereditary early onset of breast cancer, for coinheritance of breast and ovarian cancer, and for conferring increased risk also for colon and prostate cancer was cloned." BRCA1 mutations are slightly more common than BRCA2 mutations, but the impact on women with the gene mutation is more profound. Fortunately, biochemists are working on various forms of gene therapy [34–38] where drugs are delivered through some virus to inhibit growths of malignant cells.

An immunotherapeutic drug pembrolizumab has been applied to treat advanced metastatic breast cancer in patients with a BRCA mutation. So some progress has already been made. But we believe if gene therapy is added with immunotherapy results will be much better.

The other genetic factors are: abnormality of the gene ATM. Under normal conditions, this gene repairs damaged DNA. Inheritance of one mutated copy of this gene is a cause for breast cancer. Also, inheritance of a mutated copy of the gene CHEK2 enhances the risk of breast cancer. The gene PTEN regulates cell growths. Inherited mutated copies of this gene can cause both benign and malignant breast tumors. It is also a cause for tumors in the digestive tract, thyroid, uterus, and ovaries. Inherited mutations of the gene CDH1 increase the risk of invasive lobular breast cancer. A defective STK11 gene also increases the chance for breast cancer. The gene PALB2 releases proteins that interact with proteins secreted by the gene BRCA2. So, a mutated copy of PALB2 poses a serious risk for breast cancer. This could also be a cause for a male breast cancer.

In the Ref. [38], a mathematical model of gene therapy has been presented which could be used together with the immunotherapy. We have tried to mathematically discuss several methods, for a combination of gene therapy and immunotherapy concentrating our search to look into gene expression and analyze mathematically how proteins released by mRNAs could be corrected properly.

The gene replacement therapy could be quite risky. It often implies at least one of the following procedures:

(i)   Replacement of the mutated gene that caused the cancer with a healthy copy of the gene.
(ii) Inactivating the mutated gene.
(iii) Inserting a new gene into the body to deactivate the cancer process.

Although it is well known that cancer is a genetic disorder, each of the above steps involves risks which could impose some drastic results on the body. Yet, researchers in many laboratories around the world are investigating gene therapy, not only for cure of cancer but many other diseases.

Chen [37] from Harvard University presented an article on mathematical modeling of gene expression considering degradation of proteins and mRNAs. Dey [38], the first author, started with somewhat similar ideas but considered another approach of gene therapy. He made a preliminary attempt to introduce a mathematical model regarding how proteins released by proto-oncogenes and tumor suppressor genes could effectively control proteins released by oncogenes and thereby will correct the expressions of oncogenes leading to inhibition of cancer. This topic is discussed in this monograph in Chap. 7.

1.18   A Note on Surgery

Surgery at the outset of treatments could be quite dangerous. After standard "cut, burn, and poison"-type treatment, cancer may completely vanish from the primary site and reappear at a distant location in the body at a later time or at the same time. This mathematical finding, especially with respect to breast cancer, has been backed up by several oncologists. Also, the immune system gets weakened after surgery and radiation and cancer could thrive and appear at a distant location with full strength at a later time (https://breastcancerconqueror.com/why-does-surgery-spread-cancer-cells/).

"During the surgical procedure, natural barriers that contain the tumor are breached, enabling cancer cells to escape their original confinement and spread to other parts of the body. (i) Surgery induces immune suppression while initiating an inflammatory cascade that provides cancer cells to propagate. (ii) In response to the trauma, the body secretes growth factors to facilitate the healing. Unfortunately, these same growth factors also stimulate tumor cell growth. (iii) Cancer spills into the bloodstream from the surgical margins and establishes metastatic colonies in other parts of the body. (iv) Cancer cells have a 'Velcro-like' surface that allows them to stick to each other and the blood vessel walls. In one experiment that mimicked surgical conditions, the sticking and binding of cancer cells to blood vessel walls increased by 250%. (v) Surgery reduces the natural killer cell activity. NK-cells' function is to gobble up cancer cells."

So any surgery must be monitored with extreme care and caution. Most patients should go through adjuvant therapy. Even then, more chemo and more surgeries could become a necessity. The body often could not tolerate all these, and the patient does not survive. The first author witnessed that a few times. If immunotherapy is done as a neoadjuvant therapy (before other treatments) before any surgery, it could be quite effective.

1.19   The Role of Stress

Stress is a killer. It is often the root cause of hundreds of diseases including breast cancer. In the second century A.D, Greek Scientist Galen [39] "noted a greater tendency for development of breast cancer among melancholic women than more sanguine traits." In around 400 BC, Hippocrates, the famous Greek Scientist and Philosopher, believed that by keeping both internal and external environments peaceful we can retain good health [39]. Stress destroys our peace and makes the body vulnerable to sickness. Then cells misbehave. That often causes mutations. In general, women are more susceptible to stress than men. In [39], it is stated that day-to-day stress does not affect our health in general. But a period of chronic stress does. It "increases the production of growth factors that increase blood supply. This can speed the development of cancerous tumors." Recently, a Swedish Study suggested

[41] that researchers found that "women who reported being under stress had twice the risk of developing breast cancer as women who managed to stay cool, calm, and collected. This twofold risk held up even when they took into account other factors that might explain the increased risk for breast cancer such as family history of cancer, alcohol use, body weight, smoking, and factors related to reproduction, such as age when women had their first baby, and the age they were they began menopause." There are several reports like this [54].

Diamond [42] stated "The dramatic atrophy of the thymus gland in a person under-going stress is not fully understood. Within a day of severe injury or sudden illness, millions of lymphocytes are destroyed and thymus shrinks to half its size." Selle [43] reported that acute stress caused thymic atrophy and this effect was mediated by the release of glucocorticoids from the adrenal glands. Stress suppresses splenic NK (natural killer cells) activities and reduces the number of lymphocytes, which promotes tumor growth.

Authors in [39] described more scientifically how the stress hormone glucocorticoids affect our immune system. Lymphocytes and macrophages, two most powerful sets of the army of the immune system, are severely suppressed from doing their anti-cancer activities. T-cells trained in our thymus gland, just above the heart, release a sequence of cytokines to destroy malignant cells. Macrophages phagocytize cancer cells. "Glucocorticoids inhibit the production of a wide range of cytokines (kind of poisonous chemicals, released by cytotoxic T-cells) and monokines including IL-1, IL-2, IL-6, and TNF-α" [44]. This results in a reduction in a number of activities of lymphocytes and macrophages. Glucocorticoid response to stress is also enhanced by estrogen. Thus estrogens are immunosuppressive. Another important observation has been recorded by Clancy [45]: "If stress occurs before an immune response, antigen may not be able to activate immune cells that have been suppressed by corticosteroids. On the other hand, if stress occurs during an immune response, the level of corticosteroids may be higher and remain elevated for a longer period of time, resulting in prolonged immune suppression of effector cell functions." A group of scientists from Japan [31] found that "Lymphotoxins (TNF-β) and estrogen, as well as glucocorticoids induce profound thymic involution" which suppress immune response very significantly. Unfortunately, it has been statistically observed that in comparison with men, "women feel stress more often" (Newsweek, June 14, 1999, A report by Ronald Kessler of Harvard University).

Thus we may state: Stress causes immunosuppression. Stress management could do immunostimulation. In our mathematical model in Chap. 6, we did include this significant information as a part of the immunotherapy [51].

1.20 Other Well-Known Causes

Some other causes (American Cancer Society [32]) for breast cancer are the following:

- Early menarche and late menopause make estrogen work in the body for a longer period.
- Drinking alcohol is a risk factor.
- Smoking.
- Having children after the age 30.
- Hormone therapy after menopause.
- Less physical activities.
- Obesity.

1.21   Some Perspectives of Our Models

Our models revealed quite encouraging qualitative agreements with what oncologists have found in their practice. Some of these are the following:

- If the tumor is detected at an early stage and it is still in situ, which means it did not metastasize, it is curable.
- After forming the tumor, malignant cells, even at an early stage, move toward the lymph nodes.
- After standard "cut, burn, and poison"-type treatment, cancer may completely vanish from the primary site and reappear at a distant location in the body at a later time or at the same time. Also, the immune system gets weakened after surgery and radiation allowing cancer to thrive and appear at distant locations with full strength. Most patients go through adjuvant therapy. Even then, more chemo and more surgeries become a necessity. The body seldom can tolerate all these, and the patient does not survive.
- If cancer is not treated at an early stage, it does spread.
- If immunotherapy is introduced as a neoadjuvant therapy (before other treatments), such that the immune system could recognize the cancer cells as "foreign" (non-self), then the soldiers of the immune system can keep on destroying them wherever they can find them and the patient may live a normal life. Vitamin C strengthens the immune system [56].
- Targeted treatments will be most effective if not just the tumors but all cancer cells at every point in the body are attacked simultaneously while reducing the rate of growth. Use of two drugs was considered. The first one is a Smart Combat Drug #1 (SCD1) which reduces the rate of growth of cancer cells at every point where they could be, stimulate the performance of the immune system, and destroy them at those locations. The second one is Smart Combat Drug #2 (SCD2) which improves the performance of the immune system and kills cancer cells. These are targeted drugs. It has also been found that another drug Skilled Killer Drug (SKD) could work as a silver bullet. It locates all cancer cells all over the body, and while most resiliently sticking to it, it destroys the enemy. Here, the concept of application of artificial intelligence has been considered.

These models are not meant to recommend any particular medicine or quantify the dosages or state what kind of CT scan or MRIs that medical professionals should prescribe. Only physicians are authorized to do these. However, the models could make proper assessments of the procedures undertaken by doctors to treat patients.

1.22 Conclusion

Our starting point is observing nature. In nature, there is always a balance of energy when chemical interactions take place. That means if there is an interaction between two chemicals A and B, and A gives energy to B, then A loses energy and B gains an equal amount of energy. In physiology however, if biochemicals interact then if B does not gain an equal amount of energy, it indicates that the body stores that amount of energy, possibly as fat for future use. Or in other words, the body gains that energy. Like a child becoming an adult. If there is a loss of energy, the body is aging or is sick. Mathematical biochemistry has to follow this regulation while constructing many models dealing with physiology.

In this text, all our models deal with equations following rules of biochemistry. A nice explanation is given in [46].

A word of caution: Even if all the tumors start shrinking and finally vanish, and a whole-body scan reveals no sign of accumulation of any foreign mass, cancer cells can still remain hiding inside adipose tissues, under the skin or at practically any part of the body for years. Cancer is that dangerous !!! So oncologists need to check the patient regularly. We witnessed that oncologists and other doctors do follow up their treatments long after patients start living healthy lives.

So, we have developed various models to treat cancer at different stages of the disease. We hope that medical professionals dealing with cancer treatment will find these models useful. The codes that we have developed are still preliminary codes. Yet, they are practical enough to help medical professionals find better procedures to treat cancer patients. The algorithms for numerical solutions could be easily parallelized. This is discussed in the Appendix C.

The overall picture of breast cancer has been studied here. We have tried to discuss most of the fundamental aspects of this disease with some details. Readers may add more aspects and go into much more detail and modify our models.

A very strong encouragement came first from Webb [47, 48], a superstar of mathematical biology from Vanderbilt University, Tennessee, who saw our mathematical models on metastatic breast cancer treatment developed at NASA Ames Research Center, consisting of several nonlinear reaction–dispersion equations, said "it is a very comprehensive model in this area," and invited the first author to give a seminar presentations at Vanderbilt University in 2001. Such encouragement from a top scientist was very helpful.

References

1. Magiorkinis, E., Petrogiannis, N., Bissias, C., Diamantis, A.: The concept of health in ancient greek medicine. Balkan Military Med. Rev. (2013)
2. Faguet, GB.: A brief history of cancer: age-old milestones underlying our current knowledge database. Int. J. Cancer **11**, August 2014
3. Edelstein-Keshet, L.: Mathematical Models in Biology. Birkhauser Mathematics Series. McGraw-Hill (1987)
4. Murray, J.D.: Mathematical Biology. Springer (1993)
5. Keener, J., Sneyed, J.: Mathematical Physiology. Springer (1998)
6. Williams, P.L. (ed.).: Gray's Anatomy, 38th edn (1999)
7. Adam John, A., Bellomo, N.: A Survey of Models for Tumor-Immune System Dynamics. Birkhauser (1996)
8. Stoll Basil, A. (ed.): Women at High Risk to Breast Cancer, Kluwer Academic Publishers (1991)
9. McKenna, M., et al.: precision medicine with imprecise therapy: computational modeling for chemotherapy in breast cancer. Transl. Oncol. **11**(3) (2018)
10. Dr. Susan Love Research Foundation: Estrogen Receptor (ER) and Progesterone Receptor (PR) Positive Breast Cancer
11. Banerjee, S., et al.: Heat shock protein 27 differentiates tolerogenic macrophages that may support human breast cancer progression. Cancer Res. (On line) (2011)
12. Ghajar, C., et al.: Is it possible to prevent breast cancer metastasis? www.fredhutch.org
13. Forys, L., et al.: Logistic equations in tumour growth modelling. Int. J. Appl. Math. Comput. Sci (2003)
14. Williams, H.: Oncologist, San Antonio, Texas. Pers Commun (2019)
15. Rachel, R.-D., et al.: A mathematical model of breast cancer treatment with CMF and doxorubicin. Bull. Math. Biol. **73**, 2011
16. Lange, C., Yee, D.: Progesterone and breast cancer. Women's Health **4**(2) (2008)
17. https://breast-cancer.ca/mammims/#:~:text=Scar%20tissue%20can%20often%20appear,likely%20to%20be%20breast%20cancer
18. Oncolytic Virus Therapy—Cancer Research Institute (CRI). http://www.cancerresearch.orghttp://www.cancerresearch.org
19. Study Points the Way for Future Therapy to Revive a Damaged Immune, April 6, 2012. System. https://www.mskcc.org/news/study-points-way-future-therapy-revive-damaged-immune-system
20. Ventevogel, M.S., Sempowski, G.D.: Thymic rejuvenation and aging. Curr. Opin. Immunology **25**(4) (2013)
21. Garaci, E., et al.: Thymosin alpha 1 in the treatment of cancer: from basic research to clinical (2000). https://www.pubmed.ncbi.nim.nih.gov
22. Abusarah, J., et al.: Thymic rejuvenation: are we there yet? In: Chapter 3, Gerontoloy. IntechOpen (2018)
23. https://icnr.com/
24. Halberg, N.: NCMM young associate investigator. The Department of Biomedicine, University of Bergen
25. Chithra Devi, M., Audithan, S.: Analysis of different types of entropy measures for breast cancer diagnosis using ensemble classification. Res. Art.—Biomed. Res. **28**(7) (2017)
26. Messerschmidt, J.L., Prendergast, G.C., Messerschmidt, G.L.: How cancers escape immune destruction and mechanisms of action for the new significantly active immune therapies: helping non immunologists decipher recent advances. Oncologist **21**(2) (2016)
27. https://blog.dana-farber.org/insight/2017/06/catch-me-if-you-can-finding-cancer-cells-that-hide-in-plain-sight/
28. American Cancer Society: Breast Cancer Facts & Figures 2017–2018
29. Women in Balance Institute. Portland, Oregan (2019)

30. Hollstein, M., Sidransky, D., Vogelstein, B., Harris, C.C.: Science **253** (1991)
31. American Cancer Society, Breast Cancer, report 2014
32. How Stress Affects Cancer Risk, A Report. MD Anderson Cancer Center, Texas, December 2014
33. Franks, L.M., Teich, N.M.: Introduction to the Cellular and Molecular Biology of Cancer, 3rd edn. Oxford University Press, Oxford (1997)
34. McCrudden, C.M., McCarthy, H.O.: Current status of gene therapy for breast cancer: progress and challenges. Appl. Clin. Genet. **7**(2014)
35. Stoff-Khalili, M.A., Dall, P., Curiel, D.T.: Gene therapy for carcinoma of the breast. Cancer Gene Therapy **13**(7) (2006)
36. Chigvintsev, A., Marino, S., Kirschner, D.E.: A mathematical model of gene therapy for the treatment of cancer. http://mathus.micro.med.umich.edu/lab/pubs/Tsygvintsev_et_al.2012.nf.pdf
37. Chen, T., et al.: Modeling gene expression with differential equations. In: Pacific Symposium of Biocomputing (1999)
38. Dey, S.: Mathematical modeling of gene expression (invited lecture). In: International Conference on Gene Expression, Institute of Chemical Biology, Kolkata (2005)
39. Plotnikoff, M., Faith, W.: Stress & Immunity. CRC Press (1991)
40. Why fat matters. Cancer Today. https://www.mag.org/. Winter 2017–2018
41. Does Stress Cause Breast Cancer ? WebMD, September 2003
42. Diamond, J.: Behavioral Kinesiology, How to Activate Your Thymus and Increase Your Life Energy. Harper & Row, Publisher, New York (1979)
43. Selle, H.: (Plotnikoff [11], p. 438). Stress & Immunity, CRC Press (1991)
44. Hirahara, et al.: Glucocorticoid independence of acute thymic involution induced by lympho-toxin and estrogen. Cell. Immunol. **153**(2) (1994)
45. Clancy, J.: Basic Concepts in Immunology. The McGraw Hill (1998)
46. Chauviere, A.H., Hatzikirou, H., Lowengrub, J.S., Frieboes, H.B., Thompson, A.M., Cristini, V.: Mathematical oncology: how are the mathematical and physical sciences contributing to the war on breast cancer? Curr. Breast Cancer Rep. **2**, 121–129 (2010)
47. Webb, G.: Mathematics Department of Vanderbilt University. Pers. Commun. (2002)
48. Fitzgibbon, W.E., Webb, G.F.: A career in mathematics. Math. Model. Nat. Phenom. **3**(7) (2008)
49. www.breastcancer.org. U.S. Breast Cancer Statistics
50. Guyton, Hall: Textbook of medical physiology. www.amazon.com
51. Does Stress Cause Breast Cancer? WebMD, September 2003
52. Women in Balance Institute. Oregan, Portland (2019)
53. Weil, A.: Spontaneous Healing. www.publishersweekly.com
54. Smell, L., Graham, S.: Social trauma as related to cancer of the breast. Br. J. Cancer (1971)
55. Linos, E., Willett, W.C., Cho, E., Frazier, L.: Adolescent diet in relation to breast cancer risk among premenopausal women. Cancer Epidermiol Biomakers Prev. **19**(3) (2010)
56. Cameron, Pauling, L.: Cancer and Vitamin C. Linus Pauling Institute of Science and Medicine, Menlo Park (1999)
57. Kelly, C.: Forcing the immune system to attack cancer. Alliance Adv. Bio Eng. July 24, 2017
58. Rivlin, N., Brosh, R., Oren, M., Rotter, V.: Mutations in the p53 tumor suppressor gene: important milestones at the various steps of tumorigenesis. Genes Cancer **2**(4) (2011)
59. Chauviere, A.H., Hatzikirou, H., Lowengrub, J.S., Frieboes, H.B., Thompson, A.M., Cristini, V.: Mathematical oncology: how are the mathematical and physical sciences contributing to the war on breast cancer? Curr. Breast Cancer Rep. **2**(3) (2010)
60. Koujan, S.E., et al.: Matrix metalloproteinases and breast cancer. Thrita (2015)
61. https://news.cancerresearchuk.org/2020/10/20/sugar-and-cancer-what-you-need-to-know/

Chapter 2
Statistics: The Background and the Basis

Abstract Cancer cells do not always follow the same pattern of attack. They often display a very heterogeneous nature of growths and movements confusing the body and confusing even expert doctors. Uncertainty is the very nature of certainty of cancer. So, principles of statistics are often applied to study their behavior. We have discussed here some statistical principles to analyze and understand treatments of cancer. Before and after solving mathematical models computationally, medical professionals often apply statistical principles to analyze data and estimate their validity and applicability.

2.1 Rationale

Mathematical modeling works quite well in the world of physical science, because rules and regulations are rather uniform. In biological sciences, the rules that guide us today could fool us tomorrow. Scientists often say that they feel like "Alice in Wonderland." Diversities are uniformly present almost everywhere that confuse even expert scientists. Here statistics come into play.

In the most simplistic form, we often say that statistics is the study of averages and their distributions. Samples are drawn, and mathematics comes in to look into what these samples describe now and beyond. The aim of statistics is to predict population by studying samples. This means analysis and organization of huge amounts of data.

On the basis of such analysis, we have modeled our studies on cancer treatment, especially breast cancer treatment. Statistics is one of our primary guides our leading guides to conduct this research.

Breast cancer treatment is an aspect of medical science governed by biochemistry which must follow the principles of physiology. The laws of physiology are very different from the laws of physics and chemistry. The state of psychology is one of the primary guides of practically any state of physiology. So, any aspect of physiology should be guided by keen observations and experiments. A major tool that mathematicians have to use to investigate any phenomenon in physiology is statistics. When a mathematician decides to look into a model in physiology by using certain aspects of biochemistry, statistical observations and explanations are

© Springer Nature Singapore Pte Ltd. 2021
S. Dey and C. Dey, *Mathematical and Computational Studies on Progress, Prognosis, Prevention & Panacea of Breast Cancer*, Forum for Interdisciplinary Mathematics, https://doi.org/10.1007/978-981-16-6077-1_2

applied. In this chapter, we will look into some fundamental laws of statistics applicable in our studies. These could help the readers better understand many aspects of our mathematical models on the treatments of breast cancer.

The number of cancer survivors is on the rise every year. Yet much more work needs to be done. It is often said that "Cancer is Darwinian." In this regard, we may refer to: https://academic.oup.com/nsr/article/5/1/15/4815783 and https://doi.org/10.15252/embj.2021108389.

Also, most Darwinian populations may be described with some applications of logical formulas. This is practically impossible when it comes to cancer. Christopher Benz of the Buck Institute for Age Research in Novato, California, says that "no insights from evolution should be accepted until they are put to an experimental test the way any other hypothesis would be." This should be applicable to the growth of cancer cells. And that fails here because there are too many unknown factors which drive cancer. So, oncologists always change strategies to fight against this disease [Dr. Ghafoori, Austin Cancer Center]. They always wonder what comes next, which is often not easy, and could be very hard to determine.

A mathematician's insight into cancer shows that there is a revolution in an evolution and an evolution in a revolution. Body's destructive power, by virtue of mutations of cells, suddenly rises up and moves forward with some very calculated paradigms and some precise protocols to annihilate itself. Defense fails to save the body often in any orderly way. This may be attributed to Darwin's theory of "survival of the fittest." But this could happen all on a sudden, statistical scales failing for all predictions. Some scientists believe that regardless of better and better treatments cancer cells will evolve and come back with newer mutations and survive. Scientists and mathematicians are determined to stop this Darwinian process. An attempt has been made in this regard in Chap. 8.

When we find hundreds of forms of cancers with widely different rates of growth, then Darwinian concepts literally evaporate because all predictions are nullified. For instance, for a lung cancer tumor to grow large enough to be detected by X-rays, a single cancer cell must divide (i.e., double in size) at least 30 times. This will put it at just under a half an inch (or one centimeter) in dimension. While this division may not seem like much, try this. Using a calculator, multiply 1×2. Then, do 2×2. Then, multiply 4 by 2 again and keep doing this 30 times [https://kymeramedical.com/how-quickly-does-cancer-grow]. That is how the tumor is growing. Fortunately, this is not always true. We have also seen a lung cancer that grew like:

$$f(t) = a \, \exp((\alpha(t) - \beta)), \text{ where } \alpha(t) \ll \beta,$$
$$for \ 0 \le t < T_0, \text{ and } \alpha(t) \gg \beta \text{ for } t \ge T_0. \ \beta > 0, \ a > 0.$$

This means as t (time) increases, $\alpha(t)$ suddenly starts increasing fast after $t \ge T_0$.

We must note that so long a and $\alpha'(t)$ are of the same sign then $f(t)$ is an increasing function because $df/dt > 0$.

In Darwin's theory, species grow and subdivide with some uniformity. No principle of growth, no deterministic factor, no nothing can predict the strategies that different cancer cells adopt as they grow and attack the body.

Possibly, the simplest example is: Let us think of a lady coming out of her home and starts jogging. While jogging she became a gorilla. Soon after she started having babies who look like lion cubs. Finally, she looks like an angel. This is how a malignant tumor often behaves. That makes it extremely difficult to make any proper predictions regarding treatment procedure.

Not necessarily this is their strategy of survival. Sometimes, they do it because this is a game they play even though the patient is no longer resisting. All drugs become useless. This is certainly survival of the fittest but no evolution. Unpredictability is the only predictability.

Every mathematical equation that we have derived is only statistically valid. Bioscience of cancer is not an exact science like Physics. Behavior of cancer is possibly the most chaotic in the world of science. It often baffles expert oncologists and statisticians.

As an example, I would like to mention a case where at the end of August 1988 at Dayton Hospital a patient was thoroughly examined for a tumor in the brain and it showed up nothing. In early November, an MRI scan in the DePaul Hospital at St. Louis showed nothing positive. Still brain surgery was done!!! Even that surgery found nothing. But in early January, the same brain surgeon found a golf ball size tumor in the brain. It already spread all over the bones. This is cancer!!! Changes are revolutionary. Often the most non-Darwinian in nature.

Cancer carries too many unknown factors which change stochastically following no regulation. Often with no links between their sudden changes. "We have discovered that some microRNAs, a group called microRNA-200S, undergo chemical inactivation and inhibit their expression. When these cellular appearance drivers are not present, tumour cells change, stretch, stop their inhibition and thus the tumour progresses," explains Dr. Esteller, adding that "the results from research show that this is a very dynamic process" [8]. There is no evolutionary process that could predict when and how "drivers are not present." Cancer cells change their strategies continuously that often make the old techniques of treatments outdated and obsolete. Many of them can survive "poison and burn." These stark uncertainties outweigh and outperform any evolutionary process. They have made almost all mathematical models related to Darwinian predictions [13–15] totally invalid. There is no so-called smooth transition of cancer from one state to the other. No pattern could match their pattern of growth. Patterns of aggression and patterns of attack are all random and unpredictable to some extent. For instance, cancer will attack which breast first cannot be predicted nor how it will grow and metastasize. Thus, statistical search to look for protocols for better understanding of cancer and for newer treatments is always very much needed.

Statistics is possibly the most useful tool to study the behaviors of biological phenomena, applying mathematical modeling where events happening inside the body cannot be properly understood by the rules of physical sciences. In statistics, we work with stochastic events happening in stochastic spaces. They are random variables. The primary objective is to find out, on the average, underlying rules of causes and effects connected with an event under investigation. To do this, we collect data dealing with the event, analyze them, and make predictions about the effects as outcomes, even though the events and their effects are happening as random variables in a stochastic space, in a microenvironment. Mathematics comes in to shape these protocols analytically in as much detail as possible. These mathematical techniques of developing protocols have vast applications in medical science and biology.

If data are collected from a homogeneous group connected with an event under investigation, statistical inferences become quite effective. In physical sciences, engineering, and business, such homogenous groups of data are generally available. In biological sciences, however, such homogeneity is hard to find. Two human bodies are not the same. For exactly the same ailments of two persons, the exact same medications may give entirely different results, because the mechanisms of a human body are often strongly controlled by the mind which is infinitely complicated. There exists no straightforward algorithmic pattern which could connect mind with body, yet there exists a very strong rapport between mind and body in every aspect of our actions and reactions. So, statisticians often face an uphill challenge [1–3] while making statistical inferences in biology and medicine. Data collection and data organization are very intriguing tasks in this regard.

It is extremely difficult to observe the behavior of an individual cancer cell and generalize such observations. Some basic facts about cancerous tumors are: A 1-cm tumor has about *100 million* cells, a 0.5-cm tumor has about *10 million* cells, and a 1-mm tumor has about *100 thousand* cells. They group together to form tumors. They move in a group in blood and lymph so that they could fight together, in case they are detected. They release a number of proteins to confuse the body's defense. They hide in fatty tissues. They can even hide their foreign antigens. What they can and cannot do is hard to understand. However, cancer is not contagious. It is a genetic disease. It begins with mutations of one or more genes.

In any mathematical modeling of treatment, the most important parameters are estimations of the number of cancer cells, sizes and locations of tumors, the assembly of the forces of defense, their effectiveness, medications and their effectiveness, their antigens [9], their other significant characteristics [10–12] etc. All these should be estimated through statistical studies.

Doctors first do some medical tests to confirm that the disease has happened. They test the vitals, like body temperature, blood oxygen level, blood pressure, heart beats, etc., followed by several blood tests, X-rays, CT scans, MRIs, ultrasounds, angiograms, and biopsy. Some tests could be positive, and some could be negative. But it could be a false positive or a false negative. These results affect the psychology of a patient. This psychology plays a very significant role regarding the state of health of a patient. We will first study some statistical analysis of test results. Such studies begin with organization and analysis of the available data and treatments for better prognosis of the disease. Mathematically, data analysis is conducted applying some principles of the set theory. Let us discuss some basic concepts.

2.2 Some Preliminary Concepts on Sets

Let us consider a set $S = \{-1, 0, 1, 2, 3, 4\}$. If A is the set $A = \{0, 1, 3, 4\}$, then all elements of A are in S. We say that A is a subset of S. We write $A \subset S$. If B $= \{-1, 2\}$, then $B \subset S$. Obviously, A and B have no element in common. We say A and B are mutually exclusive or disjoint, meaning that they have no element in common. Two mutually exclusive events cannot occur together. Like night and day cannot occur at the same time at the same place.

Furthermore, if we join A and B together, that becomes S. We say A and B are complements of each other and write: $A \cup B = S$. Also, since A and B have nothing in common, $A \cap B = \emptyset$ (a null set).

If $C = \{0\}$, C is not null and C is not the complement of B, because $C \cup B \neq S$.

C has only one element. Also, C and B are mutually exclusive, while $C \subset A$, as well as $C \subset S$. If $D = \{\}$, then D has no element in it. D is a null set. D is a subset of all sets. We may write $D = \emptyset$.

If S is a set with m number of elements, then $n(S) =$ the number of elements of $S = m$. So from the above example, $n(A) = 4 =$ number of elements of A, $n(B) = 2$ = number of elements of B, $n(C) = 1$, and $n(S) = 6$.

2.2.1 Cardinality

In cardinality, we count only the unique elements of a set. If $A = \{1, 2, 3, 4\}$, then all elements of A are unique and none is repeated. Then, Card $(A) = 4$. If $B = \{1, 1, 1, 1, 1\}$, then only one element is unique. Others are repetitions. Here, Card $(B) = 1$. If $C = \{1, 2, 3, 1, 2, 3, 1, 2, 3, 4\}$, then Card $(C) = 4$, because only four elements are unique. Only Card (Null Set) $=$ Card$(\{\emptyset\}) = 0$, because it contains no element. $\emptyset$ is a symbol of null.

The concept of cardinality is useful for data organization. If some data on a particular event are repeated, then we deal with those data in statistics in a special way where cardinality plays an important role.

2.2.2 Union and Intersection of Sets

We will discuss now some fundamental properties of unions and intersections of sets.

(i) $A \cup B$ means the operation A. OR. B. It means if A or B, any one of them is true $A \cup B$ is true, like two blood tests. If any one of them shows an abnormality, then the test $A \cup B$ is abnormal.

(ii) $A \cap B$ means the operation A. AND. B. Here, both must be true so that $A \cap B$ will be true, like chemotherapy and some nutritional therapy. If both worked well, then we conclude that the treatment goes well.

If A and B are disjoint sets, the operation $A \cap B$ gives a null set, because both cannot occur together at the same time when sets are time dependent, like some surgery and chemo. I noticed doctors performing them at the same time.

(iii) $n(A \cup B) = n(A) + n(B) - n(A \cap B)$, because $n(A \cap B)$ is in both A and B. So it is added twice in $n(A) + n(B)$.

A and B are disjoint if no element is common between A and B. Then,

(iv) $n(A \cup B) = n(A) + n(B)$ and $n(A \cap B) = 0$, like the number of radiation therapies and the number of nutritional therapies when nutritional therapies are going on and the patient does not go through any radiation treatment.

2.2.3   Probability Using Sets

An Example:

Let A = a set of patients under treatment plan#1, and B = a set of patients under treatment plan#2. Let 50% of patients in A be under the plan #1, 50% of B be in plan#2, and 30% of the patients be under both plans, then what is the probability that if a patient is chosen arbitrarily, she belongs to either A or B.

$$P(A.OR.B) = P(A \cup B) = P(A) + P(B) - P(A \cap B) = (50 + 50 - 30)\% = 70\%.$$

For a number of treatments A, B, C, we have:

$$P(A \cdot OR \cdot B \cdot OR \cdot C) = P(A \cup B \cup C) = P(A) + P(B) + P(C) - P(B \cap C)$$
$$- P(C \cap A) - P(A \cap B) + P(A \cap B \cap C).$$

The general formula is:

$$P\left(\cup_{i=1}^{n} A_i\right) = \sum_{i=1}^{n} P(A_i) - \sum_{1 \le i < j \le n}^{n} P\left(A_i \cap A_j\right)$$
$$+ \sum_{1 \le i < j < k \le n}^{n} P\left(A_i \cap A_j \cap A_k\right) - \cdots + (-1)^{n-1} P\left(\cap_{i=1}^{n} A_i\right). \tag{2.1}$$

The notation: $\left(\cup_{i=1}^{n} A_i\right)$ means $A_1 \cup A_2 \cup A_3 \cup \ldots \cup A_n$.

A simple application is let a doctor conduct several different procedures for treatment of a patient and he is aware of the odds regarding how each treatment plan works as well as how they work in combination. But he is anxious to find out if all of these procedures are applied and how it will work. Suppose he has 4 such procedures. Let these be A_1, A_2, A_3, and A_4, respectively. Then, the probability that all will work or at least one will work is:

$$P(A_1 \cup A_2 \cup A_3 \cup A_4) = \sum_{i=1}^{4} P(A_i) - \sum_{1 \le i < j \le 4}^{4} P\left(A_i \cap A_j\right)$$
$$+ \sum_{1 \le i < j < k \le 4}^{4} P\left(A_i \cap A_j \cap A_k\right) - P\left(\cap_{i=1}^{4} A_i\right) \tag{2.2}$$

And none will work: $P(None\ will\ work) = 1 - P(A_1 \cup A_2 \cup A_3 \cup A_4))$. Sometimes two sets may be disjoint, yet there could be some set sharing information belonging to both sets. For example, there may be two entirely different diet plans A and B which have no food in common. If both plans are tried on a group of breast cancer patients G, then so far diet plans are concerned, G belongs to both A and B. We will see such cases in Chap. 9.

2.2.4 Conditional Probability

Let us now go back to our previous discussions in Sect. 2.2. The universe was the set $S = \{-1, 0, 1, 2, 3, 4\}$ and n(S) = number of elements of S = 6. Then P(S) = probability of S = 6/6 = 1. Also P(A) = n(A)/n(S) = 4/6 = 2/3. Note that the probability of any event can never exceed 1.

Also $P(A) + P(A') = 1$.

A' is a subset of S such that any element of A' is not an element of A and $A \cup A' = S$. This is the property of any complement set.

Now let us understand the concept of conditional probability. Let A and B be subsets of S, where $S = \{-2, -1, 0, 1, 3, 4, 5\}$, $A = \{0, 1, 4, 5, 3\}$, and $B = \{-1, 0, 1\}$. The elements 0 and 1 are common to both A and B. We write $A \cap B = \{0, 1\}$. Then $P(A \cap B) = 2/7$. P(B/A) = probability of B assuming that A has happened. Here, A is considered to be the universe. So, $P(B/A) = n(A \cap B)/n(A) = 2/5$. Note: When we assume A as the universe, we cannot use all the elements of B. We must use only those elements which are common to both A and B. This is also equal to $\{n(A \cap B)/n(S)\}/\{n(A)/n(S)\} = P(A \cap B)/P(A)$. Thus,

$$P(B/A) = P(A \cap B)/P(A).$$

Hence $P(A \cap B) = P(A) P(B/A)$. Similarly, $P(A \cap B) = P(B) P(A/B)$.

So, $P(A) P(B/A) = P(B) P(A/B)$.

For $P(A \cap B \cap C)$, let us assume $A \cap B = D$. So, $P(A \cap B \cap C) = P(D \cap C) = P(D)P(C/D)$.

$= P(A \cap B)P(C/(A \cap B)) = P(A)P(B/A)P(C/(A \cap B))$.

So, in general, if $A = \bigcap_{i=1}^{n} A_i$, then

$$P(A) = P(A_1)P(A_2/A_1)P(A_3/A_1 \cap A_2)P(A_4/A_1 \cap A_2 \cap A_3) \ldots P\left(A_n/ \bigcap_{i=1}^{n-1} \ldots A_i\right) \quad (2.3)$$

This is the basis of the so-called tree diagram used in probability.

When a patient goes through several diagnostic tests, the level of anxieties of her and her near and dear ones gets higher and higher.

Let us consider a simple example.

Ex. 1. Let there be three diagnostic tests $T_1, T_2, and T_3$. Let T_i^+ means the test $T_i (i = 1, 2, 3)$ is positive and T_i^- means it is negative. The doctor estimated that:
$P(T_1^+) = 0.8$, $P(T_2^+/T_1^+) = 0.85$ and $P(T_3^+/T_1^+ \cap T_2^+) = 0.9$, also $P(T_2^+/T_1^-) = 0.4$,

$$P\left(T_3^+/T_1^+ \cap T_2^-\right) = 0.6,\ P\left(T_3^+/T_1^- \cap T_2^+\right) = 0.3 \text{ and } P\left(T_3^+/T_1^- \cap T_2^-\right) = 0.2$$

With these, we can find:

$$P\left(T_1^+ \cap T_2^+ \cap T_3^+\right) = P\left(T_1^+\right)P\left(T_2^+/T_1^+\right)P\left(T_3^+/T_1^+ \cap T_2^+\right) = 0.8 \times 0.85 \times 0.9 = 0.612$$

$$P\left(T_1^+ \cap T_2^+ \cap T_3^-\right) = P\left(T_1^+\right)P\left(T_2^+/T_1^+\right)P\left(T_3^-/T_1^+ \cap T_2^+\right) = 0.8 \times 0.85 \times 0.1 = 0.068$$

$$P\left(T_1^+ \cap T_2^- \cap T_3^+\right) = P\left(T_1^+\right)P\left(T_2^-/T_1^+\right)P\left(T_3^+/T_1^+ \cap T_2^-\right) = 0.8 \times 0.15 \times 0.6 = 0.072$$

$$P\left(T_1^+ \cap T_2^- \cap T_3^-\right) = P\left(T_1^+\right)P\left(T_2^-/T_1^+\right)P\left(T_3^-/T_1^+ \cap T_2^-\right) = 0.8 \times 0.15 \times 0.4 = 0.048$$

$$P\left(T_1^- \cap T_2^+ \cap T_3^+\right) = P\left(T_1^-\right)P\left(T_2^+/T_1^-\right)P\left(T_3^+/T_1^- \cap T_2^+\right) = 0.2 \times 0.4 \times 0.3 = 0.024$$

$$P\left(T_1^- \cap T_2^+ \cap T_3^-\right) = P\left(T_1^-\right)P\left(T_2^+/T_1^-\right)P\left(T_3^-/T_1^- \cap T_2^+\right) = 0.2 \times 0.4 \times 0.7 = 0.056$$

$$P\left(T_1^- \cap T_2^- \cap T_3^+\right) = P\left(T_1^-\right)P\left(T_2^-/T_1^-\right)P\left(T_3^+/T_1^- \cap T_2^-\right) = 0.2 \times 0.6 \times 0.2 = 0.024$$

$$P\left(T_1^- \cap T_2^- \cap T_3^-\right) = P\left(T_1^-\right)P\left(T_2^-/T_1^-\right)P\left(T_3^-/T_1^- \cap T_2^-\right) = 0.2 \times 0.6 \times 0.8 = 0.096$$

An easy way to validate the results is if we add them up it will be exactly 1, because we have taken care of all the possibilities.

Ex. 2. Let there be two medications for breast cancer treatment to be taken, one after another. Let M_1 means 70% chance that some patients may take it and M_2 means 80% may take it with taking M_1. What is the probability that a patient will take (1) both and (2) neither. (Obviously these are not all the possibilities.)

Ans: (1) $0.7 \times 0.8 = 0.56$; (2) $0.3 \times 0.2 = 0.06$.

2.2.5   Independent Events

Let X= the event that "I will pass my history test" and Y= the event that "Germany will win the world cup." These two events can certainly occur at the same time. However, the occurrence of one does not affect the occurrence of the other. So they are two independent events. Or in other words, X does not affect Y and Y does not affect X. So, P(X/Y) = P(X) and P(Y/X) = P(Y). Then, P(X∩Y) = P(X) P(Y/X) = P(X)P(Y). This means that the chance that both will occur together is smaller than the chances that they occur alone independent of each other.

2.2.6   Mutually Exclusive vs. Independent Events

In this regard, we will mention about mutually exclusive events once again. They are not independent, because, for independent events, the occurrence of one event does not affect the occurrence of the other. So independent events can occur together. On the contrary mutually exclusive events cannot occur together. So, if X and Y are mutually exclusive, $P(X \cap Y) = 0$. So $P(X/Y) = P(Y/X) = 0$ (because if X happens, Y cannot happen and vice versa). If we toss a coin, occurrence of a head guarantees that a tail did not happen. They are mutually exclusive. But if we toss two coins and a head is the outcome of tossing the first coin, then it has no effect on the outcome of tossing the second coin. They are independent.

Often students also asked me difference between disjoint sets and complementary sets. Two sets are complementary if and only if their union is the universe. This is not valid for any two mutually exclusive sets unless they are complementary. For example, let S= {1,2,3,4}, A= (1,2), B= {3,4} and C = {4}. S is the universe. With regard to S, A and B are complementary sets (obviously they are disjoint), A and C are disjoint (mutually exclusive), but not complementary.

Ex. 1. If X = sky above me is overcast by dark clouds and Y= sky above me is sunny, then X and Y are mutually exclusive. They cannot occur together. They are mutually exclusive. If one occurs, the other cannot occur at the same time.

Ex. 2. In medical science, most tests are often neither mutually exclusive nor independent events. Like in the case of cancer, blood tests, angiograms, MRIs, and biopsies may all be connected with each other. However, an ultrasound test for prostate malfunction and a test for hypothyroidism are generally independent of each other, because both can happen together, independent of each other.

Similarly, if T_1^- = a test that MRI is negative for breast tumor and T_2^- = a test that shows that the patient is not diabetic, then these two tests are independent, because both can occur at the same time. They are not mutually exclusive.

Ex. 3. In a hospital, 60% of the patients having some breast ailments were tested negative for breast cancer through MRI and 70% were found to be negative through blood test for diabetes. If a patient is chosen at random what is the probability that (i) both tests will be negative, (ii) at least one will be negative, and (iii) none will be negative.

Ans. Let T_1^- = MRI is negative and T_2^- = no diabetes. Then, $P(T_1^-) = 0.6$, $P(T_2^-) = 0.7$. These two events are independent. So, (i) P(both will be negative) = $P(T_1^- \cap T_2^-) = 0.6 \times 0.7 = 0.42$.

(ii) P(At least one test will be negative) = $P(T_1^- \cup T_2^-) = P(T_1^-) + P(T_2^-) - P(T_1^- \cap T_2^-) = 0.6 + 0.7 - 0.42 = 0.88$. (iii) P(None will be negative) = 1−P(At least one will be negative) = $1 - 0.88 = 0.12$.

2.3 Bayes' Formula

A sample space S is generated by collections of samples of events $A_1, A_2, \ldots, A_n$. Let E be an event such that

$$E = (E \cap A_1) \cup (E \cap A_2) \cup (E \cap A_3) \cup \ldots \cup (E \cap A_n) \qquad (2.4)$$

This implies that all the events $(E \cap A_i)$, $i = 1, 2, \ldots n$ are mutually exclusive.

Then, $P(E) = \sum_{i=1}^{n} P(E \cap A_i). = \sum_{i=1}^{n} P(A_i)P(E/A_i)$.
This gives

$$P(A_i/E) = P(A_i)P(E/A_i)/P(E) = P(A_i)P(E/A_i)/\left(\sum_{i=1}^{n} P(A_i)P(E/A_i)\right). \quad (2.5)$$

This is Bayes' formula. In statistical models, it has vast applications.

$P(A_i)$ is called a priori probability of A_i. How it will affect E is unknown. $P(A_i/E)$ is called a posteriori probability because we have figured out how it works in our situation.

Ex. 1. John has been examined, and it is found that he has cancer. He worked at three different industries $A_1, A_2,$ and A_3 at different times under some degree of carcinogenic environments. 35% of the time he worked for A_1, 25% of the time he worked for A_2, and the rest for A_3. Statistics show that 1.02% workers at A_1, 1.08% of the workers at A_2, and 0.5% of the workers at A_3 did get cancer. What are the probabilities that John got cancer working at $A_1, A_2,$ and A_3, respectively.

Ans. Let $C =$ cancer. We have the following data: Since John worked 35% of the time at A_1 and that could have caused the disease, $P(A_1) = 0.35$. However, statistics showed that $P(C/A_1) = 1.02\% = 0.0102$. Similarly, $P(A_2) = 0.25$, $P(C/A_2) = 0.0108$ and $P(A_3) = 0.4$ and $P(C/A_3) = 0.005$.

So, $P(C) = P(C \cap A_1) + P(C \cap A_2) + P(C \cap A_3) = P(A_1)P(C/A_1) + P(A_2)P(C/A_2) + P(A_3)P(C/A_3) = 0.00357 + 0.0027 + 0.002 = 0.008427$.

Then applying Bayes' rule: $P(A_1/C) = P(C \cap A_1)/P(C) = 0.00375/0.00827 = 0.4534$

$P(A_2/C) = P(C \cap A_2)/P(C) = 0.0027/0.008427 = 0.3204$
$P(A_3/C) = P(C \cap A_3)/P(C) = 0.002/0.008427 = 0.2373$

So, there is a 45% possibility that he got it working at A_1, 32% chance that he got it at A_2, and 24% chance that he got it at A_3.

Ex. 2. A person lived 6 years in the city A_1, 5 years in A_2, 7 years in A_3, and 9 years in A_4. These cities have high rates of pollutants which are well known to be carcinogenic. In A_1, it is 2.35%, in A_2 it is 3.5%, in A_3 it is 1.5%, and in A_4 it is 1.04%. The doctors suspected that one of these cities could be the primary cause for his cancer. Which city could it be ?

Ans. In all this person lived a total of 27 years in all those 4 cities. 6 years in A_1 means 22% of time, 5 years in A_2 means 19% of time, 7 years in A_3 means 26% of time, and 9 years in A_4 means 33% of time.

$$P(C) = 0.00235 \times 0.22 + 0.0035 \times 0.19 + 0.0015 \times 0.26 + 0.00104 \times 0.33$$
$$= 0.001182$$

$$P(A_1/C) = (0.00235 \times 0.22)/0.001182 = 0.44.$$

$$P(A_2/C) = 0.56, P(A_3/C) = 0.33, P(A_4/C) = 0.29$$

2.4 Testing Tumors: Specificity and Sensitivity and Applications of Bayes' Formula

Detection of any disease is first conducted by using several medical tests. With regard to testing breast cancer the gold standard, as spoken by oncologists, is biopsy. Even then, to know the type and the stage of the disease, several tests are performed.

Let D^+ = the disease is present and D^- = the disease is not present, and T^+ = the test is positive and T^- = the test is negative.

Then, doctors call $P(T^+/D^+)$ = *Probability of a positive test assuming that disease is present* = the sensitivity test and $P(T^-/D^-)$ = *Probability of a negative test assuming that disease is absent* = specificity test.

So, sensitivity test shows the probability that a test is positive, assuming that the disease is present, and specificity shows the probability that the test is negative, assuming that the patient is free from the disease. However, a patient needs to know what are $P(D^+/T^+)$ and $P(D^-/T^-)$, or in other words, a patient would like to know if the test is positive, what is the chance that she has the disease and if the test is negative, what is the chance that she does not have this disease.

$$\text{Let } \xi = \text{the stress parameter} = P(D^+/T^+) \text{ and}$$
$$\zeta = \text{the relief parameter} = P(D^-/T^-).$$

The stress parameter measures the probability for contracting the disease when the test is positive. That is the same as the probability that the test is a true positive.

The relief parameter is the probability that the test is a true negative. However, the probability that the test is a false negative also causes stress and the probability that the test is a false positive is also a measure of relief.

Let $p = P(D^+)$, then in medical science, q = sensitivity = $P(T^+/D^+)$ and r = specificity = $P(T^-/D^-)$. Also, $P(D^-) = 1-p$, $P(T^-/D^+) = 1-q$ and $P(T^+/D^-) = 1-r$.

Following the law of conditional probability, $P(A)P(B/A) = P(A \cap B)$, we get

$$P(T^+)P(D^+/T^+) = P(D^+ \cap T^+), \text{ and } P(T^+) = \{P(D^+ \cap T^+) + P(D^- \cap T^+)\}.$$

So,

$$\begin{aligned}
\xi = P(D^+/T^+) &= P(D^+ \cap T^+)/P(T^+) = P(D^+ \cap T^+)/ \\
&\quad \{P(D^+ \cap T^+) + P(D^- \cap T^+)\} \\
&= P(D^+)P(T^+/D^+)/\{P(D^+)P(T^+/D^+) + P(D^-)P(T^+/D^-)\} \\
&= pq/\{pq + (1-p)(1-r)\} \qquad\qquad\qquad\qquad\qquad (2.6)
\end{aligned}$$

Similarly,

$$\begin{aligned}
\zeta = P(D^-/T^-) &= P(D^- \cap T^-)/P(T^-) = P(D^- \cap T^-)/ \\
&\quad \{P(D^- \cap T^-) + P(D^+ \cap T^-)\} \\
&= P(D^-)P(T^-/D^-)/\{P(D^-)P(T^-/D^-) + P(D^+)P(T^-/D^+)\} \\
&= (1-p)r/\{(1-p)r + p(1-q)\} \qquad\qquad\qquad\qquad (2.7)
\end{aligned}$$

If the values p, q, r are known (given by pathologists), a patient can easily compute these *parameters* ξ *and* ζ.

For example, if $p = 0.125$ (*because* 1 *out of* 8 *could get breast cancer*), $q = 0.98$, *and* $r = 0.87$, then $\xi = 0.52$. So, there is a 52% chance that she has the disease and 48% chance that it is a false positive.

Also, $\zeta = 0.9967$. That means if the test is negative, then there is a 100% chance that she does not have the disease.

There is another simpler way to find these results by applying the available statistics as follows:

Let us consider statistics of a group of patients on whom a test has been performed. Let T be that test.

Let $T^+ =$ test is positive and $T^- =$ test is negative. Then, T^+ could be a *true positive* or a *false positive*. Let $a =$ the number of patients with true positive and $b =$ the number of patients with false positive.

Similarly, let $T^- =$ test is negative and $c =$ number of patients with a *false negative* and $d =$ number of patients with *true negative*.

So, we have the following chart given in Table 2.1:

$$\begin{aligned}
\text{Total } T^+ &= \text{True Positive} + \text{False Positive} = no\#\{D^+ \cap T^+\} \\
&\quad + no\#\{D^- \cap T^+\} = a + b
\end{aligned}$$

$$\begin{aligned}
\text{Total } D^+ &= \text{True Positive} + \text{False Negative} = no\#\{D^+ \cap T^+\} \\
&\quad + no\#\{D^+ \cap T^-\} = a + c
\end{aligned}$$

Table 2.1   Test Results for Table 2.1

	D^+ = Disease Exists	D^- = No Disease	
T^+ = Test Positive	#of True Positive = a	#of False Positive = b	Total Positive = a + b
T^- = Test Negative	#of False Negative = c	#of True Negative = d	Total Negative = c + d
	Total D^+ = a + c	Total D^- = b + d	

$$Total\ T^- = False\ Negative\ +\ True\ Negative = no\#\{D^+ \cap T^-\}$$
$$+ no\#\{D^- \cap T^-\} = c + d$$

$$Total\ D^- = False\ Positive + True\ Negative = no\#\{D^- \cap T^+\}$$
$$+ no\#\{D^- \cap T^-\} = b + d$$

where $no\#\{S\}$ = Number of elements of a set S.
Let,

$$\alpha = Sensitivity = P(T^+/D^+) = P(D^+ \cap T^+)/$$
$$\{P(D^+ \cap T^+) + P(D^+ \cap T^-)\} = a/(a + c)$$

$$\beta = Specificity = P(T^-/D^-) = P(D^- \cap T^-)/$$
$$\{P(D^- \cap T^+) + P(D^- \cap T^-)\} = d/(b + d)$$

$$\xi_1 = Stress\ Parameter\#1 = P(D^+/T^+) = P(T^+ \cap D^+)/$$
$$\{P(T^+ \cap D^+) + P(T^+ \cap D^-)\} = a/(a + b)$$

$$\xi_2 = Stress\ Parameter\#2 = P(D^+/T^-) = P(T^- \cap D^+)/$$
$$\{P(T^- \cap D^+) + P(T^- \cap D^-)\} = c/(c + d)$$

$$\zeta_1 = Relief\ Parameter\#1 = P(D^-/T^+) = P(T^+ \cap D^-)/$$
$$\{P(T^+ \cap D^+) + P(T^+ \cap D^-)\} = b/(a + b)$$

$$\zeta_2 = Relief\ Parameter\#2 = P(D^-/T^-) = P(T^- \cap D^-)/$$
$$\{P(T^- \cap D^+) + P(T^- \cap D^-)\} = d/(c + d)$$

$$Total\ number\ of\ patients = a + b + c + d.$$

Thus, stress parameters deal with probabilities that the disease is present regardless of whether the test is positive or negative, and relief parameters deal with probabilities that the disease is not present regardless if the test is positive or negative.

We must note that $P(D^+) + P(D^-)$ *must be equal to* 1.

Here, $P(D^+) + P(D^-) = (a+c)/(a+b+c+d) + (b+d)/(a+b+c+d) = 1$

Similarly, $P(T^+) + P(T^-) = 1$. Also, from the table we may see : $P(D^+/T^+)$ $+ P(D^-/T^+) = a/(a+b) + b/(a+b) = 1$. If we consider D^+/T^+ as an event E, *then* $E^- =$ *the complement of* $E = D^-/T^+$. Then if S is another test done on the patient, then, following a table similar to the table above, we will be able *to compute:* $P(E/S^+)$ *and* $P(E^-/S^+)$, etc. Sometimes, instead of talking about the probability of an event E, we talk about the odds that E will happen. Odds that E will happen $=$ odds in favor of $E =$ (probability that E will happen)/(probability that it will not happen $) = P(E)/(1 - P(E))$. Odds get better as $P(E)$ gets better.

2.5 Confidence in the Estimation of Prognosis

If cancer is detected and/or suspected, then several tests are performed by the oncologists, to validate their findings. Let D^+ be the event that the disease is suspected to be present. Let $T_1, T_2,$ *and* T_3 be three successive tests. Then, in general, the prognosis is grim, if all tests are found to be positive. Among all the tests to detect cancer, biopsy is considered to be the gold standard. Yet it is not 100% correct.

One common question that the first author was often asked is if $T_1, T_2..., T_n$ are n number of tests and $P(D^+/T_i^+)$ for $i = 1, 2, \ldots, n$ are all approximately the same, then what is $P(D^+/\bigcap_{i=1}^n T_i^+)$. In a more simplistic form, the question is: Let us consider three tests: $T_1, T_2, and T_3$. Let $P(D^+/T_i^+) = 98\%$ for $i = 1, 2, 3$. Then what is $P(D^+/T_1^+ \cap T_2^+ \cap T_3^+)$. The answer is exactly 98%. Let us understand this using simple statistics (Table 2.2).

Then, $P(D^+/T_1^+) = a_1/(a_1 + b_1)$, $P(D^+/T_2^+) = a_2/(a_2 + b_2), P(D^+/T_3^+) = a_3/(a_3 + b_3)$. Let each equal α.

Then, $P(D^+/T_1^+) = a_1/(a_1 + b_1) = \alpha$ gives $a_1 = \alpha(a_1 + b_1)$; similarly, $a_2 = \alpha(a_2 + b_2), a_3 = \alpha(a_3 + b_3)$.

Let $A \odot B$ means $n(A) + n(B) =$ the sum of elements of A and the elements of B, then

Table 2.2 Test Results for Table 2.2

	D^+	D^-	Total
T_1^+	a_1	b_1	$a_1 + b_1$
T_2^+	a_2	b_2	$a_2 + b_2$
T_3^+	a_3	b_3	$a_3 + b_3$
Total	$a_1 + a_2 + a_3$	$b_1 + b_2 + b_3$	$a_1 + a_2 + a_3 + b_1 + b_2 + b_3$

$$P(D^+/(T_1^+ \odot T_2^+ \odot T_3^+) = (a_1 + a_2 + a_3)/((a_1 + a_2 + a_3) + (b_1 + b_2 + b_3))$$
$$= \alpha(a_1 + b_1 + a_2 + b_2 + a_3 + b_3)/$$
$$((a_1 + a_2 + a_3) + (b_1 + b_2 + b_3)) = \alpha.$$

Instead of 3, if we consider n number of such findings, that indicates that $P(D^+/T_n^+) = a_n/(a_n + b_n) = \alpha$ is true for a large set of data. This implies that if a doctor takes just one test and finds that the probability that the disease is present is α, after examining that the test is positive; then afterward if he takes a number of tests and finds that the probability of those tests are positive, the probability that the disease is present will not increase a bit. It only increases the confidence of a physician that the diagnosis is valid.

2.6 Analysis of Multiple Probabilistic Tests for Tumor Detection

Doctors do various tests because that gives them directions for methods of treatment that they should adapt, hoping that such methodical treatments will benefit the patients. In general, at least three major tests are conducted. These are MRI, angiogram, and biopsy (other than chest X-ray, blood test, urine and stool tests, bone density test, etc.). In that case, anxieties and level of stress of the patients and their well-wishers could be quite high which is understandable. So, we will now discuss this concept applying statistical data. Let three of these tests be $T_1, T_2,$ and T_3. Some may be positive and some may be negative. In all, there will be 2^3 or 8 results like: $T_1^+ \cap T_2^+ \cap T_3^+, T_1^+ \cap T_2^+ \cap T_3^-, T_1^+ \cap T_2^- \cap T_3^+$, etc., which may be computed by the *principle of counting*. Similarly, if n *number of tests are done, the total will be* 2^n.

For three tests, there will be a total number of 8 combined tests. Let $a_1 =$ *Number of patients for whom all tests are positive* (true positive), *meaning they all have the disease*. Let $b_1 =$ *Number of patients for whom all tests are positive, however, they are really disease free* (false positive). (Total number of positive cases $= a_1 + b_1$).

So $P(D^+/T_1^+ \cap T_2^+ \cap T_3^+) =$ *Probability that the patient has contracted the disease when* all *tests are positive* is $a_1/(a_1 + b_1)$. Similarly, *Probability that the patient has contracted the disease when all tests are negative* $= a_8/(a_8 + b_8)$, where a_n, b_n are all defined in the table below (Table 2.3):

Then,

Total Number of Patients $= \sum_{n=1}^{8} a_n + \sum_{n=1}^{8} b_n$. Number of cases that disease is present $= \sum_{n=1}^{8} a_n$. Number of cases that disease is not present $= \sum_{n=1}^{8} b_n$. So we have:

$$P(D^+) = \sum_{n=1}^{8} a_n/(\sum_{n=1}^{8} a_n + \sum_{n=1}^{8} b_n) \text{ , and}$$

Table 2.3 Test Results for Table 2.3

	D^+	D^-	Total
$T_1^+ \cap T_2^+ \cap T_3^+$	a_1	b_1	$a_1 + b_1$
$T_1^+ \cap T_2^+ \cap T_3^-$	a_2	b_2	$a_2 + b_2$
$T_1^+ \cap T_2^- \cap T_3^+$	a_3	b_3	$a_3 + b_3$
$T_1^+ \cap T_2^- \cap T_3^-$	a_4	b_4	$a_4 + b_4$
$T_1^- \cap T_2^+ \cap T_3^+$	a_5	b_5	$a_5 + b_5$
$T_1^- \cap T_2^+ \cap T_3^-$	a_6	b_6	$a_6 + b_6$
$T_1^- \cap T_2^- \cap T_3^+$	a_7	b_7	$a_7 + b_7$
$T_1^- \cap T_2^- \cap T_3^-$	a_8	b_8	$a_8 + b_8$

$$P(D^-) = \sum_{n=1}^{8} b_n / (\sum_{n=1}^{8} a_n + \sum_{n=1}^{8} b_n),$$

$$P(T_1^+ \cap T_2^+ \cap T_3^+ / D^+) = a_1 / \sum_{n=1}^{8} a_n,$$

Similarly,

$$P(T_1^- \cap T_2^- \cap T_3^- / D^-) = b_1 / (\sum_{n=1}^{8} b_n).$$

A point of caution !!!

We may note that:

$$P(T_1^+ \cap T_2^+ \cap T_3^+) P(D^+ / T_1^+ \cap T_2^+ \cap T_3^+) = P(D^+ \cap T_1^+ \cap T_2^+ \cap T_3^+)$$

$$= a_1 / \left(\sum_{n=1}^{8} a_n + \sum_{n=1}^{8} b_n \right)$$

And $P(T_1^+ \cap T_2^+ \cap T_3^+) = (a_1 + b_1) / \left(\sum_{n=1}^{8} a_n + \sum_{n=1}^{8} b_n \right)$.

That also gives: $P(D^+ / (T_1^+ \cap T_2^+ \cap T_3^+)) = a_1 / (a_1 + b_1)$. In this way, we can get all results like $P(D^+ / (T_1^- \cap T_2^+ \cap T_3^+))$, $P(D^+ / (T_1^+ \cap T_2^- \cap T_3^+))$, etc. This study could be extended to a large number of tests. Assuming that all the tests are negative the probability that the patient has contracted the disease is $= P(D^+ / (T_1^- \cap T_2^- \cap T_3^-)) = a_8 / (a_8 + b_8)$. If b_8 is very small, $P(D^+ / (T_1^- \cap T_2^- \cap T_3^-))$ will be close to 1.

Also assuming that the disease is present, probability that all tests will be negative is $P(T_1^- \cap T_2^- \cap T_3^- / D^+) = a_8 / \sum_{n=1}^{8} a_n$. Obviously, this is very small. Yet, it did happen as witnessed by the authors.

2.7 A Few Topics on Fundamentals of Probability Theory

To comprehend how statistics could help us in mathematical modeling, especially how various assumptions that we have to use for our models may be supported by statistical analysis, we need to recall some aspects of probability theory and statistics. Some references are [1–4]. Here, we will discuss some very briefly.

2.7.1 Uniform Distribution

This distribution is often used to estimate how soon a treatment could be effective. The (cumulative) distribution function is:

$$F(x) = (x - a)/(b - a) \, for \, a \leq x \leq b \tag{2.8}$$

with $F(x) = 0$ if $x <= a$ and $F(x) = 1$ if $x >= b$. Mean $= (a + b)/2$, variance $= (b - a)^2/12$

Let us consider a simple example (hypothetical).

Let us consider a sequence of data $\{x_1, x_2, \ldots, x_n\}$ such that frequency of each entry is k (exactly the same). If $a \leq x_i \leq b$, then $x_i's$ are uniformly distributed if

$$k(b - a) = 1.$$

Ex. 1. If a person takes carcinogenic food and drinks continuously for ten years, then what is the probability that any time within the next 15 years he/she will contract some form of cancer ?

Ans. If f(t) is the density function, f(t) is a uniform density function. So, 15 f(t) $= 1$, givingf(t) $= 1/15$. If $P(x = 5)$ means cancer will happen after 5 years, then $P(x = 5) = 5f(t) = 5/15 = 1/3 = 33$ % and $P(x = 10) = 10/15 = 2/3 = 66\%$, etc.

Ex. 2. A dietician prescribed a special diet to a patient and said that within two to five months blood report will improve. What is the probability that she is correct?

Ans. f(t) $= 1/(5–2)$. Using the distribution function, $P(x = 3) = 1/3$, $=$ in three months it will work. $P(x = 5) = (5 - 2)/(5 - 2) = 3/3 = 1$. (That means in five months it is guaranteed to work).

2.7.2 Binomial Distribution

The binomial coefficients are of the form (combination of n things taken r at a time).

$$C(n, r) = n!/(r!(n - r)!). \tag{2.9}$$

A Note: $0! = 1$

From this formula, it is easy to see that

$$C(n, r) = C(n, n - r). \tag{2.10}$$

Bernoulli Trial.

A Bernoulli trial is a random experiment with only two outcomes: "a success" or "a failure." Both are equally likely. Like a medical test is either positive or negative.

We consider a sample size n drawn with replacement from a population size N where N, being the size of the population (like a census), is much larger than n. Let $p =$ the probability of a success. Then $0 \leq p \leq 1$. If there are r number of successes, then if $q = 1 - p =$ probability of a failure, there are $(n - r) =$ the number of failures.

Then, the probability of r number of success is given by

$$f(r, n, p) = C(n, r)p^r q^{n-r}, \tag{2.11}$$

where $C(n, r) =$ the binomial coefficient=combination of n things taken r at a time.

Ex. A diet has been set up with some statistical data suggesting that it will be effective in reducing stomach discomfort and the probability is 70%. In 8 patients it will be tried. What is the probability that at least 4 patients will get the benefit?

Ans. Here $n = 8, p = 0.7, q = 1 - p = 0.3$. At least 4 means 4 or more. So the answer is

$$C(8, 4)(0.7)^4(0.3)^{8-4} + C(8, 5)(0.7)^5(0.3)^{8-5} + C(8, 6)(0.7)^6(0.3)^{8-6}$$
$$+ C(8, 7)(0.7)^7(0.3)^{8-7} + C(8, 8)(0.7)^8$$
$$= 0.1361 + 0.2541 + 0.2965 + 0.1977 + 0.0576$$
$$= 0.942$$

Here it is evident that if $n = 80$, we must do quite a bit of long computation.

A rule is: If N is large, binomial distribution $\rightarrow$ normal distribution. Possibly, the simplest example is tossing a fair coin. If we toss a coin 10 times, we may get 8 heads and 2 tails. But if we toss it 1000 times, we will get about 500 heads and 500 tails.

Normal distribution has very wide applications in medical science. We used it doing analysis of data while modeling the treatment for breast cancer. Let us briefly discuss some aspects of normal distribution.

2.7.3 Normal Distribution

Any mathematical modeling dealing with large-scale predictions especially with heterogeneous data requires some good background in statistics. In this regard, we will briefly discuss some properties of normal distribution with some practical applications.

The frequency function is:

$$y = f(x) = \Theta exp(-\theta), \tag{2.12}$$

where $\Theta = 1/\sigma\sqrt{2\pi}$ and $\theta = (x - \mu)^2/(2\sigma^2)$. $\mu = $ *mean* and $\sigma = $ *standard deviation*.

The frequency function is symmetric about the vertical line through the mean. About 68% of data are one standard deviation from the mean. 95.5% data are two standard deviations from the mean, and 99.7% data are within three standard deviations from the mean.

The total area under the curve is 1. That means

$$\int_{-\infty}^{\infty} y\,dx = 1 \tag{2.13}$$

The standardized normal frequency function is found by a transformation (Fig. 2.1)

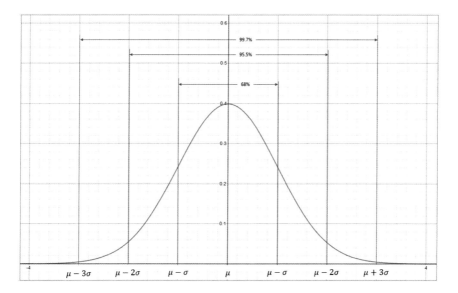

Fig. 2.1 Frequency function (probability density function) of normal distribution

$$z = (x - \mu)/\sigma \tag{2.14}$$

Then $dx = \sigma \, dz$, and at $x = \mu$, $z = 0$.
So, the standardized normal density function is:

$$f(z) = exp(-z^2/2)/\sqrt{2\pi} \tag{2.15}$$

$f(z)$ is symmetric about the $y - axis$.
Tables given in texts on statistics are for:

$$P(0 < Z \le z) = \int_0^z f(z)dz. \tag{2.16}$$

It is important to know that $\int_{-\infty}^{+\infty} f(z)dz = 1$. It is symmetrical about the origin.
This means that $P(Z \text{ is at most } z) = 0.5 + P(0 < Z \le z)$.
And $P(Z \text{ is at least } z) = 1 - P(Z \text{ is at most } z)$.
In general

$$P(a \le z \le b) = \int_a^b f(z)dz \tag{2.17}$$

Please use the standardized normal table from any book on statistics.

Ex. 1. A dietician found that effectiveness of a particular diet for breast cancer patients is normally distributed with a mean of 82% and standard deviation 15%. What is the probability that she will see (i) at least 90%, (ii) at most 95%, and (iii) exactly 70% effectiveness, respectively?
    Ans. Here, $\mu = 0.82$, and $\sigma = 0.15$, From (2.14), $z = (x - 0.82)/0.15$.

(i)   If $x = 0.9$, $z = 0.08/0.15 = 0.53$.
          $P(z) = 0.5 - 0.2019 = 0.2981 = $ about 30%.

(ii)   If $x = 0.95$, $z = (0.95 - 0.82)/0.15 = 0.13/0.15 = 0.87$. So $P(z) = 0.5 + 0.3078 = 0.8078 = about 81\%$.

Explanation: The total area under the normal frequency curve is 1. The curve is symmetrical around $X = \mu$(the mean). So, the area on each side of the mean is 0.5. The shaded area represents $P(x_1 \le X)$.
    That means probability of X which will be at least x_1, implies the area starting from x_1 or more. The area under the curve on the right side of the point, starting from that point.
    The shaded area defines $P(X \ge x_1) = P(X \text{ is at least } = x_1)$.
    Dotted vertical line is $X = 0.5 \text{ or } Z = 0$. If we assume $z = 0.53$, then we get from the table of standard normal distribution, probability of $P(z \le x_1) = $ the area under

Fig. 2.2 $P(X \geq x_1)$ [Ref.
Using the Normal
Distribution-Statistics Libre
Texts. Created by Anthony
Palmiotto]

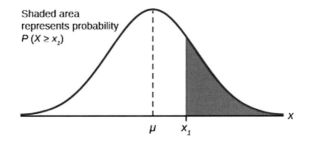

Shaded area
represents probability
$P(X \geq x_1)$

the curve between the dotted line and the shedded area = 0.2019. So, the shaded
area $= \int_{\mu}^{\infty} y dx - \int_{\mu}^{0.53} y dx = 0.5 - 0.2019 = 0.2981$ (at least means not less than
x_1, $X \geq x_1$ or $Z \geq z_1$) (Fig. 2.2). Instead of at least if we say at most, then it will be
1- 0.2981 = 0.7019.

Now let us assume that the point x_1 represents $z = 0.87$. Then,
$P(x = at\ most\ 95\%) =$ the area under the curve from the extreme left side up to
the point 0.87. $= \int_{\mu}^{\infty} y dx + \int_{\mu}^{0.87} y dx = 0.5 + 0.3078 = 0.8078$ (at most means not
more than x_1). On two sides of the mean, area under the curve is 0.5.

(iii) Exactly 70% is a bit more interesting. Here $z = (0.7 - 0.82)/0.15 = -0.8$.

Due to symmetry, we look for $z = 0.8$ and $P(0 \leq z \leq 0.8) = 0.2881$.
So, $P(z \leq -0.8) = P(z \geq 0.8) = 0.5 - 0.2881 = 0.2119$.
So, on the right side of the vertical line $z = -0.8$, the area is $0.2881 + 0.5 = 0.7881$ and on the left side of $z = -0.8$, the area is 0.2119.
We want $P(z = 0.8)$. From (2.17), here $a = b$, so $P(z = 0.8) = 0$, Meaningless
result.
We approximate this as $P(z - \varepsilon < z < z + \varepsilon) = \int_{z-\varepsilon}^{z+\varepsilon} f(z)dz = \int_0^{z+\varepsilon} f(z)dz - \int_0^{z-\varepsilon} f(z)dz = 0.2910 - 0.2852 = 0.0058$,
if $\varepsilon = 0.01$. So, it is extremely small !!! Possibly meaningless.

*So, we must move back to binomial distribution from normal distribution to resolve
this issue.*

Here $\mu = 82$, out of 100 samples. If we consider 30 samples, $\mu = 0.82 \times 30 = 24.6$. So, with $n = 30$ and $\mu = 24.6$ (still it is 82%), considering a binomial
distribution,$p = 0.82$, $q = 0.18$ (must remain for all n).
One must remember that from binomial distribution we can proceed to normal
distribution if: $n \geq 30$, $np > 5$, and $nq = n(1 - p) > 5$. In this case, we took $n = 30$.
So, normal distribution remains valid here because the inequalities are satisfied. Now
70 out of 100 means 21 out of 30. Probability of exactly 21 successes in 30 trials =
$b(21; 30, 0.82) = C(30, 21)(0.82)^{21}(0.18)^9 = 0.04398$.
So, if we consider 30 samples, the probability that exactly 21 is a success is about
4.4%.

2.7.4　Significance of μ in the Practical World

Working with medical professionals, our experience shows that most of the time they use the concept of taking averages while evaluating a procedure or any medication or even economic evaluations of treatments. It is necessary to study how to establish the validity of such applications through data analysis. We are guided by these applications in selecting/analyzing inputs while developing our mathematical models.

In reality, we can only analyze a limited number of samples and compute their statistics. To check statistics of a population is a humongous task. For instance, the cost to buy a particular medication in the world is a population. We can only collect a few samples and compute the average cost. The aim of statistics is to predict the statistics of a population by analyzing statistics of data.

2.7.5　The Central Limit Theorem (CLT)

The inner point of the central limit theorem, which has broad applications in statistics, starts with a fundamental question: Can we assess statistics of any population by studying statistics of samples drawn from this population. The answer is: yes.

This is indeed one of the major tools in studying mathematical modeling of breast cancer treatment in this monograph, because we have to estimate the average results of all treatments and extend our studies to the entire population.

We will briefly discuss this with an example: Let us consider a population with only 4 elements (1, 3, 5, and 7). Let frequency of each element is 1. So, this is a uniform distribution. Its statistics are $\mu = \text{Mean} = \sum_{i=1}^{4} x_i f_i / \left(\sum_{i=1}^{4} f_i \right) = 4$, here $f_i = 1 \ \forall i$, $\sigma^2 = \text{Variance} = \sum_{i=1}^{4} (x_i - \mu)^2 f_i / \left(\sum_{i=1}^{4} f_i \right) = (9 + 1 + 1 + 9)/4 = 5$.

Let us take 2 elements at a time with replacement from the population. Since there are 4 elements and 2 will be taken at a time with replacement, using the principle of counting, there will be a total of $4 \times 4 = 16$ sets of samples which are : the first set of samples of 2 = (1, 1), (1, 3) ,(1, 5), (1, 7), which means 1, 2, 3, 4; the second set of samples of 2 = (3, 1), (3, 3), (3, 5), (3, 7) which means 2, 3, 4, 5; the third set of samples (5, 1), (5, 3), (5, 5), (5, 7) which means 3, 4, 5, 6; and the fourth set of samples (7, 1), (7, 3), (7, 5), (7, 7) which means 4, 5, 6, 7.

This distribution of means is not uniform. Frequencies of the means 1, 2, 3, 4, 5, 6, 7 are, respectively, 1, 2, 3, 4, 3, 2, 1. So there are 7 means, and if we denote these means by m_i, $i = 1, 2, \ldots 7$, then the mean of the means is $= 4$. If we denote this as m, then $\mu = m$. If $s =$ the standard deviation of the means, then the variance of the means is given by

$$s^2 = ((1 - 4)^2 + 2 \times (2 - 4)^2 + 3 \times (3 - 4)^2 + 0 \times (4 - 4)^2 + 3 \times (5 - 4)^2$$
$$+ 2 \times (6 - 4)^2 + 1 \times (7 - 4)^2)/(1 + 2 + 3 + 4 + 3 + 2 + 1) = 2.5$$

So, $s^2 = \sigma^2/n$, *where* $n = 2$ = the size of the sample.

In fact, this is true regardless of the nature and size of the samples. As the size of the samples increases at least $n \geq 30$, the frequency function will take a bell shape and become a normal frequency curve. This is the central limit theorem which we say in brief CLT.

Mathematically, we state that if from any population of any size and any distribution with mean μ and standard deviation σ, we arbitrarily choose samples of exactly the same size n, then if m is the mean of the means m_i of the samples (all of the same size n) and if s is the standard deviation of the means of the samples, then $s^2 = \sigma^2/n$, so the standard deviation of the means is $s = \sigma/\sqrt{n}$. Also as $n \to \infty$, $z = (x - m)/(\sigma/\sqrt{n})$ is a standardized normal variable. This is valid if $n \geq 30$.

Example: (1) Oncologists found that there is a targeted treatment procedure quite successful treating breast cancer patients. Success is measured in terms of how many days they started feeling much better. We assume that it is normally distributed.

If the procedure is applied to 25 patients chosen at random, what is the probability that they will start feeling better between 95 and 120 days on the average?

(1)   In a hospital where for such treatments population mean $\mu = 90$ days and $\sigma = 18.5$ days.

Ans. (1) If $m = 95$, $z = (95 - 90)/(18.5/\sqrt{25}) = 5/3.7 = 1.3514$.

If $m = 120$, $z = (120 - 90)/3.7 = 8.11$

$P(95 \leq x \leq 120) = P(1.35 \leq z \leq 8.11) = 0.4999 - 0.4115 = 0.0884 =$ about 9%.

(2)   In another hospital, the oncologists found a very similar treatment with same mean but a standard deviation which is much larger, $\sigma = 25.37$. How this will differ from the previous case.

If $m = 95$, $z = (95 - 90)/(25.37/5) = 0.9854$, and if $m = 120$, $z = 150/25.37 = 5.91$

$P(0.99 \leq z \leq 5.91) = 0.4999 - 0.3413 = 0.1586$ which is about 16%. It is obviously better although a larger standard deviation means data are too scattered.

(3)   In a third hospital $\mu = 110$, and $\sigma = 30$. Then if $m = 95$, $z = (95 - 110)/6 = -15/6 = -2.5$.

If $m = 120$, $z = 10/6 = 1.7$, $P(-2.5 \leq z \leq 1.7) = 0.4938 + 0.4554 = 0.9492$ about 95%. Thus, a bigger mean and a larger standard deviation are much better.

So sometimes, the patient and her family have to choose their doctors and the medical facilities.

2.8   Hypotheses Testing: The p-Value

In medical treatments and trials, analyses of data are of primary concern. Often various doctors give various suggestions regarding use of exactly the same medications. In such cases, statisticians come in to analyze the data and make some validity of various claims regarding what could be statistically acceptable. This task is sometimes done by conducting statistical hypothesis testing.

First they generally look into what the manufacturer of the medication has suggested. This is their null hypothesis, meaning that this is the starting point. We denote it by H_0. Then, they look into what the medical practitioners are saying. Each statement is called an alternate hypothesis and is denoted by H_a.

But the problem is, if a hypothesis is accepted, there is a chance of making an error which is known as the Type I error, and similarly, if it is not accepted, there is a chance of making an error which is called the Type II error.

The upper limit of the Type I error is denoted by α. This must be chosen first before conducting any test. The probability of not accepting the alternate hypothesis is the $p - V value$. If $p - value < \alpha$, we reject H_0 and accept H_a. Else, we say that we do not accept H_a and also say that we have failed to reject H_0, the null hypothesis.

Ex. A pharmaceutical company stated that a medication should be effective for 72 h on the average with a standard deviation of 6 h. (i) A group of medical practitioners used 36 samples and found that the average is 70 h. (ii) Another group said they also used a different set of 36 samples and found that it should be 74 h. (iii) A third group examined another box of 36 samples, declared 72 h is inaccurate meaning that it could be more or could be less, and requested that the company must check their products more thoroughly and give a new estimate.

Let us look into what is going on by hypothesis testing.

Here $\mu_0 = 72 \, and \, \sigma = 6$, $n = 36..$ These are valid for all cases.

The null hypothesis is $H_0 : \mu_0 = 72$. This must be tested.

$$z = (\mu_a - \mu_0)/(\sigma/\sqrt{n})$$

For the case (i) $\mu_a = 70$. So, $z = -2$ (test statistic).

The alternate hypothesis is: $H_a : \mu_a = 70 < 72$ (Fig. 2.3).

Now we must choose the max level of acceptance or the max acceptable Type I error given by the value of α. That gives how much error we are ready to tolerate, if we reject H_0.

(a)   Suppose that one statistician accepts that a 5% error is tolerable, then $\alpha = 0.05$ giving, on the left side of the mean (the origin $z = 0$), the critical statistic $= z(-\alpha) = -1.645$. Since $-2 < -1.645$, we will reject H_0 and accept H_a.

(b)   Suppose another statistician decides to accept only a 1% error. Then $\alpha = 0.01$. Then the critical statistic $= z(-\alpha) = -2.33$. Now $-2 > -2.33$. So this statistician will fail to reject H_0.

These decisions often complicate serious medical studies.

Fig. 2.3 Rejection of $H_0:\mu_0$ (left tail test)

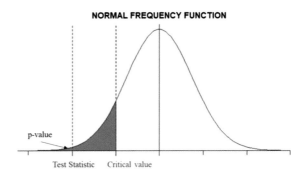

NORMAL FREQUENCY FUNCTION

p-value

Test Statistic   Critical value

For the case (ii), we use the right side of the normal frequency curve. Here, if we choose $\alpha = 0.05$, accepting a 5% error, then $z(\alpha) = 1.645$. Since $2 > 1.645$, we will reject H_0 and accept H_a. However, if again we decide to accept only a 1% error, $\alpha = 0.01$. Then, the critical statistic $= z(\alpha) = 2.33$. Now $2 < 2.33$. So this statistician will fail to reject H_0.

For the case (iii), the situation is a bit different. Here, we will test $H_0 \neq \mu_0$. In order to do that, we will look into both sides of the normal frequency curve (figures may be found in any statistics book).

Here, we look for $z(\alpha/2)$ and $z(-\alpha/2)$ on two sides of the frequency curve. If we choose $\alpha = 0.05$, then $\alpha/2 = 0.025$.. Then, the area under the curve or the critical values of z are $z(\alpha/2) = 1.96$ and $z(-\alpha/2) = -1.96$. Our test statistic is $z = -2 < -1.96$. So we reject the null hypothesis $H_0 : \mu_0 = 72$.

A short note on p-value. This simple concept has created a great deal of discussions among the users. Some of them are very confusing. So, based upon the ideas of conditional probability we have planned to discuss it now.

Let us assume a population of some event A_0 is distributed normally with a mean μ_0 and a standard deviation σ.

Exactly from the same population, a sample of size n is drawn. It has a mean μ_a, and its standard deviation is $\sigma/\sqrt{n}$. If $\mu_0 = \mu_a$, or even approximately close to μ_a, then there is no problem. But if they significantly differ, then comes hypothesis testing to choose one of the two. We have to look into which will give us less error of the Type I.

For that, we have to choose prior to our test how much error is acceptable. That is given by the value of α.

Let us consider two similar events A and B which are not independent or mutually exclusive. These are two observed events related to the same phenomenon. Question is: Which to accept?

Then from the principle of conditional probability,

$$P(A) = P(B)\,P(A/B) \text{ and } P(B) = P(A)\,P(B/A).$$
$$\text{iff } P(A/B) = P(B/A), \text{ then } P(A) = P(B).$$

This is rather impossible. So, we will look into the standardized normal frequency curve assumed to be valid for both A and B. Then, we preset how much error is tolerable, if A is rejected and we denote that error by α; then in comparison with that, we compute what will be the error, if B is rejected. This is the $p - value$ for rejecting B. Whichever is less is accepted.

We know if z_A and z_B are the standardized z values for A and B, respectively, $P(A) = \int_0^{z_A} f(z)dz$ from Eq. (2.17).

If we reject A, the error is $1 - P(A)$. That we set at the beginning. This is the critical value α.

If $p = 1 - P(B). < \alpha$, we accept B over A, etc.

Most of these statistical analyses have been applied forming the foundation of data used in our models. For more details on statistics, we refer to the references [1, 2].

2.9   A Search For Panacea for Metastatic Breast Cancer by Applying Dey–Markov Chain

After detection of a cancerous lump in the breast, the very first concern of an oncologist is whether it has spread elsewhere and/or how it could metastasize. Even though they could detect this by applying a whole-body scan using MRIs, a mathematical determination of the metastasization could be used for a backup check. Because MRIs do not reveal the mechanism of metastasization, Markov chain analysis could lead us inside that realm. Mason and Newton and several others [7] made a theoretical attempt to study this. Ours is much more practical and is in keeping with the recent mode of treatments of cancer. Theoretical understanding is certainly useful, but modeling the mode of treatments is now a crying need. Such an attempt has been conducted here by applying an algorithmic approach. After gaining some understanding of how cancer has spread, doctors may also choose some unique strategies to treat individual patients. And then comes the most fundamental question: How to cure them? For each patient, a unique strategy of treatment is often an urgent necessity. The Dey chain, somewhat similar to Markov chain analysis, could help the doctors in this respect.

In the true sense of the term, the word "cure" is still not applicable to cancer. Genes often go through several mutations before cancer begins. So, cancer is cured, if all these mutations are reversed. Cancer is indeed a genetic disease.

2.9.1   Modified Markov Chain or The Dey–Markov Chain

Let us first think of a theater hall. If we call it S, then certainly it is a bounded space. Let us assume that a time-dependent event E will take place with a number of participants $X_k's, k = 1, 2, \ldots, m.$ $X_k's$ which will change in time as participants will move arbitrarily from one position to another. Some will come in and some will leave arbitrarily. So, this is a stochastic variable and E is a stochastic event because E changes arbitrarily as participants change arbitrarily. Now if we think that as E is arbitrarily changing, S, the size of the sitting arrangements in the theater is also arbitrarily changing, then S is a bounded stochastic space. All are happening inside S though. If at t_n, the event is E^n and there are m numbers of players denoted by $X_1^n, X_2^n, \ldots, X_m^n$, then E^n is obviously dealing with all these m number of players in a regular Markov chain model when we know ahead of time the probabilities of how players will change their positions. These probabilities will remain fixed throughout the shows for the Markov chain.

Now we will consider a slightly different scenario. Let there exist a unique space $S^* \subset S$, containing one and only one event E^* which has one and only one fixed group of players $X_1^*, X_2^*, \ldots, X_m^*$, which may be presented in a vector form as $X^* = (X_1^*, X_2^*, \ldots, X_m^*)^T \in E^*$, a unique event space. This is the only one unique element of $S^* \subset S$, such that: as $n \to \infty, P(X^n) \to P(X^*) = 1$. Everything is happening at random, yet merging with a unique predetermined point in $S^* \subset S$, while the time-dependent events $E^n \to E^*$.

In our study, E^n is the state of health of a cancer patient whose malignancies have spread to m different organs which are $X_1^n, X_2^n, \ldots, X_m^n$ at t_n and the event is denoted by E^n and all are happening at random in a stochastic space S which is the body. Even though $S^* \subset S$, mathematically S^* is a virtual state for a cancer patient, which is the state of good health, free from cancer, a reality of existence, yet at the outset that state was not physically present.

Charlie (the second author) with consultation of the first author figured out an algorithmic procedure for cure of cancer by applying the basic concepts of the Markov chain. We will often mention this simply as Dey chain. Here, the existence of a virtual state $S^* \subset S$ has been transformed into a real unique state. $X_1^n, X_2^n, \ldots, X_m^n$ are the states which must all merge with $X^* = (X_1^*, X_2^*, \ldots, X_m^*)^T \in E^*$, and all $X_i^*'s$ are exactly the same $\forall i$, where there is no trace of cancer cells. In order to achieve that, the transition matrices must be redone and redefined at the outset of every new procedure. This is precisely what oncologists do by changing their strategies of treatments.

2.9.2   Definition: The Dey–Markov Chain

Let $X_1, X_2, \ldots, X_m$ be a set of random variables and $E^r(X, t)$ be a set of stochastic operators, defined in a bounded space S^r, where $S^r \subset S$, a measurable space. Let $S^* \subseteq S^r \subset S$ and as $t \to \infty, E^r(X, t) \to E^*$, a stationary virtual operator (because at $t = 0, it\ did\ not\ exist$) operating on only one element $X^* = (X_1^*, X_2^*, \ldots, X_m^*)^T \in E^*$ which is the unique point of attraction of $P(X^k) \forall k$, such that $P(X^*) = P(E^*) = 1$ and both X^* and E^* are predetermined (non-stochastic) by the condition $E^* happening\ in$

$S^* \subseteq S^r \subset S \to S^*$ when $P\left(X_j^n/X_i^n\right) = p_{ij}^n \to 0 \ \forall \ i,j$, except for $P\left(X_j^*/X_i^n\right) = p_{ij}^*$, where $p_{ij} =$ the probability that X_i^n will move into X_j^n, $\sum_{i=1}^m p_{ij}^* = 1 \forall j$ as $t \to \infty$. Also, $\forall i$, $\sum_{i=1}^m p_{ij}^n = 1$, $\forall n$.

So the Dey-Markov chain moves from a series of indeterministic positions, which are arbitrarily chosen, to a deterministic position which was assumed to exist initially, by applying a random process guided by predetermined probabilities. That ultimate destination has been mandated at the very beginning of these processes of transitions. So, it is a random process where the end of it is not stochastic. Hence, the space E^* is predefined and its properties are known meaning, $P(X^*) = P(E^*) = 1$ and $E^* : S^* \to S^*$.

Standard Markov chain analysis finds out a steady state which is not preconditioned, which defines a space E^* where all these random events finally settle down being operated by a fixed transition matrix, and this steady state is a location which is an element of the event space before the stochastic process starts. E^* is defined at the very end. This is obviously a subspace of the event space E or in other words, $E^* \in \{E^1, E^2, \ldots, E^n\}$ for $\forall n$.

In Dey-Markov chain, $E^* \notin \{E^1, E^2, \ldots, E^n\}$. It is a predetermined limiting state of X_k^0, $k = 1, 2, \ldots m$ as $p_{ij}^n = P\left(X_j^n/X_i^n\right)$, which are variables, settling down as $n \to \infty$ and $p_{ij}^n = 0$ at all i, j 's, except at one point $X^* \notin \{X_1, X_2, \ldots, X_m\}$, identified by $X^* \in S^*$, caused by the limiting event E^*.

Let at the outset, at a given time $t = t_0$, $X^0 = \{X_1^0, X_2^0, \ldots, X_N^0\}$, and $x_i^0 = P\left(X_i^0\right)$, where there are N different non-overlapping sets of data. Let $X^0 =$ the event that an individual has been diagnosed with cancer. Let $X_i^0 =$ the event that the same cancer has been detected as a metastatic state at the ith location of the body, $i = 1$, being the primary location. Let $S = \cup_{i=1}^n S_i$, where S_i 's are all disjoint, which are the sources of the events X_i 's, respectively.

In mathematical terms, the Dey–Markov chain may be expressed as: Let $\Omega_m(X, t)$ be a set of stochastic operators defined in an n dimensional stochastic bounded space Φ generated by them. Let P be an operator, and $P \in \Omega_m(X, t)$. Let X be a set of events, such that if $X \in \Phi$, $P(X) \in R^1 \subset \Phi$. ($R^1$ is a set of real numbers, and $P(X)$ satisfies $0 \leq P(X) \leq 1 \ \forall X$). Let $M(\Omega_m(X, t))$ be the measure of $\Omega_m(X, t)$., because the space is assumed to be bounded. Let E^i be the set of events generated by Ω_i.

Let at a given time $t = t_0$, $X^0 = \{X_1^0, X_2^0, \ldots, X_N^0\}$, and $x_i^0 = P\left(X_i^0\right)$. This means, there are N different non-overlapping sets of data at the beginning. Let $X^0 =$ the event that a unique patient has been diagnosed with cancer. Let $X_i^0 =$ the event that the same cancer has been detected as a metastatic state at the ith location of the body, $i = 1$, being the primary location. Let, $S = \overset{n}{\underset{i=1}{\cup}} S_i$, where S_i 's are all disjoint, which are the sources of all the events X_i 's. Let $p_{ij} = P(X_j/X_i) = $ Probability of X_i moving into X_j. That also means that assuming X_i has already happened, the probability that X_j will happen is p_{ij}. If X_i has not happened, then p_{ij} is zero. Let $x_i = P(X_i)$,

(X_i is the detection of a tumor at the ith site).

Then,

$$P(X) = \sum_{i=1}^{N} P(X_i) = \sum_{i=1}^{N} x_i = 1. \left(because\ X_i's\ are\ disjoint\ sets\right). \qquad (2.18)$$

For a *cure* of the disease, we need: (*noting that* E_i *is the event, generating* X_i).
$\lim_{t \to \infty} M(E^i) \to 0$. Thus, $P(X_i) \to 0\ \forall i\ (M(E^i)$ is the measure of (E^i),

$$i = 1, 2, ..., N - 1.$$

And

$$P(X_N) \to 1. \qquad (2.19)$$

Since, $p_{ij} = P(X_j/X_i)$, then $x_i p_{ij} = P(X_i)P(X_j/X_i) = P(X_i \cap X_j)$. *Since both* $P(X_i)$ *and* $P(X_j/X_i)$ *are less than 1,* $P(X_i \cap X_j)$ *will be much less than 1 and less than both* $P(X_i)$ *and* $P\left(\frac{X_j}{X_i}\right)P\left(\frac{X_j}{X_i}\right)$.

First, let us understand this phenomenon in cancer treatment. When a physician says that there is a 50% chance that breast cancer will metastasize in the bone, 35% chance that it will metastasize in the liver, 30% chance that it will metastasize in the lungs, etc., this means, according to medical records, out of 100 patients with metastatic breast cancer, he found that 50 patients have metastases in the bone, 35 have metastases in the liver, 30 have metastases in the lungs, etc. For the Markov chain analysis, all sites must be disjoint sets. This concept is applicable in the business world. For example, from a city A, a person can move to a city B or to a city C, but not both at the same time. Mathematically, A,B,C are disjoint sets. However, this is not valid in biological science. A cancer cell coming out of a breast may duplicate itself in the breast, one may go to the bone and other to the liver at the same time. This means, a person with bone metastasis may also have liver metastasis too at the same time (Fig. 2.4).

Or in other words, metastatic breast tumors (like an invasive ductal carcinoma) could metastasize in the bone and in the liver simultaneously.

At first, all data should be properly organized for a meaningful statistical analysis following the principles of *data normalization*. Any overlapping of data is not in conformity with the application of a Dey–Markov chain analysis. Here data must not overlap. Or in other words, data must be organized in such a way that if A_1 and A_2 are two sets, then if $a \in A_1$, then $a \notin A_2$. If A_1 and A_2 are disjoint, then there is no problem. But if they are not, then out of these two, we can find three disjoint sets, namely: $A_1 - A_1 \cap A_2, A_1 \cap A_2$, and $A_2 - A_1 \cap A_2$ ($A_1 - A_1 \cap A_2$ means an element which is in A_1 is not in A_2). Let us again consider the sets A_1 and A_2 which are disjoint. Let these two states represent two different metastatic states of cancer, then there exists one more state A_3 when treatment begins, which denotes a state of being almost cancer-free. However, there is some possibility that surviving cancer cells may come out of it to infect some organs and by the same token some cancer cells from infected organs will invade this state. If there are a total of three overlapping

Fig. 2.4 Intersections of sets

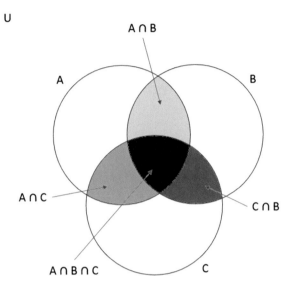

sites A_1, A_2, and A_3, then the number of disjoint sites is: $C(3,1) + C(3,2) + C(3,3) = 7$ (If we add another virtual set to these three sets, then the result will be $C(4,1) + C(4,2) + C(4,3) + C(4,4) = 15$) disjoint sets that cancer cells can move back and forth in a Markov chain, in a completely random manner. As treatment begins, these random movements will be estimated by some probabilistic pattern.

This means, in our search for the success of treatment, if there are n number of overlapping metastatic sites, and if the total number of non-overlapping sites is N, then in a Markov chain N is given by:

$$N = \sum_{r=1}^{n+1} C(n+1, r). \tag{2.20}$$

where $(n + 1)th$ state is the state S^* when the body is practically cancer-free. This is not a state in an actual Markov chain. That is an essential difference between the Dey–Markov chain and regular Markov chain. Applying this principle, researchers may approximate the probabilities of the pathways of metastatic cancer as treatments are being planned and monitored applying the principle of Markov chain.

Let us consider 100 patients with breast cancer. Some may have cancer still in situ, some may have metastasis in the other breast, and some may have metastasis in the bone, lungs, liver, and brain. These are the common sites where this cancer metastasizes. It should be noted that when cancer metastasizes, it may leave the primary site completely or any other site. Our mathematical models (to be discussed in the later chapters) also validated this finding of the oncologists. Let us consider a patient with metastatic breast cancer.

First we need to define the event space. Events are generated by the sources. If $E_1=$ the event, observing cancer in the first breast (the primary site), $E_2=$ the event of detection in the second breast, $E_3=$ cancer detected in the bone, $E_4=$ cancer detected in the liver, $E_5=$ cancer detected in the lungs, and $E_6=$ cancer detected in the brain, then the total number of events which are non-overlapping that could happen individually is (from Sect. 2.4.5):

$$N = \sum_{r=1}^{7} C(7, r) = 127. \qquad (2.21)$$

(the events like E_{ij} mean the pathways from $S_i \rightarrow S_j$ are possible *for* $i =$ 1, 2, ..., 127; *and* $j = 1, 2, ..., 127$).

Let us denote the elements of E_i mathematically by X_i, $i = 1 ..., 2, ..., N$. They all are changing arbitrarily at every moment. So, $P(X_i)$'s are changing arbitrarily. Estimates of these probabilities are to be given by pathologists, radiologists, and oncologists. The matrix

$$T_{N \times N} = \begin{pmatrix} p_{11} & p_{12} & \cdots & p_{1N} \\ p_{21} & p_{22} & \cdots & p_{2N} \\ \cdots & \cdots & \cdots & \cdots \\ p_{N1} & p_{N2} & \cdots & p_{NN} \end{pmatrix}$$

is called the transition matrix. Here, $p_{ij} = P\left(X_j/X_i\right)$, changing at each t_n. such that at each n, $\sum_{n=1}^{K} p_{ij}^n = 1$, $K =$ number of cancer infected sites. In the standard Markov chain, the transition matrix remains the same during all computations. But here it will change as treatment goes from one phase to the next as determined by the attending doctor, depending on the changes of a patient's physical conditions. So with the initial probability distribution, satisfying Markov chain conditions we start our computations and after reaching the steady state, we record the values of x_i's. Then, start a new round of treatment with a new transition matrix and steady-state values are noted again. This process goes on till at a particular steady state the oncologists are satisfied with their treatment plan. Obviously, at that point $p_{ij} \rightarrow 0$, $\forall i$ and $j = 1, 2, ... J - 1$ and $p_{iJ} \rightarrow 1$ $\forall i$.

This entire process will be: $(x^1)^T = (x^0)^T T^0$, giving, $(x^k)^T = (x^{k-1})^T T^{k-1}$, where x^k is the vector

$$x^k = (x_1^k, x_2^k, ..., x_n^k)^T \in \Omega, \ \forall k$$

obtained by a Markov process at the k *th time* and T^{k-1} is the transition matrix at the $(k - 1)$ *th time*.

Let us assume a group of metastatic breast cancer patients. Let $x_i^0 (i = 1, 2, \ldots, 7)$ be the initial estimates of the probability of metastatic states.

If we first consider 100 patients, that means 100% of them have breast cancer and for some it has metastasized. In all, there can be 127 groups. The method of treatment for those having only bone metastasis and those having only liver metastasis is different in general than those who have both. So, these groups cannot be put together. They must be separated forming three groups: (1) $A_1 =$ *the set who have only bone metastasis*, (2) $A_2 =$ *the set who have only liver metastasis*, and (3) $A_3 =$ *the set who have both.*

For an individual patient, having several metastatic states is now our primary concern.

Let us *assume* that there are in all 7 states, left breast, right breast, bone, liver, lungs, and brain that cancer cells can move from one to the other, the 7th state being the state where a patient is almost cancer-free. Let us assume that if B and C are two metastatic sites and A is the site from where cancer is moving toward those two metastatic sites, then cancer is moving only to B or to C, but not to B and C. Or in other words, B and C are mutually exclusive.

As treatment continues and physical conditions of the patient change, p_{ij}'s must change.

So we have

$$\sum_{j=1}^{7} p_{ij} = 1, \ for \ \forall i = 1, 2, \ldots 7; \forall t. \tag{2.22}$$

The metastatic process is happening through blood vessels and lymph vessels. Unfortunately, lymph nodes which are castles of the immune system protecting the breasts are often overrun by cancer cells and become their castles.

These malignant cells release several varieties of proteins. One of them is Hsp27 (heat shock protein 27). That makes the immune system inactive and unresponsive. It also increases blood flow feeding cancer cells and fueling their energy [Asit De, Cancer Research, University of Rochester]. Furthermore, those women who take too much alcohol deprive their body of energy. Immune system gets weaker. Breast cancer cells take full advantage of that and thrive.

Let us assume that the sites of cancer are denoted by: Site 1 is the primary site. The secondary sites are 2, 3, 4, 5, and 6 representing the second breast, bone, liver, lungs, and brain, respectively. Site #7 is the cancer-free site. Let:

$p_{ij} =$ Probability of the pattern of metastasis from $i \rightarrow j, i = 1, 2, \ldots 7; j = 1, 2, \ldots, 7.$

And, $x_i =$ Probability that cancer is detected at the *i*th metastatic state. (The treatment is to find under what conditions $x_7 \rightarrow 1$ *and others will tend to* 0.)

At the outset: $\sum_{i=1}^{7} x_i = 1$ and $\sum_{j=1}^{7} p_{ij} = 1 \ \forall i, j = 1, 2, \ldots, 7.$

As treatment proceeds, oncologists analyze the prognosis and adjust treatment plans which in turn change the transition matrix. This is a tailored individual care. This is where there is a major difference between Dey chain and Markov chain.

Our algorithm is:

Assume : $X_1 = Breast \#1, X_2 = Breast \#2, X_3 = Bone, X_4 = Liver, X_5 = Lungs, X_6 = Brain, and X_7 = (Almost)Cancer - Free State.$ X_1 is the primary site. $X_2, X_3, X_4, X_5,$ *and* X_6 are the metastatic sites. X_7 *is a virtual site where cancer cells should die when they metastasize at this site.* Body becomes *almost* cancer-free when all cells reach X_7 statistically.

Step #1. Choose a mode of treatment with medications, radiation, etc. Following the state of condition of a patient, oncologists determine at the outset, the values of x_i's by analyzing the statistics supplied by the pharmaceutical companies, onco-researchers, oncologists, and FDA and, of course, their own experience. Then, they set up a scheme of treatment. Following this schedule, they estimate the elements of the initial transition matrix:

$[p_{ij}^0]_{i,j}, i = 1, 2, \ldots, 7; j = 1, 2, \ldots, 7$ *for each i*; and values of $P(X_i^0) = x_i^0, i = 1, 2, \ldots, 7$ initial probability distributions regarding the growth of tumors at the probable metastatic sites. We have assumed that cancer cells move from one location to only one other location, not two locations simultaneously which may not be true. Often doctors set up their goal after analyzing their methods of treatments, when to make some adjustments of treatments, under consideration, and when treatment should stop and adjuvant therapy should resume. Then, the patient returns to performing her daily routine.

Dey–Markov chain begins: Compute

$$[x_1^k, x_2^k, \ldots, x_7^k] = [x_1^{k-1}, x_2^{k-1}, \ldots, x_7^{k-1}][p_{ij}^{k-1}]_{i=1,2,\ldots,7; :j=1,2,\ldots,7}$$

$k = 1, 2, \ldots$ at each state $\sum_{j=1}^{7} p_{ij}^{k-1} = 1.$

When the steady state is found, the values of x_j^k reveal the state of conditions of the new metastatic sites. At that level, a *new* transition matrix is set up for a modified treatment. And the steady-state values of x_j 's are computed again applying Dey–Markov chain again, etc. This process continues until the doctors are satisfied with the values of x_j's.

Convergence Principle:

If $\lim_{t \to \infty} max(x_j) \to 0, \forall j = 1, 2, \ldots, 6$ *and* $\lim_{t \to \infty} x_7 = P(X_7) \to 1$, *the treatment is a success.*

The transition matrix is as follows:

$$
\begin{array}{ll}
\textit{First row} & \left[p_{11}\ p_{12}\ p_{13}\ p_{14}\ p_{15}\ p_{16}\ p_{17} \right] \\
\textit{Second row} & \left[p_{21}\ p_{22}\ p_{23}\ p_{24}\ p_{25}\ p_{26}\ p_{27} \right] \\
\textit{Third row} & \left[p_{31}\ p_{32}\ p_{33}\ p_{34}\ p_{35}\ p_{36}\ p_{37} \right] \\
\textit{Fourth row} & \left[p_{41}\ p_{42}\ p_{43}\ p_{44}\ p_{45}\ p_{46}\ p_{47} \right] \\
\textit{Fifth row} & \left[p_{51}\ p_{52}\ p_{53}\ p_{54}\ p_{55}\ p_{56}\ p_{57} \right] \\
\textit{Sixth row} & \left[p_{61}\ p_{62}\ p_{63}\ p_{64}\ p_{65}\ p_{66}\ p_{67} \right] \\
\textit{Seventh row} & \left[p_{71}\ p_{72}\ p_{73}\ p_{74}\ p_{75}\ p_{76}\ p_{77} \right]
\end{array}
$$

The objective of the treatment is to transport the initial vector x^0 to the final vector x^F, where

$$
x^F = \{ \gamma_1^F, \gamma_2^F, \gamma_3^F, \gamma_4^F, \gamma_5^F, \gamma_6^F, 1 - (\gamma_1^F + \gamma_2^F + \gamma_3^F + \gamma_4^F + \gamma_5^F + \gamma_6^F) \},
$$

(where $\gamma_j^F s$ should be small in comparison with the previous values of $x_j's$. At that point a patient starts her adjuvant therapy.). Theoretically, we would like to see how an initial vector $x^0 = (1, 0, 0, 0, 0, 0, 0)^T$ could be transformed into the final vector $x^F = (0, 0, 0, 0, 0, 0, 1)^T$.

Let treatment begin with the following initial entries: $x_1 = 1.0, x_2 = 0.0, x_3 = 0.0, x_4 = 0.0, x_5 = 0.0, x_6 = 0.0$, and $x_7 = 0.0$. That means the cancer has been detected in the breast and it has not yet metastasized elsewhere. So the initial vector is $(x^0)^T = (1.0, 0.0, 0.0, 0.0, 0.0, 0.0, 0.0)$

Let the transition matrix, at the outset be chosen, be determined by the first protocol of treatment as follows:

0.05, 0.05, 0.02, 0.02, 0.01, 0.05, 0.35 *** DATA #0 * **
0.05, 0.05, 0.02, 0.15, 0.01, 0.01, 0.35
0.05, 0.05, 0.02, 0.15, 0.01, 0.01, 0.35
0.05, 0.05, 0.02, 0.15, 0.01, 0.01, 0.35
0.05, 0.05, 0.02, 0.15, 0.01, 0.01, 0.35
0.05, 0.05, 0.02, 0.15, 0.01, 0.01, 0.35
0.05, 0.05, 0.01, 0.01, 0.05, 0.05, 0.60

This means that during the initial treatment, oncologists are estimating that from the primary site, 5% chance that cells will come back to the primary site, 5% moving to the second breast, 20% into the bone, 20% into the liver, 10% into the lungs, 5% into the brain, 35% into remission, etc. Applying Markov chain analysis, the steady-state result is $x^1 = (0.05, 0.05, 0.15, 0.13, 0.075, 0.075, 0.47)^T$. We note that $P(X_7) = x_7 = 0.47$. That means 47% of metastases have been cured or 47% chance that cancer has gone to a state of remission. *We certainly want more of them to move to the state of remission. So, treatment should be modified for the better and oncologists do figure that out with a new strategy.*

So, the second round of treatment begins with the initial vector x^1 as the starting point.

The transition matrix has been chosen according to a new mode of treatment as follows:

The second transition matrix is chosen as:

```
0.05, 0.05, 0.15, 0.15,   0.01,    0.05,  0.45 *** DATA #1  * **
0.05, 0.05, 0.15, 0.15,   0.01,    0.05,  0.45
0.05, 0.05, 0.15, 0.15,   0.01,    0.05,  0.45
0.05, 0.05, 0.15, 0.15,   0.01,    0.05,  0.45
0.05, 0.05, 0.15, 0.15,   0.01,    0.05,  0.45
0.05, 0.05, 0.15, 0.15,   0.01,    0.05,  0.45
0.05, 0.05, 0.10, 0.025, 0.0125, 0.0125, 0.75
```

The first row implies that there is a 5% chance that cancer will grow in the first breast, 5% chance that it will move from the primary site to the second breast, 15% chance that it will move from the primary site to the bone, 15% chance to the liver, 10% chance into the lungs, 5% chance into the brain, 45% chance to the state of remission, etc. Doctors estimate that once cancer cells enter into the state of remission, there is a 75% chance that cancer cells will move into remission (as seen in the last row).

One oncologist at Carle Clinic, Champaign, Illinois, being asked why blood tests of a cancer patient are done every day, told the first author that blood tests give some indication regarding the status of a treatment, its success or failure. And accordingly, procedures of treatments are modified everyday if needed.

The steady-state values are now:
$x_1 = 0.05$, $x_2 = 0.05$, $x_3 = 0.1178$, $x_4 = 0.07$, $x_5 = 0.0438$, $x_6 = 0.0255$, $x_7 = 0.6429$ (so, over 64% remission of cancer is obtained; treatment is going well).

The next transition matrix is:

```
0.025, 0.025, 0.1,  0.05, 0.05, 0.05, 0.7 *** DATA #2  * **
0.025, 0.025, 0.1,  0.05, 0.05, 0.05, 0.7
0.025, 0.025, 0.1,  0.05, 0.05, 0.05, 0.7
0.025, 0.025, 0.1,  0.05, 0.05, 0.05, 0.7
0.025, 0.025, 0.1,  0.05, 0.05, 0.05, 0.7
0.025, 0.025, 0.1,  0.05, 0.05, 0.05, 0.7
0.005, 0.005, 0.08, 0.02, 0.02, 0.02, 0.85
```

Notice: The sum of each row is 1. This is true at all iterations.

The steady-state values are: $x_1 = 0.025$, $x_2 = 0.025$, $x_3 = 0.09$, $x_4 = x_5 = x_6 = 0.03$, $x_7 = 0.77$.

Here $p_{77} = 0.795$ means that out of 100, 79.5% is the chance to be cancer-free, cancer moving into the state of remission, and only 2.5% will grow in the primary site.

At this point, the transition matrix is selected as follows:

0.005, 0.005, 0.08, 0.05, 0.02, 0.02, 0.85 *** DATA #3 ***
0.005, 0.005, 0.08, 0.05, 0.02, 0.02, 0.85
0.005, 0.005, 0.08, 0.05, 0.02, 0.02, 0.85
0.005, 0.005, 0.08, 0.05, 0.02, 0.02, 0.85
0.005, 0.005, 0.08, 0.05, 0.02, 0.02, 0.85
0.005, 0.005, 0.08, 0.05, 0.02, 0.02, 0.85
0.001, 0.001, 0.01, 0.02, 0.005, 0.005, 0.973

The results are the following: $x_1 = 0.0011, x_2 = 0.0011, x_3 = 0.0122, x_4 = x_5 = x_6 = 0.0255, x_7 = 0.9691$.

These show that the treatment for this patient is quite successful. She is 97% cancer-free. At this point, adjuvant therapy could begin.

These are obviously all theoretical works. Only oncologists will judge their validity. As we have mentioned it earlier, the first author consulted with several doctors to develop these data with care and caution.

2.10 Conclusion

Many statistical studies on cancer are available online. However, no details of data about any treatment procedure are available. This monograph is meant primarily for researchers, graduate students, and advanced undergraduates majoring in bioengineering who have a good background in mathematics. However, for all medical practitioners, doctors, nurses, interns, and also patients several other books are available. We prefer to mention only two references that we find quite valuable in this respect. The first one has been published by the National Cancer Institute of USA [5], and the second one is [6]. In Miller et al. [5], a huge collection of data on various kinds of cancer and the survival rates have been recorded. In Shambaugh et al. [6], authors applied simple statistical methodologies to explain how medical practitioners and patients could analyze the success rates of methods of treatments and evaluations of the state of health of patients. In Dey [2], all the fundamental statistical concepts have been elucidated. Medical students and practitioners may find it handy.

Now we will discuss how treatments could be conducted following some mathematical models.

References

1. Feinstein, A.R.: Principles of Medical Statistics. Chapman & Hall/CRC (2002)
2. Dey, S.K.: A Practical Guide to Applied Statistics. Paradigm Publisher. Austin, Tx (2012)
3. Armitage, P.: Statistical Methods in Medical Research. John Wiley, New York (1971)
4. Fisher, R.A.: Statistical Methods and Scientific Inferences. Oliver & Boyd, Edinburgh (1959)
5. Miller, K., et al.: Cancer Treatment and Survivorship Statistics, 2019. The National Cancer Institute, USA (2019)
6. Shambaugh, E.M., et al.: Self Instructional Manual for Cancer Registrars. Book#7. Statistics and Epidemiology for Cancer Registrars. Cancer Statistics Branch. The National Cancer Institute, NIH Publication # 94–3766 (1992)
7. Mason, J., Newton, P.K.: Markov Model of Cancer Metastasis. BioRxiv, Cold Spring Harbor Lab (2018)
8. Davalos, V., Moutinho, C., Villanueva, A., Boque, R., Silva, P., Carneiro, F., Esteller, M.: Dynamic epigenetic regulation of the microRNA-200 family mediates epithelial and mesenchymal transitions in human tumorigenesis. Oncogene (2011). https://doi.org/10.1038/onc.2011.383
9. Coulie, P.G., Hanagiri, T., Takanoyama, M.: From tumor antigens to immunotherapy. Int J Clin Oncol **6**, 163 (2001)
10. Meissner, T.B., Li, A., Biswas, A., Lee, K., Liu, Y., Bayir, E., Iliopoulos, D., van den Elsen, P.J., Kobayashi, K.S.: NLR family member NLRC5 is a transcriptional regulator of MHC class I genes. Proc. Natl. Acad. Sci. **107**(31), 13794–13799 (2010)
11. Bloch, M., Klein, W.M., Hesse, B.W., et al.: Behavioral research in cancer prevention and control: a look to the future. Am. J. Prev. Med. **46**(3), 303–311 2014. https://doi.org/10.1016/j.amepre.2013.10.004
12. https://www.scientificamerican.com/article/evolved-for-cancer-2008-07/
13. https://www.tbi.univie.ac.at/~pks/Preprints/pks_349.pdf
14. Vernon Williams–vernvlw81@gmail.com:Mathematical Demonstration Darwinian Theory of Evolution
15. Price's equation made clear, Andy Gardner. Philosophical Transaction of the Royal Society B (2020)

Chapter 3
Attacker and Defender Model: The Dynamics of the Immune System

Abstract We will introduce the attacker/defender (or activator/inhibitor model). Cancer means a warzone inside the body. The attackers are the malignant cells carrying foreign antigens. They activate the war. Defenders of the body are our leukocytes, the white blood cells and medications used to contain/cure cancer. They are the inhibitors. We will study the first part of this war applying mathematical models.

3.1 Rationale

Nature is dynamic. Behind all bounties and beauties, each element of nature is battling for its survival constantly. That is the picture we will observe now through mathematical modeling on cancer treatment.

Many details happening in the microenvironment inside the body, especially inside cancerous tumors and tissues, were estimated statistically for the past twenty years. We begin with the following note.

Biochemicals and chemicals are entirely different. When two chemicals interact, if the ambient conditions are identical, the results should be identical. However, if exactly the same chemicals interact inside a human body, results could be widely and wildly different. That is why in vivo experiments conducted within a living organism and in vitro experiments conducted outside a living organism often produce very different results. Even in vivo experiments conducted in two similar living organisms could be entirely different. Body is built to defend itself from foreign aggressions by bacteria, viruses, pathogens, allergens, cancer, etc. When these invaders attack the body, the immune system (our leukocytes, which are the white cells) attacks back. Spleen is an organ that filters the blood. Abodes of leukocytes. It sits in the upper left of the abdomen. Bone marrow is primarily a hematopoietic organ. It is at the center for the entire immune system. From here cells which move to thymus become T-cells. Thymus is an endocrine gland between the lungs and just below the neck. Lymph nodes are the castles of the soldiers of our immune system. They are camps of hundreds of T-cells and B-cells. These are some of the most powerful fighters.

© Springer Nature Singapore Pte Ltd. 2021

S. Dey and C. Dey, *Mathematical and Computational Studies on Progress, Prognosis, Prevention & Panacea of Breast Cancer*, Forum for Interdisciplinary Mathematics, https://doi.org/10.1007/978-981-16-6077-1_3

There are about 500–700 lymph nodes in our body. Near the armpit and groin, there are about 100 lymph nodes. They are linked by lymphatic vessels where the fluid lymph flows. Around the breast areas and nearby, there are plenty of lymph nodes connected by hundreds of lymph vessels. Each lymph node is a castle of T-cells and B-cells. So, these organs are guarded by the body's defense mechanism.

When the foreign invaders carrying foreign antigens are identified, the T-cells and the B-cells launch very antigen-specific attacks. There are several varieties of T-cells. Cytotoxic T-cells (CD8+) are one of them. Together with B-cells, they form some ruthless army of fighters. CD8+ T-cells release their granules, kill an infected cell, then move to a new target, and kill again. They act like a serial killer. They form our defense against tumors. They secrete interleukin 2 (IL-2) which are cytokine molecules. They do cell-to-cell communication and stimulate strong movements of T-cells toward the sites of infections, inflammations, or tumors. They also secrete interferon gamma (IFN-γ) and tumor necrosis factors TNF-β, IL-4, IL-5, and IL-10. Wang and Lin [1] noticed that "TNF is a double-dealer. On one hand, TNF could be an endogenous tumor promoter, because TNF stimulates cancer cells' growth, proliferation, invasion and metastasis, and tumor angiogenesis. On the other hand, TNF could be a cancer killer. The property of TNF in inducing cancer cell death renders it a potential cancer therapeutic, although much work is needed to reduce its toxicity for systematic TNF administration."

Any promotion of TNF should be conducted with extreme caution.

"B-cells also have proteins called recombination-activating genes (RAGs) that enable them to rearrange the components of antibodies and customize their shapes." With the help of these RAGs, which were first discovered by Susumu Tonegawa (cancer research.org), it has been estimated that B-cells have the potential to synthesize over one trillion distinct antibodies. Unfortunately, abnormal RAG activity can also cause cancer-promoting mutations in B-cells and is believed to contribute to the development of lymphoma. This connection was discovered by Frederick W. Alt, Ph.D., and Klaus Rajewsky, M.D. (Cancer Research Institute, New York).

Recently, in 2011, Prof. Asit De from the University of Rochester Medical Center discovered a protein heat shock protein 27 (Hsp27) which is released by breast cancer cells that causes immune system unresponsive to the presence of cancer cells and causes more blood flow toward themselves so that they can get more food and oxygen and grow rapidly.

At the very early stages of cancer, in general, the immune system does give a good fight to destroy cancer cells. "However, if the rate of tumour growth begins to match the activity of our immune system, then we enter a stage of equilibrium… Some tumours can actually get fairly big but still be kept in check by our immune cells," says Elliott. "This behaviour can sometimes last for several years" [Cancer Research Center, UK, Feb. 2019].

In an adult body, there are about 10^{14} (one hundred trillion) cells, and at any time, some possibly disobey the rules of homeostasis (the law of dynamic equilibrium in the body) and mutate. They start growing abnormally, and the immune system, generally, destroys them. If this fact is not true, possibly 99% of the world population would have had cancer. Also, it takes several mutations of genes before a cell becomes cancerous. To escape the clutches of the immune system, sometimes these cells hide in adipose tissues or under the skin tissues, for years.

With the passing of time, cancer cells develop strong genetic changes and could adopt very ingenious methods to confuse the immune system. Then, T-cells and B-cells fail to launch an attack against them and cancer thrives. Practically, cancer cells have the ability to turn off the body's natural immune response and suppress the activity of local immune cells.

At this point, immunotherapy could revitalize the immune system and detect the villains, and body fights back. When monoclonal drugs come to assist them, cancer could be defeated. We will discuss more on these through mathematical modeling later.

In this chapter, it has been assumed that the immune system has detected the "non-self" antigen and is able to launch an attack. This may be called an activator/inhibitor model. An inhibitor in pathology usually means one who is employed to prevent an activator from being activated. Cancer activates itself. As it attacks the neighboring normal cells, it has the ability to transform its very genetic structure in its favor. When the immune system recognizes them, they literally fight back. It is a direct fight, not just an inhibition. Body is really a battlefield—a warzone. Metastasis means a newly opened fronts. The model could be called the attacker/defender model.

3.2 A Preliminary Model

In the very first preliminary model, we have considered how the immune system fights when a tumor is being formed and the malignant cells are not very active elsewhere in the body. Doctors often call it cancer in situ (local, not metastasized). It has been assumed that the antigen of a malignant cell has been detected as "foreign" (non-self) by the immune system and as such it triggers an attack by the immune system. Among trillions of cells in an adult human body, there is a strong probability that at any time one of these cells mutates signaling for an attack by the immune system, else cancer could start. Since everybody, fortunately, does not get cancer, it is evident that in general, the mechanism of the immune response is wide awake in a human body. Leukocytes rush to the sites of foreign aggression and grow in number. This is the naturally built-in chemotaxis in a human body. This idea is incorporated in our models. Researchers did find that sometimes the immune system fails to recognize that some mutated cells are "foreign." Then, doctors use medications to unmask the real chemistry of these truly "foreign" cells, thereby triggering attacks by the immune system.

Ductal carcinoma in situ (DCIS) is a noninvasive or preinvasive breast cancer. DCIS means the cells that line the ducts (the pipelines through which milk flows) have mutated and become cancerous. However, they have not spread through the walls of the ducts affecting the nearby breast tissues. This is stage zero cancer and very much curable. But early detection is a necessity. However, ductal carcinoma could be very invasive. The American Cancer Society predicts that every year, over

180,000 women, here in the USA, are diagnosed with invasive breast cancer and a majority of them have invasive ductal carcinoma.

Lobular carcinoma in situ (LCIS) is a lobular neoplasia. These cancer cells grow in lobules of the milk-producing glands of breasts. But they do not grow through the walls of the lobules. Oncologists do not call it cancer. They cannot be noticed in a mammogram. Yet we believe any such abnormality should be duly treated medically. No abnormality could be trusted.

Let us model mathematically some of the underlying principles of the growth of cancer and the fight between cancer and the immune system.

Let u represent a population of malignant cells. It starts growing uncontrolled. A simple mathematical model for that is:

$$du/dt = au, a > 0 \qquad (3.1)$$

If at $t = 0$, $u = u_0$, the solution is:

$$u(t) = u_0 exp(at) \qquad (3.2)$$

This clearly shows that a being positive, u will keep on growing. However, since the equation is only time dependent (changes inside the three-dimensional body are not considered), this implies that the tumor is still being formed and may not have started spreading to other parts of the body using blood vessels and lymphatic vessels. So, it could still be *in situ* (in place). Still, it must be destroyed. However, in general, cancer cells never stay at a particular location in the body, in a three-dimensional configuration, we should consider u as $u(x, y, z, t)$ moving with time at various locations of the body and that is not given by (3.2). For a benign tumor the model is valid. Not for a malignant tumor in general. These cells start spreading from the very beginning. Also most unfortunately, not all cancer cells at different locations of the body behave in the same way. So, a three-dimensional time-dependent model is necessary. We will study this in the next chapter. The time-dependent model in (3.1) will reveal how an individual cancer in situ is behaving as time changes.

We need to say a bit more on the a, the rate of growth of u. If in case $a = a(t)$ is a decreasing function of time, that does not mean u is decreasing. It means rate of growth of u is decreasing and yet u may still increase. Let us look into this phenomenon:

Let $du/dt = a(t)u$, $a(t) > 0 \ \forall t$. Let $a(t)=1/t$. Obviously $a(t)$ is a decreasing function of t. Mistakenly, one may think that cancer is decreasing. That is wrong. Mathematically, so long $du/dt > 0$, u is increasing. Integrating we get:

$u(t) = (t/t_0)u_0 = (u_0/t_0)t$ (representing a line on the u_t plane with a constant slope (u_0/t_0)).

So $du/dt = (u_0/t_0)$ is a constant, while $u(t)$ is increasing linearly with time. At a time $t < < t_0$, u could be unnoticeably small, yet growing linearly. This could pose a life-threatening condition in the long run. Sometimes, we must watch and analyze these situations very keenly.

If an attack is launched to control this growth of u, then (3.1) may be written as: $du/dt = a_1 u - a_2 u$, **both a_2 and a_1 are positive** (a_1 and a_2 are called rate constants). The second term is an inhibitor.

If $a_2 > a_1$, the solution is: $u(t) = u_0 exp((a_1 - a_2)t)$ and $u \to 0$ as $t \to \infty$.

If $a_1 > a_2$, $u(t)$ keeps on increasing as $t \to \infty$. If $a_1 = a_2$, u will remain a constant.

If a_1 and a_2 are time dependent, the solution is: $u = u_0 exp(-\int_{t_0}^t (a_2(t) - a_1(t))dt)$.

So, if $\int_{t_0}^t (a_2(t) - a_1(t))dt > 0$ $\forall t > 0$, then the treatment will work and the tumor must regress.

With this in mind, we will proceed to develop a mathematical treatment of cancer in situ (like DCIS) or a cell in motion in time. There are some protagonists, which help cancer to grow (like stress, certain foods, carcinogens, etc.) and some antagonists (like our immune system, medications, etc.) which fight back to stop any uncontrolled growth of tumors.

3.3 First Line of Defense of the Body

If we add a natural antagonist, the immune system, having a fixed value V_0 in (3.1), then we get the following equation:

$$du/dt = a_1 u - a_2 u V_0 \tag{3.3}$$

Here, the first question is to look into the dimensional analysis for the equation. u and V_0 are treated like biochemical masses. So, their dimensions will be represented by mass (M). $a_1 =$ the growth of u per unit time, and $a_2 =$ destruction of u per unit time, per unit of V_0. So dimensionally, the dimension of the right side is: $M/T - (1/(T*M))*(M*M) = M/T$. Obviously, the dimension of the left side of (3.3) is also M/T, where T is time. So, the equation is dimensionally matched.

Subject to the previous initial condition in (3.1), the solution of (3.3) is given by:

$$u(t) = u_0 exp((a_1 - a_2 V_0)t) \tag{3.4}$$

If $a_1 < a_2 V_0$, then as $t \to \infty$ $u(t) \to 0$. Logically what V_0 is doing is: It is reducing the rate of growth of u. Thus, cancer in situ is cured by reducing the rate of growth of cancer. This is not true in the real world.

If $a_1 = a_2 V_0$, then $u(t) = u_0$, $\forall t$, however if $a_1 > a_2 V_0$, then u will grow without bounds. At this point, there are two choices: (1) introduce another antagonist W to inhibit the growth u, or (2) make V_0 vary as time progresses, such that V_0 may adjust itself in order to damp out u, thereby curing the disease. In general, in the real world cancer may never be fully cured.

3.4 Immune System Gets A Helper Drug

If we introduce another antagonist W_0, which is a chemo, then the model will be:

$$du/dt = a_1 u - a_2 u V_0 - a_3 u W_0 \qquad (3.5)$$

With a solution:

$$u(t) = u_0 exp((a_1 - a_2 V_0 - a_3 W_0)t) \qquad (3.6)$$

If $a_1 < (a_2 V_0 + a_3 W_0)$, then as $t \to \infty$ $u(t) \to 0$; if $a_1 = (a_2 V_0 + a_3 W_0)$, then $u(t) = u_0$ $\forall t$; however, if $a_1 > (a_2 V_0 + a_3 W_0)$, then u will grow without bounds.

These findings point out very clearly that to wipe out u, a_1, the rate of growth of u, should be a primary target of the inhibitors.

In general, while conducting any treatment, values of W are kept fixed.

When the patient's immune system is weak, that happens to many patients suffering from HIV, some allergies, alcohol, poor nutrition, even flu virus that could weaken the immune system, cancer cells take full advantage of them and thrive. So V_0 is often a drug given at a fixed dosage to patients to improve the quality of their immune resistance.

Since this is a simple linear differential equation, quite elementary, there is no need to show any computational results. With various inputs for a_0, a_1, a_2, and V_0 and W_0, one can use the formula in a scientific calculator and see the results.

3.5 Immune System Changing Strategies to Fight Cancer

Cancer cells are dangerously clever. They often change their strategies of fighting. They can hide their antigens and pretend that they are normal cells. Immunotherapy renders the immune system more precise and powerful, unmasking the antigens of the cancer cells and attacking their pathways. Here, we will consider a variable v (representing our immune response) which will constantly adjust itself during the fight. We consider here only a time dependent model. In that case, the model will consist of two coupled nonlinear ordinary nonlinear differential equations.

$$du/dt = a_1 u - a_2 u v \qquad (3.7)$$

$$dv/dt = b_1 u - b_2 u v \qquad (3.8)$$

Here, we should observe that on the basis of our assumption, v grows in response to the growth of u which is given by the term $b_1 u$. We may think that in a three-dimensional scenario, u is really u_{ijk}, meaning that cancer cells vary from point to point in the body, which is three-dimensional and whenever and wherever, it is detected by the immune cells, they reach that location and proliferate as necessitated by the presence of u_{ijk} and fight. So, b_1u denotes the growth of v corresponding to the growth of u.

If in case $b_2 = 0$, $v = v_0 + b_1 \int_{t_0}^{t} u dt$. That means the present population of v equals the initial population of v plus the increment matching the total population of u at that location in the time interval (t_0, t). This happens due to chemotaxis. This is the natural response of the body. This means when bacteria, virus, or any pathogen attacks the body, our immune system fights back. Then, our T-cells, B-cells, natural killer cells, etc., move swiftly toward the attackers and multiply themselves as needed, at the locations of the assembly of foreign attackers. Body gets warm. Blood count goes up. The basic concept is in order to launch a fight against an incredibly tough attacker, v, our defenders have to outnumber and overpower the enemy. That is given by the second term of the right side of (3.7). The term "$-b_2uv$" indicates that the attacker u is killing some army of the defenders. The term b_2= destruction rate of v by each unit of u. The fight is externally expressed through fever. Temperature of the body increases. Fever is a symptom of a disease. Even if it is a low-grade fever, an increase of blood count takes place. There is inflammation at that point. So entropy must increase at those locations.

Thus if the microenvironments of all sites of the flux of heat inside the body could be observed by some microscopic MRI, some form of immunotherapy could be administered and cancerous cells could be detected and destroyed at the earliest stage of the onslaught of cancer.

In reality, whereas u may grow uncontrolled, the growth of v must remain within certain predetermined values. Because blood count in our body has an upper limit. For example, v being the number of leukocytes, it should not grow beyond a certain point. If it is a drug, then the body can tolerate only up to a certain amount of this drug. These are some restrictions that should be taken into consideration in modeling. In all of our codes, we considered these primary aspects of treatment.

Both analytical and numerical solutions have been discussed in [2]. Here we will briefly consider them:

From Eq. (3.7), it is quite clear that if $\forall t$, $v > a_1/a_2$, then $du/dt < 0$, so u must decrease with time. If at the same time $v = b_1/b_2$, v is a constant. So a fixed dosage of v should keep the body fit and kill cancer in situ. If $a_1 < b_1$ and $a_2 > b_2$, then it implies that v will outgrow u and will destroy u at a faster rate than u destroying v, thereby leading to a cure. Then, we get

$$(a_1/b_1) < 1 < (a_2/b_2) \tag{3.9}$$

In Fig. 3.1, this has been validated computationally. This system has a condition number $= 51.5032$. So it is not very stiff (discussed in the next section).

So the system is mildly stiff.

With $v = 25$, $a_1 = 0.01$, and $a_2 = 0.0075$, $du/dt = (a_1 - 25a_2)u = -0.00875u < 0$. (Note $u \geq 0$. So, u must be decreasing.) It may also be noted that while u is continuously decreasing, v has reached a steady state. That means if this steady-state value is maintained by the immune system, cancer will be cured.

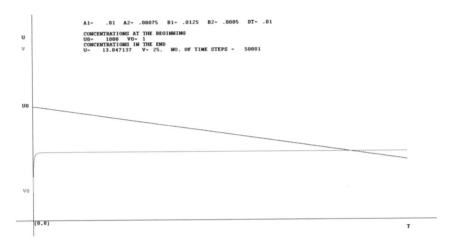

A1= .01 A2= .00075 B1= .0125 B2= .0005 DT= .01

CONCENTRATIONS AT THE BEGINNING
U0=   1000   V0= 1
CONCENTRATIONS IN THE END
U=   13.047137   V= 25.   NO. OF TIME STEPS =   50001

Fig. 3.1 B1 > A1, B2 < A2, cancer sharply decreased

If $a_1 = b_1$, then also progress of u will be inhibited if $b_2 < a_2$. This could be verified computationally. From (3.7) and (3.8), we get:

$$du/dv = (a_1 - a_2 v)/(b_1 - b_2 v) \tag{3.10}$$

If $a_1 = b_1$ and $a_2 = b_2$, then $du/dv = 1$, giving

$$u = v + u_0 - v_0 \tag{3.11}$$

So even if v_0 is significantly small, the treatment will keep u under control, provided v is kept under control, because the immune system must not exceed a certain limit. If $u_0 > v_0$, $u > v$, v being always bounded, u could be unbounded. The disease could be fatal. So care should be taken. If $u_0 < v_0$, $u < v$ and if $u_0 = v_0$, $u = v$. Depending upon the state of health of the patient, the patient will live on with a tumor. Interestingly, if the immune system is relatively strong (meaning that the patient is young and healthy) and cancer is in situ, still the method of treatment should be such that $a_1 < b_1$ and $a_2 > b_2$ else the tumor will grow and metastasize. This is represented by Fig. 3.2a.

If at $t = 0$, $u = u_0$, and $v = v_0$, then integrating (3.10) we get:

$$u = u_0 + \alpha(v - v_0) + \beta \ln((b_1 - b_2 v)/(b_1 - b_2 v_0)) \tag{3.12}$$

$$\text{where } \alpha = a_2/b_2 \text{ and } \beta = (b_1/b_2)(a_2/b_2 - a_1/b_1) \tag{3.13}$$

If $v = a_1/a_2$ (which means v has a fixed value $= V_0$ then, from (3.7)), $du = 0$, then $u = u_0$, $\forall t \geq 0$. If a_1 and a_2 are known and if V_0 is a drug for treatment, the tumor *in situ* will not grow and remain confined at the same location. So, the patient will keep on taking the drug and live on like a diabetic patient taking regular drugs and staying well.

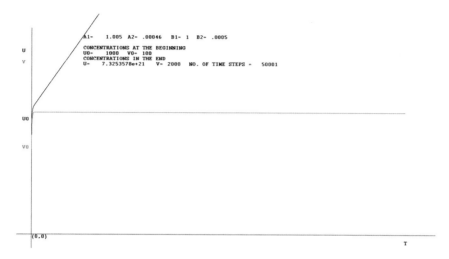

Fig. 3.2 Cancer increasing fast

3.6 Stiff Computations

In chemical biology, mathematical models are often very stiff [3]. When systems of equations are stiff, not just any numerical method of solution works. Special methods are used. So before solving the model numerically, first we should study how stiff it is and then apply a special numerical technique for computational studies. In this case, we will apply a numerical method of our own. It was developed by the second author in 1982 at NASA Ames Research Center, California [4].

Let us briefly consider how to measure stiffness of a nonlinear system. Let a nonlinear system may be represented by:

$$dZ/dt = F(u, v) \qquad (3.14)$$

where $Z = (u, v)^T \epsilon R^2$, and $F(u, v) = (f(u, v)g(u, v))^T. R^2 =$ real two-dimensional space.

Subject to the initial conditions: at $t = t_0$, $u = u_0$ and $v = v_0$.

First, we will linearize the system at (u_0, v_0). This gives:

$$dZ/dt = F(u_0, v_0) + F'(u_0, v_0) \begin{pmatrix} u - u_0 \\ v - v_0 \end{pmatrix}$$

where $F'(u_0, v_0)$ is the Jacobian matrix $\begin{pmatrix} \partial f/\partial u & \partial f/\partial v \\ \partial g/\partial u & \partial g/\partial v \end{pmatrix}$ evaluated at (u_0, v_0). The matrix is expressed rowwise.

Here, a new notation of matrix has been used: Rows have been arranged one after another.

If the eigenvalues of $F'(u_0, v_0)$ are λ_i $i = 1, 2$, the condition number may be given by

$K = max|\lambda_i|/min|\lambda_i|$. If $min|\lambda_i| = 0$, then the next least eigenvalue is chosen. If $K >> 1$, then the system is stiff.

Our numerical method is a predictor–corrector method, often called "the Charlie method" [5]. The predictor is Euler forward, and the corrector is a convex-type operation. It moves forward with time. It is an explicit differential equation solver.

3.7   Computational Studies Applying CDey (Charlie's) Algorithm [4, 5]

Let us consider a system of ordinary differential equations:

$$dU/dt = F(U) \tag{3.15}$$

where $U = (u_1, u_2, ...u_I)^T \epsilon R^I$, $R^I = I - dimensional\ real\ space$
And $F(U) = f_i((u_1, u_2, ...u_I), i = 1, 2, ..., I$.
Let U^n be known at some time step t_n. Then the algorithm is:

$$U^{n+1/2} = U^n + hF(U^n) \quad \text{(The Predictor)}$$
$$U^{n+1} = (I - \Gamma)U^n + \Gamma\{U^n + hF(U^{n+1/2})\} \quad \text{(The Corrector)} \tag{3.16}$$

where $U^0 =$ initial conditions of U, $h = \Delta t =$ the time step, $n = 0, 1, 2, ..., NT$, $NT =$ number of time steps to be computed;
$\Gamma = diag(\gamma_1, \gamma_2, ..., \gamma_I) =$ a diagonal matrix with $0 < \gamma_i < 1 \forall i$.
Lomax [7] analyzed the stability properties which could be used to choose the values of $\gamma_i's$. If we choose $\gamma_i = 1/2$, this numerical method becomes the second-order Runge–Kutta method.

Here,
$f(u, v) = a_1u - a_2uv$ and $g(u, v) = b_1u - b_2uv$,
Let $U_1 = u$, and $U_2 = v$. Then the equations may be written as:

$$dU_1/dt = f_1(U_1, U_2) = a_1U_1 - a_2U_1U_2$$
$$dU_2/dt = f_2(U_1, U_2) = b_1U_1 - b_2U_1U_2 \tag{3.17}$$

The Jacobian matrix is a t matrix given below to be evaluated at the

$$F'(U_1, U_2) = \begin{pmatrix} \partial F_1/\partial U_1 & \partial F_1/\partial U_2 \\ \partial F_2/\partial U_1 & \partial F_2/\partial U_2 \end{pmatrix}$$

initial state.

Charlie's method (CDey algorithm) is:

$$U_1^{n+1/2} = U_1^n + h(a_1 U_1^n - a_2 U_1^n U_2^n) \tag{3.18}$$

$$U_2^{n+1/2} = U_2^n + h(b_1 U_1^n - b_2 U_1^n U_2^n) \text{ (the predictor)}$$

$$U_1^{n+1} = (1 - \gamma_1)U_1^{n+\frac{1}{2}} + \gamma_1(U_1^n + h(a_1 U_1^{n+\frac{1}{2}} - a_2 U_1^{n+\frac{1}{2}} U_2^{n+\frac{1}{2}})) \tag{3.19}$$

$$U_2^{n+1} = (1 - \gamma_2)U_2^{n+1/2} + \gamma_2(U_2^n + h(b_1 U_1^{n+1/2} - b_2 U_1^{n+1/2} U_2^{n+1/2})) \text{ (the corrector).}$$

$n = 0, 1, 2, ..., NT$. ($NT =$ number of time steps)

The parameters γ_1 and γ_2 satisfy $0 < \gamma_1, \gamma_2 < 1$. They are convex parameters.

Other than Lomax [7], some authors [5] have also checked the numerical stability of Charlie's method. We have discussed this in Appendix A.

Let us consider the following data $a_1 = 1.25, a_2 = 0.00056, b_1 = 1.116, b_2 = 0.00055, u_0 = 1000, v_0 = 1$.

Here $du/dt = 113.7 > 0$ at $t = 0$. So at the outset u must sharply increase. And it did.

Very close to the origin, both u and v have increased sharply.

If $v = \frac{b_1}{b_2}$, then from (3.8), $dv/dt = 0$; hence $v = v_0 \ \forall t \geq 0$ and $v_0 = b_1/b_2$. Then (3.7) reduces to: $du/dt = (a_1 - a_2 b_1/b_2)u$, giving a solution: $u(t) = u_0 exp(a_1 - a_2 V_0)t$, where $V_0 = b_1/b_2$.

So, if $a_1 b_2 > a_2 b_1$, *then the tumor will keep on growing, if* $a_1 b_2 = a_2 b_1$, *tumor will remain the same, and if* $a_1 b_2 < a_2 b_1$, *the tumor will start shrinking.*

An interesting result is given by Fig. 3.4.

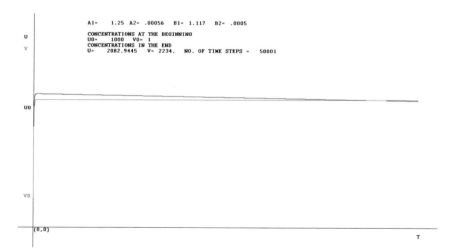

Fig. 3.3 Cancer decaying slowly

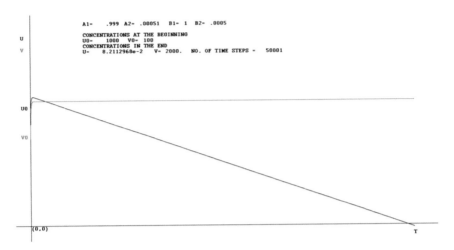

Fig. 3.4 Cancer decreasing

Here, b_1 barely exceeds a_1 and a_2 barely exceeds b_2; the model is a success. It is stiff.

First we should check the stiffness of the model represented by (3.7) and (3.8). If we express this as

$du/dt = f(u, v)$, and $dv/dt = g(u, v)$, then $\partial f/\partial u = a_1 - a_2 v$, $\partial f/\partial v = -a_2 u$, $\partial g/\partial u = b_1 - b_2 v$, and

$\partial g/\partial v = -b_2 u$. With the following inputs, $a_1 = 10^{-2}$, $a_2 = 7.5 * 10^{-4}$, $b_1 = 5 * 10^{-4}$, $b_2 = 5 * 10^{-5}$, $u_0 = v_0 = 10^3$, the Jacobian at $t = 0$ is: (written rowwise)

$[-0.74 \quad -0.75].$
$[-0.0495 \ -0.05].$

The eigenvalues are -0.7902 and 0.0001582. So, the condition number is about $5 * 10^3$. Therefore, the system is moderately stiff.

Numerical solutions are given by the three figures, namely Figs. 3.5 and 3.6. Let us briefly analyze these numerical results. Since, at $t = 0$, both $f(u_0, v_0) < 0$ and $g(u_0, v_0) < 0$, that means at $t = 0$, $du/dt < 0$, and $dv/dt < 0$. So both u and v must be initially decreasing functions of t. From the graphs, it is evident that, at some $t = t_N$, v has gone to a steady state, and u is increasing. At that point $v = b_1/b_2$, giving $du/dt = ((a_1 b_2 - a_2 b_1)/b_2)u$. Substituting the values of the parameters a_1, a_2, b_1, b_2, we find $du/dt = -0.65 < 0$. So u must be slowly decreasing first. However, later $du/dt > 0$. So the graphs are mathematically correct.

For computational studies, such validations are very much needed. For all time-dependent models, researchers need to do these checkings routinely.

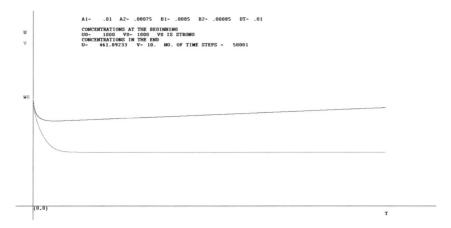

Fig. 3.5 B1 < A1, B2 > A2, cancer sharply increased

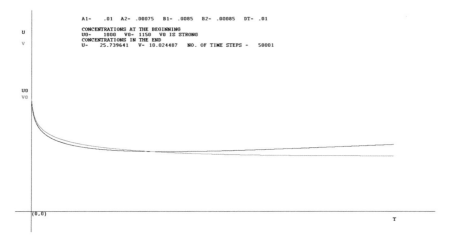

Fig. 3.6 V0 > U0, but B1 < A1, B2 > A2, treatment failed, first tumor growth is checked but then it sharply increases

3.8   Use of an Additional Drug

Oncologists examine the strength of the immune system very frequently and often apply an additional variable drug to combat a tumor in situ. So, a second drug w has to be used and the new model becomes:

$$du/dt = a_1 u - a_2 uv - a_3 uw \tag{3.20}$$

$$dv/dt = b_1 u - b_2 uv - b_3 vw \tag{3.21}$$

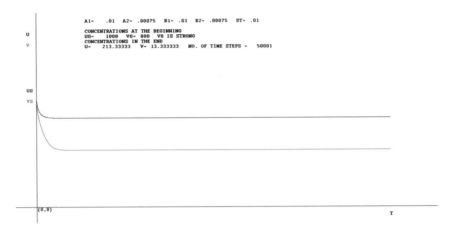

Fig. 3.7 B1 = A1, B2 = A2; Tumor growth is under control

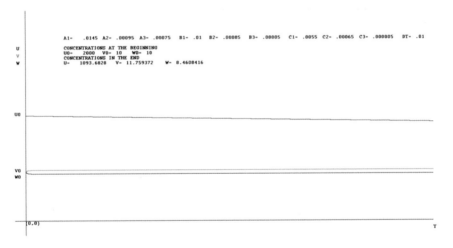

Fig. 3.8 A1 < B1 + C1, A2 > B2, A3 > C2, Tumor is large and both v & w are relatively small at to. Tumor shrinking slowly

$$dw/dt = c_1 u - c_2 uw - c_3 vw \tag{3.22}$$

We first chose: $a_1 = 0.0145$, $a_2 = 0.00095$, $a_3 = 0.00075$, $b_1 = 0.01$, $b_2 = 0.00005$, $b_3 = 0.00005$, $c_1 = 0.0055$, $c_2 = 0.00065$, $c_3 = 0.000005$. $u_0 = 2000$, $v_0 = w_0 = 10$.

If $f(u, v, w) = a_1 u - a_2 uv - a_3 uw$, then at $t = 0$, $f(u_0, v_0, w_0) = -5$. That means u is a decreasing function.

$dv/dt = 18.995$ at $t = 0$, so v is increasing, and $dw/dt = -2.0005$, so w is decreasing. The graph shows these results.

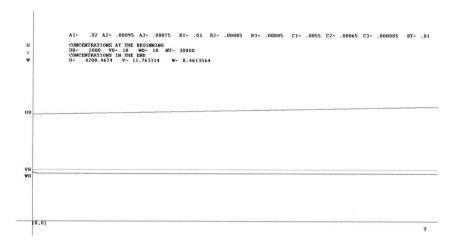

Fig. 3.9 A1 + B1 + C1, A2 > B2, A3 > C2, Tumor is aggressive, growing fast and both v & w are relatively small at to. Treatment failed

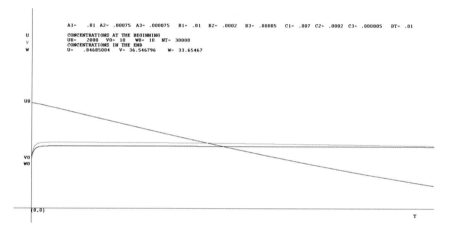

Fig. 3.10 A1 = B1 > C1, A2 > B2 = C2, A3 > B3 > C3, Tumor shrinks

Now we need to see how stiff the system is. The Jacobian matrix of the system evaluated at $t = 0$ is: (expressed rowwise)

$$[a_1 - a_2v_0 - a_3w_0 \quad -a_2u_0 \qquad\qquad -a_3u_0] = [-0.0025 \quad -1.9 \qquad -1.5 \qquad]$$
$$[b_1 - b_2v_0 \qquad\qquad -b_2u_0 - b_3w_0 \quad -b_3v_0] \quad [0.0015 \quad -1.7005 \quad -0.0005]$$
$$[c_1 - c_2w_0 \quad -c_3w_0 \quad -c_2u_0 \qquad\qquad -c_3v_0] \quad [-0.001 \quad -0.00005 \quad 1.30005]$$

The eigenvalues are: -0.003022851794615, -1.6988174597168635, -1.3012096884930575. Condition number $= 1.6988/0.003023 = 562$. So the system is relatively stiff. When the immune system fails to destroy the tumor, an extra adjustable drug w is brought in and the model is described by (3.20), (3.21), and (3.22). Results are given by Figs. 3.7, 3.8, 3.9, and 3.10. Numerical solutions consistent with our mathematical discussions are given in Fig. 3.8.

3.9   Modeling with Logistic Equation

Some researchers have discussed [6] a logistic equation to model the growth of a tumor. This is certainly not a very practical model. Because as a tumor grows in the primary site, other tumors at some secondary sites start growing too. Sometimes they may be hard to detect.

The logistic model is given by the equation:

$$du/dt = au(1 - u/Q), \text{ at } t = 0, u = u_0 \tag{3.23}$$

$u(t)$ = concentration of cancer cells in the blood.
a = rate of growth of cancer cells.
Q = the capacity of carrying cancer cells at a steady state. (We need to remember that cancer cells move in a group in the bloodstream and lymphatic vessels before forming a tumor.)

At the steady state, $du/dt = 0$, and $u = Q$. This is just speculative information. In the real world, cancer may never stop growing unless it could not get a supply of oxygen and nutrients through blood. Surprisingly they could be hypoxic. Also while metastasizing, one tumor may remain the same or even shrink but another one will grow. Such cases do not follow logistic equations.

The analytical solution of (3.23) is:

$$u = (u_0 Q)/\{u_0 + (Q - u_0)exp(-at)\} \tag{3.24}$$

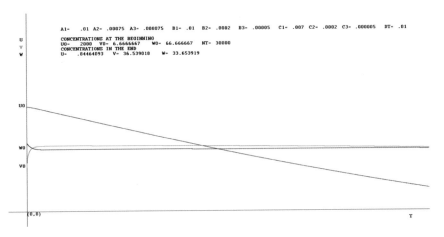

Fig. 3.11   A1 = B1 > C1, A2 > B2 = C2, A3 > B3 > C3, Tumor shrinks

From (3.24) if $a > 0$, then as $t \to \infty$, $u \to Q$. If $a = 0$, u remains u_0. If $a < 0$, then $t \to \infty$, $u \to 0$.

The first result is seldom applicable to cancer growth. Many tumors keep on growing elsewhere in the body. This shows that if the rate of growth is zero, u will remain what it was and if the tumor starts shrinking it will vanish. This is valid if and only if it is benign or strictly in situ.

Let us assume that there are three different kinds of medicines applied to inhibit the growth of cancer. Then (3.23) becomes:

$$du/dt = au(1 - u/Q) - b_1 u v_1 - b_2 u v_2 - b_3 u v_3, \text{ at } t = 0, \ u = u_0 \qquad (3.25)$$

where b_i = the rate at which u is destroyed per unit of time per unit of v_i ($i = 1, 2, 3$). These are the rate constants as discussed earlier. Equation (3.25) may be expressed as:

$$du/dt = (a - b_1 v_1 - b_2 v_2 - b_3 v_3)u - au^2/Q \qquad (3.26)$$

a = rate of growth of cancer and it is always positive. Also $Q > 0$.

If $\alpha = (a - b_1 v_1 - b_2 v_2 - b_3 v_3) \leq 0$, then u is a decreasing function of time. This is not a three-dimensional model. So, it is a realistic treatment only if the tumor is benign. A notable aspect is that from (3.23) we could notice that if with some treatment, we could make a = the rate of growth = 0, $u = u_0$ = a constant. Tumor will not die. Also, from (3.26) if $\alpha > 0$, u will start decreasing, when it will exceed $(\alpha Q/a)$. However, at the same time if a medication is used such that $a \to 0$, then $u \to 0$ as $t \to \infty$.

3.10 Conclusion

In this chapter, we have developed time-dependent models depicting the war between the defense system and cancer. These are applicable only to cancer in situ. But such cases are relatively rare. Oncologists often say that these are curable. Even though we may think that we can follow a particular cancer cell and its interaction with a particular element of the immune system and of medications, that gives very incomplete information, because even being at the same location cancer cells are able to change their antigens, their rates of growth, their strategy of fighting. Therefore, oncologists examine blood tests very often and do MRIs.

We think that even in a whole-body scan when no tumor other than one at the primary site is located, doctors should keep on searching for other tumors which most doctors do during follow-ups. Cancer is a whole-body issue inside the body, not a local one. So only time-dependent models can reveal very little information about this dangerous disease. Furthermore, cancer cells at different locations of the body could behave very differently. Accordingly, to be effective, the immune system must adjust and medications must do the same. Oncologists say they do this routinely [3, 8]. We will see these and more in Chap. 4 by using a three-dimensional model.

One primary aspect of cancer that we have not been able to discuss in a time-dependent one-dimensional model is that cancer never dies. A dead cancer cell may not be truly dead. It can release proteins to create new malignant cells. And they come out from necrotic tumors. That could be the reason why surgery, radiation, and chemotherapy (cut–burn and poison policy) often do not work. A three-dimensional model is very much needed.

References

1. Wang X., Lin, Y.: Tumor necrosis factor and cancer, buddies or foes? Acta Pharmacol. Sin. **29** (2008)
2. Dey, S.K., Dey, S.C.: Mathematical modeling of breast cancer treatment. In: Sarkar, S., Uma, B., De, S. (eds.) Applied Mathematics. Springer, Berlin (2014)
3. Chatterjee, A., Magee, J.L., Dey, S.K.: The role of homogeneous reactions in the radiolysis of water. Radiat. Res. **90**(1) (1983)
4. Dey, S., Dey, C.: An Explicit Predictor-Corrector Solver with Applications to Burgers' Equation. NASA Technical Memorandum, pp. 84402 (1983)
5. Osman, H., Wood, A.S.: Stability of Charlie's method on linear heat conduction method. Matematika, Jilkl.17, Bil.1, UTM (2001)
6. Murray, J.D.: Mathematical Biology. Springer, Berlin (1993)
7. Lomax, H.: Private Communication. NASA-Ames Research Center, Moffett Field, California (1983)
8. Ghafoori, A.: Radiation Oncologist in Austin (private communication)
9. Dey, S.K., Dey, S.C.: Mathematical modeling of breast cancer treatment. In: Special Booth Lecture Series. Eastern Illinois University (2003)
10. Dey, S.K., Dey C.: An Explicit Finite Difference Solver by Parameter Estimation. Engineering Software IV. In: Proceedings of the 4th International Conference of Kensington Exhibition Center, London, England, June 1985. Springer, Berlin (1985) (Edited by R. A. Adey)
11. Wiseman, C.: Oncologist in Beverly Hills, California (private communication)

Chapter 4
Mathematical Modeling of Metastatic Cancer

Abstract Cancer cells do not stay at one spot. Using blood vessels and lymphatic vessels, they move from the primary site to secondary sites. There they attempt to transform normal cells into malignant cells, defeat the immune systems or "trick" them such that they disregard them or even help them to spread, and deprive the normal cells from having nutrients and oxygen. They totally dominate the microenvironment around them in the tissues and start forming new colonies. This is metastasis. Then they try to create more blood vessels (angiogenesis), consume more nutrients, and grow. So now our objective is to model this scenario mathematically and look for methods to inhibit this process.

4.1 Rationale

"Mutations in an oncogene, a tumor suppressor gene, or a gene that controls the cell cycle can generate a clonal cell population with a distinct advantage in proliferation. Many such events, broadly divided into the stages of initiation, promotion, and progression, which may occur over a long period of time and transpire in the context of chronic exposure to carcinogens, can lead to the induction of human cancer. This is exemplified in the long-term use of tobacco being responsible for an increased risk of lung cancer" [1]. As the tumors grow, the body gets emaciated, loses all strength, and finally the patient dies. This is the nature of metastatic cancer. That is why cancer is so deadly. But there is more to this story. Cancer cells do not die. The p-53 gene which turns on natural apoptosis, causing death to normal cells in due time, is inactive in a cancer cell. So a cancer cell must be killed, and dead cells must be phagocytized by macrophages (a kind of white blood cells). By depriving cancer cells of nutrients like oxygen and glucose, they may be killed although they could be hypoxic.

One main point is if the rate of growth of cancer is zeroed, they still can grow. Necrotic malignant cells can do that. This could be modeled mathematically by using a flux term. For that, a three-dimensional model is needed.

© Springer Nature Singapore Pte Ltd. 2021   87
S. Dey and C. Dey, *Mathematical and Computational Studies on Progress, Prognosis, Prevention & Panacea of Breast Cancer*, Forum for Interdisciplinary Mathematics, https://doi.org/10.1007/978-981-16-6077-1_4

Thus, we need to see the growth, the reaction (between cancer cells and the immune system), and dispersion of cancer cells. This brings the pictures of both reaction and dispersion of biological cells (in the form of interactions of various proteins) which should be presented mathematically by reaction–dispersion equations.

We have two primary objectives: (i) to model a fight against each and every cancer cell wherever and whenever it can be detected in the blood, in the lymph or in a tissue and destroy it and (ii) to model the use of appropriate medications which will help the body in this regard.

The bottom line is as follows: By all means, cancer must be defeated, and life must prevail. There are over 200 genes in the body which have the potential to cause cancer, practically at any age. There are over 100 oncogenes that have been identified [1] and over 40 proto-oncogenes. Some scientists think that there are 723 oncogenes. While oncogenes cause cancer, proto-oncogenes do not. But if they mutate, they cause cancer. Those who get cancer at a very young age, they must have inherited the genes. It takes several mutations for a gene to cause cancer. Cancer should not be considered a disease for the old.

The game of a defensive war is happening in the body which is three dimensional in space and changing in time. So the corresponding mathematical model must be time dependent and three dimensional.

If and when tumors regress, all the agencies of defense must maintain the same uniform state of vigilance. That means whereas the tumors must continuously shrink, the treatments must continue till a steady state is found. Arrival of steady states means what mode of treatment should continue.

All cells, good and bad, have antigens (proteins) on their surfaces. Red blood cells have millions of antigens (self-antigens) which our immune system ignores (unless there is some blood transfusion) and our leukocytes (white blood cells) have the antigens human leukocyte antigen (HLA).

When leukocytes notice the presence of "foreign" antigens, carried by bacteria, virus, cancer cells, etc., they launch an attack. To avoid this attack, cancer cells often mask their antigens and appear to be "non-foreign or self." So with targeted drugs, they should be unmasked first.

To develop our models, we have considered all substances in the body acting like biochemicals and "the war" means action and reaction between them. In this monograph, we have used the word "antigens" to refer to "cancer cells." They are "foreign" or "non-self." All of these biochemicals react with each other and move to other locations or come in from other locations. So the model should be simulated by reaction–dispersion (diffusion) equations. Mathematically, these reactions bring in nonlinearity, and diffusion makes the models parabolic. Since, to cover a large area of the body, the models should be large scale, their finite difference analogs are stiff (Appendix B). We will discuss more on this later. This is a big challenge in computational studies.

Now, regarding treatment, there are three overall stages: (i) neoadjuvant therapy, (ii) therapy with various medications, and (iii) adjuvant therapy.

Neoadjuvant therapy is meant to reduce the size of the cancerous tumor along with its spread. For example, applying neoadjuvant therapy, a breast cancer patient's tumor is reduced, its spread is impeded, and the patient often goes through a lumpectomy instead of mastectomy, keeping a part of her breast which is not affected by cancer.

In this regard, according to the American Cancer Society, breast cancer is caused more by lifestyle than inheritance of defective genes. However, if it happens at a very young age, inheritance is more likely to be the cause [2].

Regarding all the graphs, in general, we have used $log_{10}(u)$ versus t or $log_{10}(10*u)$ $versus$ t. There are two primary reasons: (1) At the beginning, we often considered large values for the antigens and planned to see their exact changes as time moves on. We also considered a large number of time steps so that we can notice when $u(t)$ gets very large, how medications need to be adjusted. (2) We continuously check the computational stability of the numerical method which is marching forward with time with no internal iterative loop. This time-dependent solution for a large number of time steps reveals the numerical strength of the algorithm. However, the stability of this numerical algorithm has been studied in the Appendix A. This should give more encouragement and impetus to the users of computational methods to apply this numerical scheme for large-scale vector computations in modern supercomputers. In the future, attempts should be made to make the software for this algorithm available.

The graphs are drawn using colored lines. These give quite clearly the pattern of the progress of cancer and the corresponding treatments as time goes by.

At the end of the chapter, we considered intralesional or intratumoral chemotherapy, a rather new procedure that many doctors are adopting. Here a chemo is directly injected into the tumors. Mathematically, we will study such treatments. They have been found to be not very effective unless a chemo is added to destroy the runaway cancer cells moving through blood and lymphatic vessels. Dr. Blackman stated (QUORA, Feb.18, 2016) the following: "There are rare instances where intratumoral injection is used, but a large part of the reason systematic (whole body) chemotherapy is used is that you're trying to kill the cancer cells you CANNOT see—those tumor cells that are beyond the margin of the tumor. You cannot guarantee that you're going to get adequate concentrations of drugs to the periphery of the tumor or into the surrounding tissues that may have been invited (should be invaded by cancer cells) with tumor cells. In addition, patients with multiple sites of disease in hard-to-reach places are not amenable to intratumor injection."

We have validated this mathematically in this chapter and in the next chapter. Here we have to note a few aspects of the drug delivery system when a drug is injected intravenously and/or intratumoral.

A drug delivery system is a formulation or a device that enables applying one or more therapeutic substances into the body safely with a purpose to improve the quality of life of the receiver. Tumors release proteins in the blood which are called the tumors' markers which could be found through blood tests. The efficiency of a drug could be found by counting these markers. Lower these markers, more efficient is the drug. By lowering these markers, the drugs lose their strengths. So they are administered after certain intervals prescribed by the oncologists. The efficacy of the delivery of a drug is its efficacy regarding killing of cancer cells.

4.2 Mathematical Derivation of Reaction-Dispersion (Diffusion) Equation

Cancerous cells do not stay at the same location. We have already studied that. The malignant cells move from one spot to a different spot to form tumors. This is metastasis. This also means that they do not diffuse and vanish. They disperse and move on through blood and lymph circulation. "In order for cells to move through the body, they must first climb over/around neighboring cells. They do this by rearranging their cytoskeleton and attaching to the other cells and the extracellular matrix via proteins on the outside of their plasma membranes. By extending part of the cell forward and letting go at the back end, the cells can migrate forward. The cells can crawl until they hit a blockage which cannot be bypassed. Often this block is a thick layer of proteins and glycoproteins surrounding the tissues, called the basal lamina or basement membrane. In order to cross this layer, cancer cells secrete a mixture of digestive enzymes that degrade the proteins in the basal lamina and allow them to crawl through. The proteins secreted by cancer cells contain a group of enzymes called matrix metalloproteinases (MMP). These enzymes act as "molecular scissors" to cut through the proteins that inhibit the movement of the migrating cancer cells. Once the cells have traversed the basal lamina, they can spread through the body in several ways. They can enter the bloodstream by squeezing between the cells that make up the blood vessels." [https://www.cancerquest.org/cancer-biology/met astasis].

This evidently means that these malignant cells are highly organized and extremely cunning, insidious, stealthy, guileful, and what not. So it will not be easy to fight against such a dangerous enemy. But we will leave no stone unturned to defeat them. We will name our model: The Attacker–Defender model.

One fundamental assumption is as follows: Cancer cells move in a group, so do the immune cells. So here:

$u(t) = foreign\ antigen = cancer\ cells = a\ collection\ of\ malignant\ mass$;

$v(t) = immune\ cell = a\ collection\ of\ mass,\ fighting\ malignant\ mass$.

This is true for most sections in this monograph.

4.2.1 The Attacker–Defender Model [11]

Let us consider a volume bounded by a surface. Let n be the outward unit normal to S.

Let $J =$ inward diffusion. Here biochemicals are coming from outside, entering inside V.

So *inward* diffusion, $J = -\kappa\ div(q)$.

Applying Gauss' theorem, considering the limit as the control volume $V \to 0$ and then by replacing q by u, we get [11]

Fig. 4.1 V is the control volume

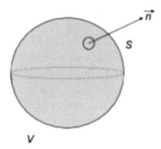

$$\partial u/\partial t = f(u) - g(u)v + \kappa \nabla^2 u. \tag{4.1}$$

Assuming $f(u)$, $g(u)$ are linear, and the rate of destruction of u is caused by each v at each unit of time, we get

$$\partial u/\partial t = a_1 u - a_2 uv + \kappa \nabla^2 u \tag{4.2}$$

where $a_1 =$ the rate of growth of u, $a_2 =$ the rate of death of u by a unit element of v. So $f(u)$ is assumed to be linear and $g(u)$ is also assumed to be linear. We considered $-a_2 uv$ to be the reaction of the immune system or a medication to reduce u. ∇^2 is the Laplacian in three dimensions and $\kappa =$ coefficient of dispersion of u. This equation may be expressed as follows (Fig. 4.1):

$$\frac{\partial u}{\partial t} = \sigma u + \kappa \nabla^2 u \tag{4.3}$$

where $\sigma = (a_1 - a_2 v)$. In all our models, we need to know both the rate of growth and the rate of annihilation of cancer.

Theoretically, Eq. (4.3) is valid at every point in the entire body. Computationally, it is valid at every grid point. Flux represented by the second term of the right side of (4.3) is valid everywhere in the body, meaning cancer cells are in motion not just in the blood, lymph, interstitial fluid, etc., and they are also in motion inside the tumors, lymph nodes, bones, etc. If by applying medications we reduce this flux, it means these motions of cancer cells are being restricted.

An interesting observation is as follows: We can imagine that the process of metastasis is cells coming inside a control volume as they attempt to form a tumor.

4.2.2 Rates of Growth of Cancer

According to Dr. Robert W. Franz, at Cancer Research Center at Providence Portland Medical Center (blog.providence.org), breast cancer cells have to divide 30 times before it can be felt. Up to 28th cell division, it cannot be felt even by a doctor. A patient stated that she had no lump in her breasts and after just one month, by self-examination, she detected that there was a lump in her breast. So within a month, the tumor cell divisions possibly have gone up from 28 to at least 30 times.

It means that for this patient, we may assume that in one month 28 cells possibly became 30 cells, at the least. So if one unit of time is one month, then in one month 100 cells become 107.1429 cells. So the rate of growth is about 7.1%. SCLC (Small cell lung carcinoma is very fast spreading) doubles in 54 to 214 days (Ref: NIH). This is definitely just an approximate estimate. For a particular patient, a patholo-gist, after doing a biopsy, could approximate this rate. And it varies from patient to patient and from one form of cancer to the other, and from one stage of breast cancer to the other. Regarding the rate of growth of breast cancer cells, according to https://www.breastcancer.org/symptoms/diagnosis/rate_grade, "pathology report may include information about the rate of cell growth — what proportion of the cancer cells within the tumor are growing and dividing to form new cancer cells. A higher percentage suggests a faster-growing, more aggressive cancer, rather than a slower, "laid back" one. Tests that can measure the rate of growth include: S-phase fraction: This number tells you what percentage of cells in the sample are in the process of copying their genetic information, or DNA. This S-phase, short for "syn-thesis phase," happens just before a cell divides into two new cells. A result of less than 6% is considered low, 6–10% intermediate, and more than 10% is considered high Ki-67 (is a protein in cells that increases as they prepare to divide into new cells). A staining process can measure the percentage of tumor cells that are positive for Ki-67. The more positive cells there are, the more quickly they are dividing and forming new cells. In breast cancer, a result of less than 10% is considered low, 10–20% borderline, and high if more than 20%." Initially, we consider rates lower than these using a different scaling.

4.2.3   Dimensional Analysis

In dimensional analysis, κ has the same dimension as the kinematic coefficient of viscosity [3].

We should check that dimensionally the two sides of (4.2) are equiva-lent. If dimensions are denoted by: $M = mass, L = length, and \ T = time$, then applying them to (4.2), we get.

Dimension of the left side is M/T. The dimension of the first term of the right side is M/T, because it represents the rate of change in u. Dimension of the second term is $(1/(M \times T)) \times (M) \times (M) = M/T$; the dimension of the flux term on the right side is $(L^2/T) \times (M/L^2) = M/T$.

Therefore, both sides have the same dimensions.

4.2.4   A Condition for Successful Treatment

Our complete model is

$\partial u / \partial t = \sigma u + \kappa \nabla^2 u$

$0 \leq x \leq L, 0 \leq y \leq M, 0 \leq z \leq N$ *and* $t \geq 0$. If we accept the validity of the following initial/boundary conditions,

$u(x, y, z, 0) = f(x, y, z)$, (initial condition) and

$$u(0, y, z, t) = u(L, y, z, t) = 0;$$
$$u(x, 0, z, t) = u(x, M, z, t) = 0;$$
$$u(x, y, 0, t) = u(x, y, N, t) = 0. \text{ (boundary conditions)} \qquad (4.4)$$

Then a necessary and sufficient condition that this model will represent a successful treatment is

$$\sigma \leq 0.$$

Since it will be very rare to make $\sigma = 0$, or in case by medications metastasis is inhibited, we will consider

$$\sigma < 0.$$

as a necessary and sufficient condition for success under the present boundary conditions. Let us look into that.

However, cancer does not stay fixed at the point of origin. Mathematically, it goes from one site to the next to form a tumor through the process of dispersion. It does go to the boundaries. It also does not go into apoptosis. It must be killed. In fact, it could move everywhere in the body. So the above boundary conditions do not really hold true in most cases in the real world. However, this is our first step.

At the outset before beginning a treatment, oncologists analyze all the test results, like blood tests, engiograms, MRIs, etc., and make a determination of the initial state of cancer. Also, regarding the boundary conditions, by considering a large field of computation, we assume that cancer is not present on the boundary. However, for our computational studies, we mostly considered boundary conditions to be variable.

4.3   Analytical Solution Revealing How Cancer Spreads

First we will consider a one-dimensional model and then generalize the solution for the 3D model.

$\partial u / \partial t = \sigma u + \partial^2 u / \partial x^2$

This equation could be solved by separation of variables as follows:

$$\text{Let } u = X(x)T(t), \tag{4.5}$$

where $X(x)=$ a function of x alone and $T(t)=$ a function of t alone.

Then, substituting in the model, we get $X(x)T'(t) = \alpha X(x). T(t) + \kappa X''(x)T(t)$. Rearranging we get

$$\frac{T'(t)}{T(t)} - \alpha = \kappa X''(x)/X(x) = -\lambda^2 \ (a \ constant \ ) \tag{4.6}$$

Since the left side is a function of t alone and the right side is a function of x alone, both sides must be equal to a constant. Also we assume that $u(x,0) = f(x)$, *the initial condition* and $u(0,t) = u(L,t) = 0$. We will explain later why $-\lambda^2$ has been chosen.

Now we have two equations to solve

$$T'(t)/T(t) - \alpha = -\lambda^2 \tag{4.7}$$

giving a solution, $T(t) =T_0(t)exp(\alpha - \lambda^2)$, where $T_0(t) =$ initial condition in time and $X''(x)+ \mu^2 X(x) = 0$, $\mu^2= \lambda^2/\kappa$. Here the solution is $X(x) = A \ sin(\mu x)+B \ cos(\mu x)$. From a practical point of view, $X(x)$ should include a description of the initial size of the tumor. This is an elementary case, because this is a one-dimensional, time dependent model. Later it will be generalized into a three-dimensional, time dependent model.

The boundary conditions (4.4) are not acceptable in a realistic model unless the tumor is truly in situ and malignant cells did not travel anywhere inside the body which Dr. Charles Roper, a Thoracic Surgeon from Barnes Hospital, St .Louis called a *myth* (1989) to the first author. He said that cancer cells almost always travel through blood vessels and lymphatic vessels and that is their very nature although they may not be forming a tumor elsewhere. Also "they keep on hiding in adipose tissues to avoid an attack from the immune system and could feed on fat for as many as 25 years, until the body gets weak and they thrive with full strength" (Dr.Roper). So cancer in situ is truly very rare unless it is detected at a very early stage.

In the breasts, there are plenty of blood vessels (built-in angiogenesis), plenty of lymphatic vessels, and plenty of fat. That is why breast cancers are so invasive in general. While being there, cancer cells secrete many chemicals (proteins) to confuse the immune system and avoid any attack from them.

They even take a ride on free radicals to metastasize. There they cause mutagenesis (induced mutations of genes) [5, 6].

The boundaries are also somewhat too simplified in most models. In engineering, for most physical models, boundary conditions are given by experimental data. But in biophysics, bioelements on the boundaries and within the field may behave in the same way and may behave entirely differently. So modeling in biophysics is sometimes very challenging. These conditions (4.4) appear when breast cancer is ductal or lobular in situ. They are mostly too close to the surface of the breasts. However, for phyllodes tumor, which develops in the breasts' soft tissues like muscles inside, such boundary conditions like $u = 0$ may not be applicable. For ductal or lobular carcinomas which are most common, if boundaries are too close to the tumor, the above boundary conditions are impractical. Here, in our models in the next chapters, boundaries are free boundaries, and biochemicals come in and move out freely. This is included in the computer codes.

Subject to the initial/boundary conditions (4.4), a closed form solution is

$$u(x,t) = \sum_{p=1}^{\infty} A_p sin(p\pi x/L)exp((\sigma - \kappa(p\pi/L)^2)t) \qquad (4.8)$$

where $A_p = (2/L)\int_0^L f(x)sin(p\pi x/L)dx$ $p = 1, 2, 3...$

Clearly, if $\sigma < 0$, as $t \to \infty$, $u(x,t) \to 0$. And conversely, if as $t \to \infty$, $u(x,t) \to 0$, then $\sigma < 0$. So $\sigma < 0$ is a sufficient condition as well as a necessary condition for $u(x,t) \to 0$, as $t \to \infty$.

In (4.7) instead of $-\lambda^2$, if we replace it by λ^2, solutions become unbounded.

This very simplistic model explains a deeper truth regarding the treatment of cancer. Let us examine the properties of $\sigma = (a_1 - a_2v)$. If $a_1 = a_1(t)$, and a medication is found such that $a_1(t) \to 0$, as t increases, then $u(x,t) \to 0$, as t increases, regardless of the values of v which may be applied intravenously or intratumoral. This has been validated computationally in the next chapter.

Let us consider solution of a three-dimensional model like

$$\partial u/\partial t = \sigma u + \kappa\left(\partial^2 u/\partial x^2 + \partial^2 u/\partial y^2 + \partial^2 u/\partial z^2\right) \qquad (4.9)$$

within a rectangle $0 \leq x \leq L; 0 \leq y \leq M; 0 \leq z \leq N$ with initial/boundary conditions as

$u(x, y, z, 0) = f(x, y, z)$ and $u = 0$ on all the six faces of the rectangle.

In actual computations, we did not consider that on the boundaries $u = 0$. Because as cancer cells reach the outermost boundary which should be the skin, they can hide under the skin. So the analytical solution is not telling the complete story regarding what is happening on the boundary. We assumed that the inward flux of cancer cells stops. And that will provide the value of u on the boundary.

Then the general solution is

$$u(x, y, z, t) = \Sigma_{p=1}^{\infty}\Sigma_{q=1}^{\infty}\Sigma_{r=1}^{\infty} A_{pqr} sin\left(\frac{p\pi x}{L}\right)sin\left(\frac{q\pi y}{M}\right)sin\left(\frac{r\pi z}{N}\right)$$
$$\times exp((\sigma - \kappa((p\pi/L)^2 + (q\pi/M)^2 + (r\pi/N)^2)t) \qquad (4.10)$$

$$A_{pqr} = (8/LMN) \int_0^L \int_0^M \int_0^N f(x, y, z) sin(p\pi x/L) sin(q\pi y/M) sin(r\pi x/N) dx dy dz$$

$$(4.11)$$

$p = 1, 2, 3...L$, $q = 1, 2, 3...M$, $r = 1, 2, 3, ...N$ Obviously A_{pqr} is bounded.

If we consider the term $\kappa((p\pi/L)^2 + (q\pi/M)^2 + (r\pi/N)^2)$, its least value is $\kappa(\pi/L)^2 + (\pi/M)^2 + (\pi/N)^2)$.

Then from (4.10), if $\sigma < \kappa((\pi/L)^2 + (\pi/M)^2 + (\pi/N)^2)$, u will decay and as $t \to \infty$, $u(x, t) \to 0$. Thus, if $\sigma < 0$, it becomes a necessary and sufficient condition so that $u(x, y, z, t)$ will be damped out.

So, we may make a concluding statement that in the treatment of cancer where a number of drugs with fixed dosages V_j, $j = 1, 2..J$ have been applied, the model for treatment is

$$\partial u/\partial t = a_1 u - (\sum_{j=0}^{J} a_{j+2} V_j)u + \kappa(\partial^2 u/\partial x^2 + \partial^2 u/\partial y^2 + \partial^2 u/\partial z^2) \quad (4.12)$$

From (4.12), the necessary condition that the treatment shall shrink the tumor, if

$$a_1 < \left(\sum_{j=0}^{J} a_{j+2} V_j \right). \quad (4.13)$$

This model is a realistic model. It is three dimensional, and it follows the mode of monitoring the infusion of drugs, generally through the veins, such that a fixed quantity of the drug V_0 stays always in the body. For each dosage of this drug, the code could quickly display the mode of treatment. Variable amounts of V_j 's will be considered in the next chapter.

In general, neither the patients nor their oncologists pay much attention to look into a nice looking closed form mathematical solution like (4.10). They want to see how cancer treatment is progressing, giving better and better prognosis, by using graphs, tables, and dynamic simulations. That will be considered in this monograph too.

We will now study a modified version of Charlie's algorithm [9] which has been used to solve our models numerically and analyze the results through graphs.

4.4 Algorithm of CDey-Simpson. A Difference-Integro Method

This is the numerical method that we will be using in all our computations in this book. The algorithm of Charlie's numerical method [9], to solve our models, we consider a simple equation:

$du/dt = f(u)$, subject to $u(t_0) = u_0$ is: $U_{Euler} = U_n + hf(U_n)$, $h = \Delta t = timestep$. This is the Euler forward scheme. Charlie used it as a predictor and corrected it by using his own corrector:

$$U_{n+1} = (1 - \gamma)U_{Euler} + \gamma(U_n + f(U_{Euler})), \; 0 < \gamma < 1; \qquad (4.14)$$

γ is a convex parameter. It will be chosen. U_n *is the value of the net function U at* t_n.

The Euler forward explicit finite difference representation of (4.9) is

$$\left(U_{ijk}^{n+1} - U_{ijk}^n\right)/h = \sigma U_{ijk}^n + \kappa\Big\{(U_{i+1jk}^n - 2U_{ijk}^n + U_{i-1jk}^n)/\Delta x^2$$
$$+(U_{ij+1k}^n - 2U_{ijk}^n + U_{ij-1k}^n)/\Delta y^2 + (U_{ijk+1}^n - 2U_{ijk}^n + U_{ijk-1}^n)/\Delta z^2\Big\}$$
$$(4.15)$$

$i = 1, 2, ..., I; j = 1, 2, ..., J; k = 1, 2, ..., K; h = \Delta t = time step$; $\Delta x, \Delta y, \Delta z$ *are the step sizes.*

$U_{ijk}^n = $ the net function corresponding to the value of u at x_i, y_j, z_k *at the nth time step* t_n.

Time derivative is approximated by Euler's forward formula, and space derivatives are approximated by central differences.

In the computational field, there are IJK number of points. In general, I, J, K are large. So, numerical solutions require large-scale computations. Equation (4.9) is a parabolic partial differential equation. So, the finite difference analog (4.9) will be stiff. (Appendix B).

Euler's explicit finite difference method has a very poor stability property, and as such, in order that the algorithm may work, time step Δt must be very small. Also when the system of equations is stiff, the Euler forward scheme fails in general. Charlie's method has a much better stability property (Appendix A). So, we have applied that in the code using larger Δt, even larger than the step sizes $\Delta x, \Delta y$ and Δz. Here, Charlie's method has been further modified by adding Simpson's corrector. In this text, it will be referred to as the CDey-Simpson method. Let us discuss its algorithm.

CDey-Simpson is a difference-integro algorithm for numerical solution of differential equations. First we apply Euler's forward difference formula as a predictor, then correct it with Charlie's convex corrector. This is then used as a predictor, and Simpson's rule for numerical integration is used as a corrector. This is the first form of the algorithm. A more advanced form is used in the code.

Let us consider a differential equation: $du/dt = f(u)$. At $t = t_0, u = u_0$. In Chap. 3, we saw the Euler forward finite difference scheme is $U_{n+1} = U_n + hf(U_n)$, $n = 0, 1, 2...$ where $U_n = $ the net function corresponding to $u(t_n)$, $t_n = $ *the nth time step.* $h = \Delta t = $ *time step.*

$U_{Euler} = U_n + hf(U_n)$ Euler forward.

$U_{n+1} = (1 - \gamma)U_{Euler} + \gamma(U_n + hf(U_{Euler}))$ Charlie's convex corrector

$$0 < \gamma < 1.$$

We did not stop here to solve our model. We went further to correct these results by introducing Simpson's rule both as a predictor and as a corrector. This new numerical scheme is the CDey-Simpson algorithm and has an excellent stability property. We need this very much, because the system is stiff, and we have to run our codes for a large number of time steps.

Let us study Simpson's rule. The equation $du/dt = f(u)$, $at\ t = t_0$, $u = u_0$ may be expressed in an integral form as follows: $du = f(u)dt$, $so\ u = u_0 + \int_{t_0}^{t} f(u)dt$. This is an integral equation. If $t = t_{n+1}$, and $t_0 = t_n$, then if $U_n =$ the net function corresponding to $u(t_n)$, and $h = t_{n+1} - t_n$, $U_{n+1} = U_n + (h/6)(f_n + 4f_{n+1/2} + f_{n+1})$, where $f_n = f(U_n)$, $f_{n+1/2} = f(U_{n+1/2})$, and $f_{n+1} = f(U_{n+1})$. This is Simpson's rule. Note: $h/2 = t_{n+1/2} - t_n = t_{n+1} - t_{n+1/2}$. In order to apply this rule, we need to evaluate U at $t = t_{n+1/2}$ or $U_{n+1/2}$, U at t_n or U_n and U at t_{n+1}. So we need a numerical method to compute $U(t_{n+1/2})$ a midpoint value and a predictor to predict U at t_{n+1}. Let us use Euler's forward algorithm as a predictor on Euler forward to move half-step forward: We will now consider the equation (4.15).

Step#1. Moving ½ step in time with Euler forward: (computing the value of U at the midpoint),

$$U_{ijk}^{n+1/2} = U_{ijk}^n + (h/2)G(U_{ijk}^n) \tag{4.16}$$

where

$$G(U_{ijk}^n) = \sigma U_{ijk}^n + \kappa((U_{i+1jk}^n - 2U_{ijk}^n + U_{i-1jk}^n)/\Delta x^2$$
$$+ (U_{ij+1k}^n - 2U_{ijk}^n + U_{ij-1k}^n)/\Delta y^2 + (U_{ijk+1}^n - 2U_{ijk}^n + U_{ijk-1}^n)/\Delta z^2) \tag{4.17}$$

$$i = 1, 2, ..., I;\ j = 1, 2, ..., J;\ k = 1, 2, ..., K$$

Step#2. Charlie's Corrector at that half-step:

$$U_{ijk}^{CH} = (1 - \gamma_1)U_{ijk}^{n+\frac{1}{2}} + \gamma_1\left(U_{ijk}^n + (h/2)G(U_{ijk}^{n+1/2})\right), \text{ where } 0 < \gamma_1 < 1 \tag{4.18}$$

$G(U_{ijk}^{n+1/2})$ is defined according to the Eq. (4.17)

$$i = 1, 2, ..., I;\ j = 1, 2, ..., J;\ k = 1, 2, ..., K$$

Step#3. Next half-step by Euler forward as a predictor:

$$EU_{ijk}^{n+1} = U_{ijk}^{CH} + (h/2)G(U_{ijk}^{CH}), \tag{4.19}$$

$$i = 1, 2, ..., I; \, j = 1, 2, ..., J; \, k = 1, 2, ..., K$$

Step#4. Correcting the values by Charlie's corrector for the second half-step computation:

$$CU_{ijk}^{n+1} = (1 - \gamma_2)EU_{ijk}^{n+1} + \gamma_2\left\{U_{ijk}^{CH} + (h/2)G\left(EU_{ijk}^{n+1}\right)\right\}, 0 < \gamma_2 < 1 \quad (4.20)$$

$$i = 1, 2, ..., I; \, j = 1, 2, ..., J; \, k = 1, 2, ..., K$$

Step#5. Simpson's rule as a second predictor:

$$SU_{ijk}^{n+1} = U_{ijk}^n + (h/6)\{G(U_{ijk}^n) + 4G(U_{ijk}^{n+\frac{1}{2}}) + G(CU_{ijk}^{n+1})\} \quad (4.21)$$

$$i = 1, 2, ..., I; \, j = 1, 2, ..., J; \, k = 1, 2, ..., K$$

where $G(U_{ijk}^n)$, $G(U_{ijk}^{n+1/2})$, and $G(CU_{ijk}^{n+1})$ are all evaluated by formula (4.17).

Step#6. Simpson's rule as a final corrector:

$$U_{ijk}^{n+1} = U_{ijk}^n + (h/6)\{G(U_{ijk}^n) + 4G(U_{ijk}^{n+1/2}) + G(SU_{ijk}^{n+1})\} \quad (4.22)$$

where $G(SU_{ijk}^{n+1}) = SU_{ijk}^{n+1} + \kappa(((SU_{i+1jk}^{n+1}) - 2(SU_{ijk}^{n+1}) + (SU_{i-1jk}^{n+1}))/\Delta x^2 + ((SU_{ij+1k}^{n+1}) - 2(SU_{ijk}^{n+1}) + (SU_{ij-1k}^{n+1}))/\Delta y^2 + ((SU_{ijk+1}^{n+1}) - 2(SU_{ijk}^{n+1}) + (SU_{ijk-1}^{n+1}))/\Delta z^2)$

$$i = 1, 2, ..., I; \, j = 1, 2, ..., J; \, k = 1, 2, ..., K$$

This is marching forward with time, an explicit forward difference method applied on a large computational field. In a super computer, it should not take a significant amount of computational time. The stability analysis is given in Appendix A.

It is a vectorized algorithm. Let us look into this briefly.

If we consider the first operation in (4.15) as follows: $U^{n+1/2} = F_1(U^n)$, then $F_1 = I + (h/2)G^n : R^n \rightarrow R^{n+1/2}$.

where R^n is a real (IJK) dimensional space at $t = t_n$.

Similarly, $U^{CH} = F_2(U^{n+1/2})$, where $F_2 = (I - \Gamma)U^{n+1/2} + \Gamma(U^n + (h/2)G(U^{n+1/2})): R^{n+1/2} \rightarrow R^{CH}$

where the components of the vector U^n may be collected from $CSM(Common \, Shared \, Memory)$.

R^{CH} is the $CDey - Convex \, Real \, Space$.

Similarly, the next operator is defined by the Eq. (4.15) as $F_3 : R^{CH} \rightarrow R^{n+1}$, $F_4 : R^{n+1} \rightarrow R^{CH}$

$F_5 : R^{CH} \to R^{SI}, F_6 : R^{SI} \to R^{SI}, F_7 : R^{SI} \to R^{n+1}$, where R^{SI} may be called *Simpson* $-$ *RealSpace*, all the spaces are real vector spaces and of the dimension (IJK) changing with time.

Thus, all the operators $F_m, m = 1, 2, ..., 7$ are vector operators, and the algorithm is a vectorized algorithm.

It may be parallelized with ease. However, when multiple processors are used, these operators become overlapping. For example, if the Nth processor solves $I_N J_N K_N$ number of equations, then the $(N + 1)th$ processor will need all information at $I_N \forall j$ *and* k, at $J_N \forall i$ *and* k, and at $K_n \forall i$ and j and so on [7]. These could be supplied by the master processor or by the common shared memory.

4.4.1 Definition: CDey-Simpson Operator

A CDey-Simpson operator F_t is a convex time-dependent operator defined on a time-dependent n dimensional real space R^n. It is defined as follows:

Let S be a normed vector space. Let $F_t : V_t \subset R^n \to S \subset R^n$, be a time-dependent difference operator operating on a time-dependent space V_t be such that

$$F_t(\gamma_t x(t) + (1 - \gamma_t)y(t)) < \gamma_t F_t(x(t)) + (1 - \gamma_t)F_t(y(t)), \text{ for all } 0 < \gamma_t < 1,$$
$$\text{and } x(t), y(t) \in V_t \ \forall t$$
$$(4.23)$$

Let $G_t : V_t \to S$ be an integral operator, such that

$$G_t(F_t) \in V_t \ \forall t$$

then S is a CDey-Simpson space, if G_t is a Simpson integral operator for F_t and $\forall t$.

4.5 Time-Dependent Extrapolated Boundary Conditions

In most problems in physical science, boundary conditions are predetermined. That is how we analyze the well posedness of a mathematical model. We do not have this property here. Here what cancer cells are doing on the boundaries depend on what they are doing inside the field in a hostile biological environment. The T-cells, B-cells, natural killer cells, and macrophages should travel anywhere in the body to attack the non-self-cancer cells when they recognize them. So, here the computational boundaries should be dependent on what they are doing inside their field of activities. As such, these must be extrapolated computationally by applying appropriate mathematical formulas. Let us now derive such formulas.

Let us assume that $f(x)$ is an analytic function on (a, b). By Taylor's theorem,

$$f(x + h) = f(x) + hf'(x) + (h^2/2!) f''(x)$$
$$+ (h^3/3!) f'''(\xi), where, x < \xi < x + h. \qquad (4.24)$$

And

$$f(x - h) = f(x) - hf'(x) + (h^2/2!) f''(x)$$
$$- (h^3/3!) f'''(\xi), where, x - h < \xi < x. \qquad (4.25)$$

Let the points $x_0, x_1, x_2, ...x_{NX+1}$ be a set of equispaced points on the x-axis, such that $x_{i+1} - x_i = h \, \forall i$.

Then,

$f(x_0) = f(x_3 - 3h) = f(x_3) - 3hf'(x_3) + 9h^2 f''(x_3)/2!$, approximately.

Assuming h is very small, terms of the order $O(h^3)$ are neglected. Noting that $x_4 - h = x_3$ and $x_4 - 2h = x_2$ and approximating the derivatives by backward difference formulas, we get

$$f(x_0) = f(x_3) - 3h(f(x_3) - f(x_2))/h + 4.5h^2(f(x_3) - 2f(x_2) + f(x_1))/h^2$$
$$= 2.5f(x_3) - 6f(x_2) + 4.5f(x_1) \qquad (4.26)$$

Since we are solving the equations in a cubical computational field (IJK), the boundaries consist of six rectangular faces. Let the first face be defined by $i = 1; j = 1, 2, ...J+1; k = 1, 2, ..., K+1$ perpendicular to the $x-axis$. At the two boundaries perpendicular to the x-axis, let x_i at $i = 1$ and $x = I + 1 \, \forall y, z$. Let $h = \Delta x$,

(i) A three point backward difference formula at the $face$ at x_i at $i = 1, \forall y, z$

$$f(1, y, z) = f(4 - 3h, y, z)$$
$$= f(4, y, z) - 3hf'(4, y, z) + (9h^2/2!) f''(4, y, z)$$
$$= f(4, y, z) - 3h(f(4, y, z) - f(3, y, z))/h$$
$$+ 4.5h^2(f(4, y, z) - 2f(3, y, z) + f(2, y, z))/h^2$$
$$= 2.5f(4, y, z) - 6f(3, y, z) + 4.5f(2, y, z). \qquad (4.27)$$

Note that the points $(2, y, z), (3, y, z), and (4, y, z)$ are points which should be inside the domain of computations. Similarly,

(ii) A three point forward difference formula for the $face$ x_i at $i = I + 1, \forall y, z$

$$f(I + 1, y, z) = f(I - 2 + 3h, y, z)$$
$$= f(I - 2, y, z) + 3hf'(I - 2, y, z) + (9h^2/2!)f''(I - 2, y, z)$$
$$= f(I - 2, y, z) + 3h(f(I - 1, y, z) - f(I - 2, y, z))/h + 4.5h^2 \qquad (4.28)$$
$$(f(I, y, z) - 2f(I - 1, y, z) + f(I - 2, y, z))/h^2$$
$$= 2.5f(I - 2, y, z) - 6f(I - 1, y, z) + 4.5f(I, y, z)$$

Here I stands for NX. The points $(NX - 2, y, z), (NX - 1, y, z)$, and (NX, y, z) are points inside the domain of computations.

Similar formulas have been applied for the four sets of boundary conditions: $f(x, 1, z), f(x, NY + 1, z)$,

$$f(x, y, 1) \; and \; f(x, y, NZ + 1).$$

However, $f(1, 1, 1) = 2.5f(4, 1, 1) - 6f(3, 1, 1) + 4.5f(2, 1, 1)$ cannot be computed by using (4.21), because they are not points where we did any computation. Similarly, $f(1, 1, 2), f(1, 1, 3)....f(1, 1, K + 1), f(1, 2, 1)$, etc., cannot be computed by using (4.21) and (4.22). In fact at all the points on the eight borderlines of the cube $(I + 1) \times (J + 1) \times (K + 1)$, these types of formulas are not applicable on the borders of the field of computation. There are 12 such borders. So in all there are $4(I + 1) + 4(J + 1) + 4(K + 1) = 4(I + J + K + 3)$ points, in which eight end points (corners) are repeated 3 times. They should be used only one time. So in all there are $4(I + J + K - 1)$ points where the formulas like (4.21) and (4.22) are not applicable.

In most flow models, we estimate boundary values of the dependent variable by considering the averages. For instance,

$f(1, 1, 1) = (f(1, 2, 2) + f(2, 1, 2) + f(2, 2, 1) + f(2, 2, 2))/4$ etc.

But, such an attempt has not been done intentionally in these models. The rationale is the very nature of cancer cells is to metastasize. Especially, when treatments begin, they disperse and look for some safe havens to hide. Certain organs, like skin, and adipose tissues are often those safe havens. Here we have tried to simulate these phenomena at the borderlines.

There are other formulas too, which we also used in some of our early codes. Let us consider them too.

If $f(x)$ is analytic on (a, b), then at any $x \epsilon (a, b)$, assuming h is small and neglecting terms of the order h^3, we get

$$f(x - h) = f(x) - hf'(x) + (h^2/2!)f''(x).$$

Approximating the derivatives by forward difference formulas,

$$f(x - h) = f(x) - h(f(x + h) - f(x))/h$$
$$+ (h^2/2!)(f(x + 2h) - 2f(x + h) + f(x))/h^2$$

This gives

$$f(x_0) = 2.5f(x_1) - 2f(x_2) + 0.5f(x_3) \tag{4.29}$$

$$f(x_2) = f(x_1 + h), \quad f(x_3) = f(x_1 + 2h)$$

Although Eqs. (4.27) and (4.29) are derived quite differently, they are exactly the same when $f(x)$ is analytic (no discontinuities) on its domain of definition, assuming h is sufficiently small. Let us prove this proposition:
$2.5f(x_1) - 2f(x_2) + 0.5f(x_3) = 2.5f(x_3) - 6f(x_2) + 4.5f(x_1)$, is True, if $2f(x_2) = f(x_1) + f(x_3)$
Giving,

$$f(x_2) - f(x_1) = f(x_3) - f(x_2), \text{ or } (f(x_2) - f(x_2 - h))/h$$
$$= (f(x_2 + h) - f(x_2))/h$$

As $h \rightarrow 0$, both limits must be equal to $f'(x_2)$. So, if the condition $f(x)$ is analytic, then (4.27) and (4.29) will give identical results.

These formulas, when applied to our models, show that the values of U_{ijk}^n on the boundaries are constantly changing with time according to the computed values inside the domain of computations, except at certain points on the boundary which will represent the skin or where cancer cells can hide. It will be discussed in the next chapter.

These are just extrapolations using values close to the boundaries. They may not be easily justified from the standpoint of physiology. So we may look into some other boundary conditions which may be justifiable physiologically.

If the field is large, and we reach the skin of the body, then from outside nothing can come in. So we may assume that there is no inward flux on the boundary. So, $|\partial^2 u/\partial x^2|, |\partial^2 u/\partial y^2|, |\partial^2 u/\partial z^2|$, etc., all of them should be zero on the boundaries. This should be a correct assumption physiologically. If we consider the direction of the normal to the outer face of the cube (our computational field) defined by $z = a \ positive \ constant$ at $k = NZ + 4$, as positive, then the direction of the normal to the bottom face $z = 0$, $given \ by \ k = 1$, will be negative. But these signs will not matter if we assume that there is no flux inward or outward on the boundaries. Therefore, we may assume:
$U_{0,j,k}^n - 2U_{1,j,k}^n + U_{2,j,k}^n = 0$, and $U_{NX+4,,j,k}^n - 2U_{NX+3,,j,k}^n + U_{NX+2,,j,k}^n = 0$
$\forall j, k$ at $\forall n$. It must be noticed that if $2U_{1,j,k}^n < U_{2,j,k}^n$, then $U_{0,j,k}^n < 0$. But the mass cannot be negative. So we have considered in the code $U_{0,j,k}^n = |2U_{1,j,k}^n - U_{2,j,k}^n|$ and similarly, $U_{NX+4,j,k}^n = |2U_{NX+3,j,k}^n - U_{NX+2,j,k}^n|$.

$$\text{So, } U_{0,j,k}^n - \left|2U_{1,j,k}^n - U_{2,j,k}^n\right| = 0 \text{ (because } U_{0,j,k}^n > 0). \tag{4.30}$$

$U_{0,j,k}^n = 2U_{1,j,k}^n - U_{2,j,k}^n$, if $2U_{1,j,k}^n > U_{2,j,k}^n$ and $U_{0,j,k}^n = U_{2,j,k}^n - 2U_{1,j,k}^n$, if $2U_{1,j,k}^n < U_{2,j,k}^n$.

Thus, in the code, we have used $U_{0,j,k}^n = |2U_{1,j,k}^n - U_{2,j,k}^n|$, $U_{NX+4,j,k}^n = |2U_{NX+3,j,k}^n - U_{NX+2,j,k}^n|$ and similar expressions for other variables.

Another simple explanation is as follows: Both outward normals to the planes, $k = 0$ and $k = K$ have opposite signs. So inward fluxes on the cube, our field of numerical solutions on these two planes, are from opposite directions. However, the numerical values of these fluxes must be positive. That is true for all boundary planes.

Another much simpler approach is assuming mesh size sufficiently small,

$$i \neq 0 \text{ and } j \neq 0, U_{i,j,0} = U_{i,j,1} \text{ and } i \neq I + 1, j \neq J + 1 \, U_{i,j,K+1} = U_{i,j,K}.$$
(4.31)

$$i \neq 0 \text{ and } k \neq 0, \; U_{i,0,k} = U_{i,1,k} \text{ and } i \neq I + 1, k \neq K + 1 \, U_{i,J+1,k} = U_{i,J,k}$$
(4.32)

$$j \neq 0 \text{ and } k \neq 0, \; U_{0,j,k} = U_{I,j,k} \text{ and } j \neq J + 1, k \neq K + 1 \, U_{I+1,j,k} = U_{I,j,k}.$$
(4.33)

A Special Note

It ought to be noted that if at the boundaries $U_{ijk} \neq 0$, then in the Eq. (4.3), the condition

$\sigma < 0$ is just a necessary, not a sufficient condition for damping out cancer, unless metastasis has been inhibited.

This may be achieved, if the values of $U_{i,j,k}$ on the boundary planes are assumed to be the same as those on the planes adjacent to them. We tried not to use these conditions in many cases, because we wanted to observe the process of metastasization as treatments and the process of hiding of the runaway cancer cells continue. It is very essential to know that, being chased by the immune cells and/or medications as time changes, how malignant cells start leaving the primary site and moving toward other organs. Later we will see these phenomena through dynamic simulations.

At the present time, doctors apply MRI scanners, with contrast, to locate and observe details of the anatomy of tumors inside the body. (MRI stands for magnetic resonance imaging. They use powerful computers, magnets, and radio waves for imaging.) Doctors not only require scanning tumors already detected, in general, they order for whole-body scanning. In this way, they may find almost all the locations related to the primary tumor as well as where it has metastasized (in case it did), and accordingly, treatments are conducted.

There is a dire need for mathematical simulations, because such models are some additional help predicting what could happen to the patient as time moves on.

4.6 Modeling General Therapy Targeting All Cancer Cells

Chemotherapy, which travels all over the body and poisons and often kills indiscriminately all fast growing cells (including some normal fast growing cells like blood cells growing in bone marrow, digestive tracts, hair follicles, reproductive cells), has been used as a standard treatment for almost all cancers. It kills tumors and runaway cancer cells. Often it is applied as neoadjuvant therapy to reduce the size of tumors before any surgical procedure is done to take the tumor out. It can destroy cancer cells and lessen their chances to return. However, it could damage normal cells causing life-threatening side effects.

We have developed a model where a medicine will destroy only cancer cells, tumors and the runaways, leaving normal cells unharmed. We have assumed here that a fixed amount of the medication will always stay in the bloodstream. This is targeted therapy.

According to the American Cancer Society, targeted therapy may (i) block or turn off chemical signals that orchestrate rapid growths and divisions of malignant cells. Then cancer cells are totally disabled and fail to do any harm, (ii) change or remove the proteins released by the damaged DNA (in oncogenes and mutated tumor suppressive genes like *BRCA1, BRCA2, p53*, etc.) and/or proteins that keep cancer cells alive and active, (iii) prevent cancer cells from doing angiogenesis or creating new blood vessels to draw oxygen, glucose, and other nutrients, (iv) trigger immune system to destroy cancer cells, and even (v) carry toxins to kill cancer cells without killing normal cells.

Targeted therapy does not mean that the drugs target only the tumors, it targets all cancer cells in the body. Killing tumors is not curing cancer.

Following Eq. (4.3), we first consider one such drug V_0 (with a fixed amount) which reduces the growth and destroys cancer cells wherever they travel by considering:

$$\partial u/\partial t = (a_1 - a_2 V_0)u + \kappa(\partial^2 u/\partial x^2 + \partial^2 u/\partial y^2 + \partial^2 u/\partial z^2) \qquad (4.34)$$

After locating the tumors, V_0 is a fixed amount of drug which is being administered in such a way that it moves through blood flow and surrounds all the tumors and starts penetrating inside the tumors to destroy them completely. Codes were written accordingly, and at each computational time step, the values of u at each location were recorded.

Comparing (4.34) with (4.9), *here*, $\sigma = (a_1 - a_2 V_0)$. Then $u(x, y, z, t) \to 0$ as $t \to \infty$, if

$$a_1 < a_2 V_0 + \kappa\pi^2(1/L^2 + 1/M^2 + 1/N^2) \qquad (4.35)$$

cancer will die where $a_1 =$ *rate of growth of u and* a_2; *is the rate of destruction of u by each unit of* V_0. This inequality implies that V_0 is a drug that reduces the rate of growth of cancer. Here one principle of chemotaxis, namely wherever u is present,

V_0 should be right there, has been applied, but V_0 does not grow as u grows. Values of the rate constants a_1 and a_2 should be supplied primarily by the pharmaceutical companies, the maker of the drug V_0. Mathematically, the drug will be effective if as discussed in Chap. 3:

$$a_2 V_0 > a_1, \text{ or } V_0 > \frac{a_1}{a_2}. \tag{4.36}$$

In the finite difference equation, this term $(a_1 - a_2 V_0)U_{ijk}^n$ means that at each location denoted by i, j, k and $\forall n$, wherever the antigen U is present, V_0 is right there. So V_0 is a targeted drug in a broader sense. We may also think that V_0 attempts to prevent malignancy wherever it appears in the body. If the drug V_0 is a variable, it will mean V_0 is being adjusted at each location and at each time where U is found.

Certainly, the same dosage of the drug cannot be administered on all patients throughout the treatment because of physiological and/or medical constraints. However, if and when application of this drug is possible, we need to predict how effective it will be. We will assume that for some patients, this drug being used as a targeted treatment could work along with the immune system and help them use more cytokines to destroy cancer cells. Targeted drugs have less side effects in general. For some, when cancer, unfortunately, has spread to a number of organs, and chemotherapy is needed, then computationally, V_0 could be supplemented by another drug, representing chemotherapy. We will discuss that too.

First we will consider a model with two tumors, a large one at a primary site and another at an axillary lymph node, and V_0 is monitored such that this fixed amount always stays in the body and works as a targeted medication.

In Fig. 4.2, we considered a computational field $I = J = K = 101$, (including the boundary points there are $105 \times 105 \times 105 = 1157625$ number of points) so there are over one million points, with mesh sizes, $\Delta x = \Delta y = \Delta z = 0.02$, $\Delta t = 0.0125$, and $NT = $ the number of time steps $= 302$. There is a tumor at the primary site, and one lymph node is also affected. We chose $a_1 = 2.5$ (growing at a rate of 150%), $a_2 = 0.0001$ and $V_0 = 1.1(a_1/a_2) = 27500$. So according to our mathematical findings, tumors should be contained. At the primary site, tumor is cubical spreading from (x_1^*, y_1^*, z_1^*) to (x_2^*, y_2^*, z_2^*), $U_0 = $ Initial value (representing population of cancer cells) $= 1000$ at each grid point. At a lymph node (x_{l1}, y_{l1}, z_{l1}), $U_0 = 200$. The results given in graph as shown in Fig. 4.2 show that at the primary site UPMAX (the largest value of the antigen) is reduced from 1000 to 624.96 in 302 time steps, and the tumor is slowly shrinking, whereas from the lymph node (LNODE1), tumor vanished. MAXU is the largest value of U in the entire field, and this value is the same as UPMAX (the largest value of the tumor at the primary site). NMAXU $= 4 = $ No. of MAXU in the entire field. This is of some concern.

That *could* mean that the cancer cells are escaping the primary site and metastasizing elsewhere.

Also, at some point in the field of computation consisting of 1,157,625 points (105 X 105 X 105), the value of U exceeded its previous value at some time.

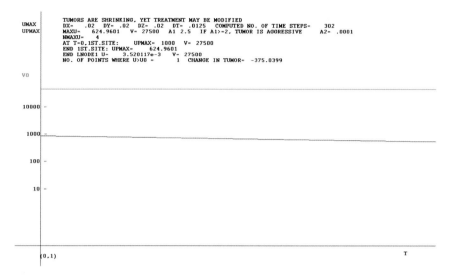

```
                TUMORS ARE SHRINKING, YET TREATMENT MAY BE MODIFIED
UMAX            DX=   .02  DY= .02  DZ= .02  DT= .0125   COMPUTED NO. OF TIME STEPS=   302
UPMAX           MAXU=   624.9601   V= 27500   A1 2.5  IF A1>=2, TUMOR IS AGGRESSIVE   A2= .0001
                NMAXU=   4
                AT T=0,1ST.SITE:    UPMAX=  1000   V= 27500
                END 1ST.SITE: UPMAX=   624.9601
                END LNODE1 U=   3.520117e-3   V= 27500
                NO. OF POINTS WHERE U>U0 =    1   CHANGE IN TUMOR=  -375.0399

VD


10000  -


1000   -


100   -


10   -



      (0,1)                                                                              T
```

Fig. 4.2 Tumor shrinking, targeted therapy. V = 1.1 *A1/A2 = fixed at the tumors

From Fig. 4.2 (drawn in a logarithmic scale like all other graphs), the redline indicates that the value of UMAX (max value of the antigen in the field) is the same as UPMAX. So, the encouraging news is that targeted treatment is working, preventing cancer from growing any further.

The redline represents the slow shrinking of the tumor. The green line shows that the amount of drug V_0 remained constant at $V_0 = 27500$. Its slope is zero that indicates the treatment has gone to a steady state (will be continued with this value of V_0). The largest values of U at the primary site and in the computational field are exactly the same. This implies the growth of cancer is under control. Yet, there is one point where value of U did increase from the previous time step, ($U_{ijk}^{302} > U_{ijk}^{301}$ *at one point in the field of computation*).

Our mathematical derivation is not restricted by the size of the primary tumor or even by the number of tumors, because it strikes every foreign antigen at every location. So, we now consider a primary tumor ten times bigger than the one discussed in Fig. 4.2

In Fig. 4.3, initial accumulation at each point in the mesh, within the tumor, is given by $U_0 = 10000$, $a_1 = 1.5$ (tumor is less aggressive growing at the rate of 50%. In the previous case where $a_1 = 2.5$, it was growing at the rate of 150%), $a_2 = 0.0001$ as before, $V_0 = 16500 > a_1/a_2 = 15000$, as required by the mathematical model in order that the tumor may shrink. Results show that the tumor is shrinking; however, at 97 points the number of cancer cells is larger than their values at the previous time step (which implies that cancer still maintains its tendency to grow and metastasize or doing both), and there are eight points in the field where it remains a maximum. Fortunately, nowhere inside the large computational field (consisting of over a million

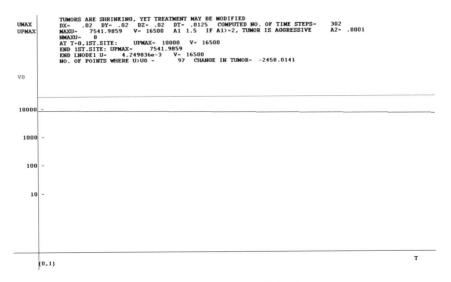

Fig. 4.3 Tumor shrinking, targeted therapy. V = 1.1 *A1/A2 = fixed at the tumors

points) UMAX (max value of U in the entire field) exceeded UPMAX (max value at the primary site). However, activities of the cancer cells appear to be slowing down.

Next we considered in Fig. 4.4, three tumors at three different points in the computational field, namely at (13,13,13) extending to (16,16,16), at (26,26,26) extending to (29,29,29) and at (47,47,47) extending to (50,50,50), where $U_0 = 10000, 1000, \ and \ 400$ at each grid point, respectively. There are 64 points in each

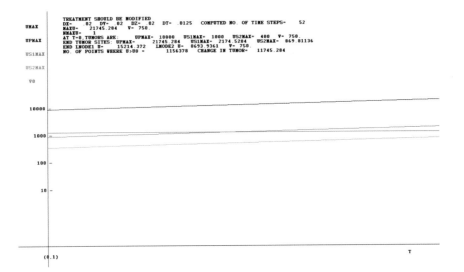

Fig. 4.4 Tumor growing, targeted therapy failed. V = S *A1/A2. S = 0.03 at every location of U

box. So the number of cancer cells in the three sites is, respectively, 640,000, 64,000, and 25,600. That means tumors exist with 729,600 numbers of cancer cells in the field of computations. The total points in the field are $105 \times 105 \times 105$, $DX = DY = DZ = 0.02$, and $DT = 0.0125$. When we chose $V = V_0 = 0.03(a_1/a_2) = 750$, as expected the number of cancer cells at each location increases. Not only that, at 1,156,378 points in the field, the number of cancer cells is greater than their values at the preceding time step. That definitely means, tumors are appearing at other sites too. The code was supposed to run for a total of 250 time steps. But the code stopped itself at 52 time steps, because cancer has been growing rampantly.

This implies that, if there is a patient who cannot tolerate more than $V_0 = 750$, this treatment does not work for her. In such cases, an oncologist tries, whenever it is possible, to apply an additional chemo, based upon the state of health of the patient.

In Fig. 4.5, we considered, $V_0 = 1.1(a_1/a_2) > (a_1/a_2)$. Other parameters remain the same. So tumors must shrink. And that has been found computationally.

Other inputs remain the same. However, regardless of the fact that tumors are shrinking, there are six points in the field of computations where $U_{ijk}^{251} > U_{ijk}^{250}$.

For Fig. 4.6, we considered $V_0 = a_1/a_2$. All other inputs remain the same, except the number of time steps which is now 502. All tumors stay virtually the same, decreasing very very slowly though. This implies that it could be possible for a patient to stay alive on a particular pattern of treatment having several tumors in the body. There are several cancer patients in such a state, according to the American Cancer Society, who are living with cancer as a chronic disease which qualitatively validates our computational findings.

Our next investigation deals with the same targeted drug with an additional drug which reduces the rate of growth of malignant cells at each point. This additional

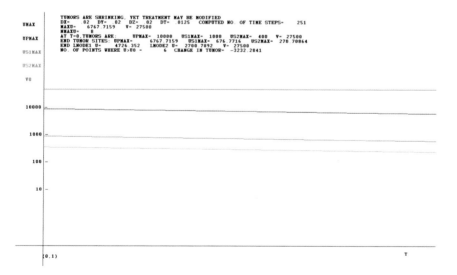

Fig. 4.5 Tumor shrinking, targeted therapy worked. $V = S * A1/A2$. $S = 1.1$ at every location of U

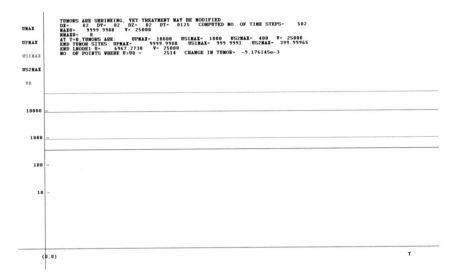

Fig. 4.6 Growths of all tumors stopped by targeted therapy. They diffused slowly, V = A1/A2, at every location

drug is decreasing a_1, the rate of growth of cancer cells at each time step. The formula that has been applied is as follows: $a_1 = a_1 \, exp \, (-\Delta t)$.

There are several drugs like alkylating agents, anti-metabolites, (Anti-metabolites kill cancer cells by acting as false building blocks in a cancer cell's genes, causing the cancer cell to die as it gets ready to divide [Breastcancer.org]) used for breast cancer patients which modify the DNA to reduce the rate of growth of cancer. However, like most chemos, they have side effects. That is why we used a very slow reduction $exp(-\Delta t)$ of a_1, meaning, a low dosage of this drug which might work.

The code ran for 377 number of time steps, and Fig. 4.7 shows that tumors are shrinking at all locations. Here, $a_1 = 2.5$, $a_2 = 0.0001$, $V_0 = (a_1/a_2) = 25000$ is fixed. MAXU = the largest accumulation of cancer cells over the entire field, UPMAX = the largest accumulation of cancer cells in the primary site, UPS1MAX = the largest accumulation of cancer cells at the first secondary site, and UPS2MAX = the largest accumulation of cancer cells at the second secondary site. All are decreasing. At no point in the field, values of U exceeded its value in the past time step.

$V_0 = (a_1/a_2) = 25000$ seems to be awfully large, if we strictly consider V_0 representing the amount of leukocytes. In fact, V_0 represents the entire regiment of defenders, all leukocytes and vitamins and medications.

Here change in tumor $U > U_0 = 0$ means, at the time step 377, at nowhere in the computational field, $U_{ijk}^n > U_{ijk}^{n-1}$ happened. This is very encouraging information. All three graphs representing the progress of tumors have negative slopes meaning that all of them are shrinking. The green line at the top is line presenting the fixed value of V_0. That implies the successful treatment by the oncologist.

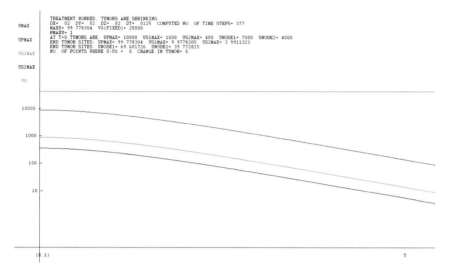

Fig. 4.7   Growths of all tumors stopped by targeted therapy and chemo. V = A1/A2 at every location

Now we will consider a case Fig. 4.8 where a patient can tolerate up to $V_0 = 0.1(a_1/a_2) = 2500$.

The same code with an extra chemo was used and the code ran for 751 time steps. Results are quite encouraging.

Here $\Delta t = 0.0125$, $\Delta x = \Delta y = \Delta z = 0.02$, *the computational field is* $105 \times 105 \times 105$. $a_1 = 2.5$ and $a_2 = 0.001$. Since a_2 is increased 10 times, V_0 is decreased 10 times. Lower dosage amount of V_0, yet more powerful. It did work. There are three tumors (resembling multifocal breast cancer) as before. First malignant cells start increasing. Watching the output at every time step, it has been recorded that from the actual computer output, we got at n = 2, there were 1,156,378 points where $U_{ijk}^n > U_{ijk}^{n-1}$, and there were eight points where $U = max\left(U_{ijk}^n\right)$. At n = 365, there were only six points where $U_{ijk}^n > U_{ijk}^{n-1}$ and at only one point where $U = max\left(U_{ijk}^n\right)$. At n = 724, there is no point where $U_{ijk}^n > U_{ijk}^{n-1}$.

So, cancer cells quit growing all over the body, and the body is moving toward a *cancer-free state* (Fig. 4.8).

Even though the results appear to be qualitatively excellent, we need to know some of the limitations of targeted therapies. These drugs, in general, target proteins which make cancer cells grow fast abnormally and metastasize. But sometimes, cancer cells become immune to such drugs and become drug resistant. Therefore, oncologists look for other similar drugs or apply some extra effective chemotherapy that a patient will be able to tolerate.

At this point, we wanted to do a reduction of the growth rate of cancer and look into the results.

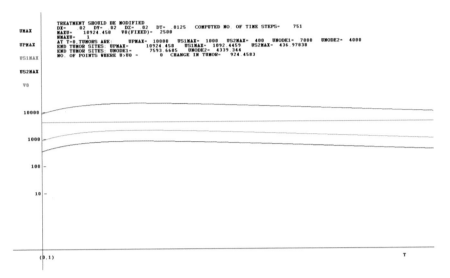

Fig. 4.8 Targeted therapy to destroy and to stop rate of growth. V = 0.1 * A1/A2 at every location

We will now consider a case where the rate of growth is being continuously reduced by the drug V_0. The formula that is used is

$$d(a_1(n))/dt = -a_1(n).V_0 \qquad (4.37)$$

where $a_1(n) = $ *value of a_1 at t_n*.

That gives $\quad a_1(n) = a_1(n-1)\exp(-\Delta t V_0). \qquad (4.38)$

The formula for V_0 is $V_0 = 0.05(a_1/a_2)$, and it remains constant throughout our computations. This is done because we need $(a_1 - a_2 V_0) < 0$ which is a necessary condition so that the treatment should work. So V_0 must satisfy $V_0 > a_1/a_2$. If a_2 is very small, V_0 becomes too large. So we are making attempts such that V_0 will continuously decrease a_1.

Inputs for Fig. 4.9

```
*********************************************************************************
! CHAPTER 4, FIG 4.9
! THERE ARE THREE TUMORS , TWO LYMPH NODES ARE AFFECTED
! DU/DT = A1*U-A2*U*VO  + (NU1)*DELSQ(U)                           ! ANTIGEN
! AT T=0, U0(OUTSIDE) IS = 20  AT THE PRIMARY SITE, U = 10000  SECOND TUMOR= 1000  3RD = 400  VO-FIXED = 2250
! DX= .02   DY= .02  DZ= .02  DT= .02  NX= 101  NY= 101  NZ = 101  NT= 1000
! A1(0)= 4.5  A2= .0001  S= .05  VO= S*A1/A2 2250 ! A1 = EXP(-LAMBDA*VO)*A1 BEING CONTINUOUSLY REDUCED.
!
! NU1= .0000001
! GAMMA1= .015  LAMBDA= .00005
! TWO LYMPH NODES ARE LOCATED AT: ( 3 , 3 , 3 ) AND  ( 35 , 35 , 35 ) BOTH ARE AFFECTED
! PRIMARY TUMOR IS LOCATED AT:   ( 13 , 13 , 13 ) UPRIMARY= 10000
! SECOND TUMOR IS LOCATED AT:  ( 26 , 26 , 26 ) US1= 1000
! THIRD TUMOR IS LOCATED AT:   ( 47 , 47 , 47 ) US2= 400
! AT THE LYMPH NODE(L11,L12,L13) U0= 7000  AT THE LYMPH NODE(L21,L22,L23) U0= 4000
*********************************************************************************
```

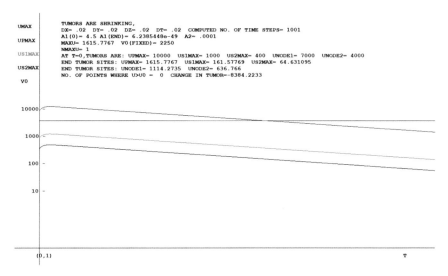

Fig. 4.9 V0 cuts the rate of growth and destroys the cancer

The results are shown in Fig. 4.9.

The inputs for Fig. 4.10 are very similar to Fig. 4.9. Only here, we basically validated the fact that if the strength is from $a_2 = 0.0001$ to 0.00065 (an increment of 6.5%) and reduced V_0 from 2250 to 103.84615, a drastic reduction, the method of treatment worked again. But it worked more slowly as could be noticed in the graphs.

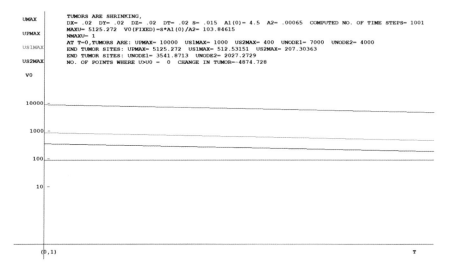

Fig. 4.10 Targeted therapy stopping the rate of growth of cancer

Inputs for Fig. 4.10

```
***************************************************************************************************
! CHAPTER 4, FIG 4.10
! THERE ARE THREE TUMORS , TWO LYMPH NODES ARE AFFECTED, S=0.015
! DU/DT = A1(N)*U-A2*U*V0  + (NU1)*DELSQ(U)                               ! ANTIGEN
! U0 AT T=0,U0 IS = 10000  AT THE PRIMARY SITE,U = 10000  SECOND TUMOR= 1000  3RD = 400  V0=FIXED =  103.84615
! DX= .02  DY= .02  DZ= .02  DT= .02  NX= 101  NY= 101  NZ = 101  NT= 1000
!
! A1(1)= 4.5  A2= .00065            S= .015  V0= S*A1/A2! A1(N) = EXP(-DT*V0)*A1(N-1) BEING CONTINUOUSLY REDUCED.
!
! NU1= .0000001 GAMMA1= .015
!
! TWO LYMPH NODES ARE LOCATED AT: ( 3 , 3 , 3 ) AND  ( 35 , 35 , 35 ) BOTH ARE AFFECTED
! PRIMARY TUMOR IS LOCATED AT:     ( 13 , 13 , 13 ) UPRIMARY= 10000
! SECOND TUMOR IS LOCATED AT:    ( 26 , 26 , 26 ) US1= 1000
! THIRD TUMOR IS LOCATED AT:     ( 47 , 47 , 47 ) US2= 400
! AT THE LYMPH NODE(L11,L12,L13) U0= 7000  AT THE LYMPH NODE(L21,L22,L23) U0= 4000
!
***************************************************************************************************
```

4.7   A Fast Growing Fast Spreading Cancer. Use of a Second Chemo

At the request of a reviewer, we have considered a case where cancer cells are dispersing faster in the model, the tumors are growing faster and spreading faster. Once cancer cells move into the interstitial fluid, which is a thin layer of fluid that surrounds the body's cells, they could metastasize faster. "Interstitial fluid (IF) flow can alter the tumor microenvironment and may therefore play a crucial role in tumor cell progression and malignancy. For instance, IF flow can create an asymmetric pericellular gradient of chemotactic proteins which the cells migrate toward through chemotaxis" [15]. This leads to a logical conclusion that a drug must be applied to increase the immune response and decrease the rate of growth. That is accomplished by applying the drug V_0. (We have used the formula (4.37)). Furthermore, a second drug W_0 is now required to halt the progress of cancer. That has been done in Fig. 4.11. The equation that we solve is

$$\partial u/\partial t = (a_1(n) - a_2 V_0 - a_3 W_0)u + \kappa(\partial^2 u/\partial x^2 + \partial^2 u/\partial y^2 + \partial^2 u/\partial z^2) \quad (4.39)$$

Since $\forall n, a_1(n) < a_1(n-1)$, a decreasing rate of growth of cancer cells, so eventually, the necessary condition for the success of the treatment, $\sigma = (a_1(n)u - a_2 V_0 - a_3 W_0) < 0$ will be satisfied. The input values have been listed in detail in the inputs for Fig. 4.11, and results have been given in Fig. 4.11.

Inputs for Fig. 4.11.

```
*****************************************************************************
! CHAPTER 4, FIG 4.6
! THERE ARE THREE TUMORS DISPERSING FASTER, NU1 (> 0.0000001)= 0.0001
! DU/DT = A1(N)*U*V0-A3=U*W0 + NU1=DELSQ(U)              ! ANTIGEN
! U0 AT T=0,U0 IS =  10000   AT THE PRIMARY SITE,U =  10000  SECOND TUMOR= 1000  3RD = 400  V0=FIXED =  103.84615
! DX= .02  DY= .02  DZ= .02  DT= .02  NX= 101  NY= 101  NZ = 101  NT= 1000
!
! A1(1)=  4.5  A2= .00065  , A3= .00075 , W0 (=5*V0)=  519.23077  ,S= .015 ,V0= S*A1(1)/A2=  103.84615
! NU1=  .0001 GAMMA1=  .015 ! A1(N) = A(N-1)*EXP(-DT*V0) BEING CONTINUOUSLY REDUCED.
!
! TWO LYMPH NODES ARE LOCATED AT:       ( 3 , 3 , 3 ) AND  ( 35 , 35 , 35 ) BOTH ARE AFFECTED
! PRIMARY TUMOR IS LOCATED AT:     ( 13 , 13 , 13 ) UPRIMARY=  10000
! SECOND TUMOR IS LOCATED AT:    ( 26 , 26 , 26 ) US1= 1000
! THIRD TUMOR IS LOCATED AT:     ( 47 , 47 , 47 ) US2= 400
! AT THE LYMPH NODE(L11,L12,L13) U0=   7000  AT THE LYMPH NODE(L21,L22,L23) U0=  4000
!
*****************************************************************************
```

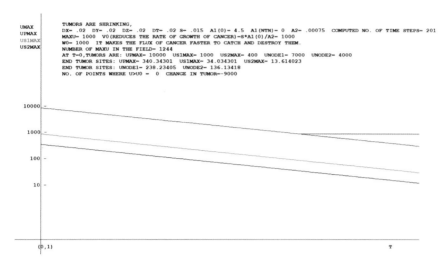

Fig. 4.11   Targeted therapy. V0 and W0 versus cancer cells

The graph reveals the success of the treatment.

The graph representing UMAX at some time during the treatment is given by a line parallel to the T-axis toward the end. It is because as UMAX started decreasing, it reached a point where it is equal to its value at the boundary line which is 1000. We have set up the delivery of medications such that it did not affect UMAX at the boundary. Cancer cells always look for a site in the body where they could stay safe. That is what we tried to see.

4.8   Reductions of Both the Rate of Growth and Dispersion of Cancer Cells

We have noticed that the necessary and sufficient condition that the cancer treatment will work, for some cases, if both $\sigma \to 0$. If $\kappa \to 0$, there will be no dispersion (formula wise this idea could be incorporated in the term representing dispersion). In this regard, we will pick up a case where the rate of growth of cancer will be reduced continuously by one drug and dispersion will be reduced continuously by another drug. As before the drug V_0 has been used to reduce the rate of growth continuously and W_0 has been used to reduce dispersion.

The formula for that is $\theta = exp(-\lambda W_0)$, $v_1 = \theta \times v_0$, where $v_0 =$ the initial value of the kinematic coefficient of viscosity (of blood). All other inputs are the same as before. (In the equation $\kappa = v$, Eq. (4.39).

Inputs for Fig. 4.12

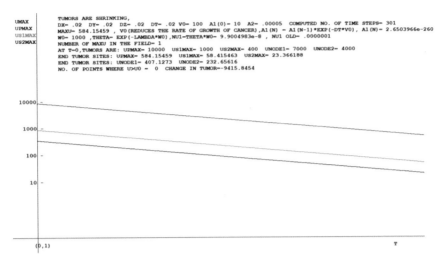

Fig. 4.12 Targeted therapy. V0 cuts growth of cancer and W0 slows flux of cancer cells

```
**********************************************************************************************
! CHAPTER 4, FIG 4.12
! THERE ARE THREE TUMORS .
! DU/DT - A1(N)*U-A2*U*VO-A3*U*WO + (NU1 * THETA)*DELSQ(U)  ,! ANTIGEN
! U0 AT T-0,U0 IS - 10000  AT THE PRIMARY SITE,U - 10000  SECOND TUMOR- 1000  3RD - 400  VO-FIXED - 100
! DX- .02  DY- .02  DZ- .02  DT- .02  NX- 101  NY- 101  NZ - 101  NT- 300
!
! A1(1)- 10  A2- .00005  A3- .00095  WO- 1000  VO- 100
! LAMBDA -  .00001     THETA - EXP(-LAMBDA*WO)
! NU1(0)- .0000001  NU1- THETA*N1(0)= 9.9004983e-8  GAMMA1- .015 ! A1(N) - A(N-1)*EXP(-DT*VO) BEING CONTINUOUSLY REDUCEI
! GAMMA1- .015
!
! TWO LYMPH NODES ARE LOCATED AT: ( 3 , 3 , 3 ) AND  ( 35 , 35 , 35 ) BOTH ARE AFFECTED
! PRIMARY TUMOR IS LOCATED AT:   ( 13 , 13 , 13 ) UPRIMARY- 10000
! SECOND TUMOR IS LOCATED AT:    ( 26 , 26 , 26 ) US1- 1000
! THIRD TUMOR IS LOCATED AT:     ( 47 , 47 , 47 ) US2- 400
! AT THE LYMPH NODE(L11,L12,L13) U0- 7000  AT THE LYMPH NODE(L21,L22,L23) U0- 4000
!
**********************************************************************************************
```

The results show that all the tumors are shrinking quite fast.
There exists not a single point in the field, where $U_{ijk}^n > U_{ijk}^{n-1}$.

4.9   Slow Growing Tumor with No Therapy. (Tubular Breast Cancer)

It is important to know how some slow growing tumors will keep on growing when no treatment is done. It may grow practically anywhere when it sees little or no resistance. Or in other words, when the immune system cannot recognize malignant cells as "foreign" or is too weak to fight (which happens at an old age, for example), cancer may grow wherever its growth is possible. Metastasis is automatic in some cases. The patient often slowly gets completely emaciated at the end with only bone and skin.

In several cases, cancer grows without showing any apparent sign of presence. In fact, this is not uncommon. So, patients see no symptoms, notice nothing unusual and as such feel no necessity to see a doctor. When symptoms appear, they often appear with some gigantic gruesome forms, and prognosis becomes equally gruesome.

Tubular breast cancer grows very very slowly. It is quite common among women over 50. Prognosis is very good when it is detected early. However, it is generally invasive.

Anyway, we have decided to look into two such cases. It has some similarities with several kinds of prostate cancer. "Prostate cancer occurs when a normal prostate cell begins to grow out of control. In many cases, prostate cancer is a slow-growing cancer that does not spread beyond the prostate gland before the time of diagnosis. Once prostate cancer forms, it feeds on androgens and uses them as fuel for growth. This is why one of the backbones of treatment for men, especially with advanced prostate cancer, is to lower a man's androgen levels with drugs collectively termed "hormone therapy." [https://www.pcf.org/about-prostate-cancer/what-is-pro state-cancer/how-it-grows/].

We considered a general case of a slow growing breast tumor growing silently with metastasis at several different locations. At each unit of time, its rate of growth remains the same ($a_1 = 0.25$). It is growing without any resistance. So, $V = V_0 = 0$ and $a_2 = 0$. Also it is barely dispersing. $v_1 = 1.0E - 10$. Yet it is a cancer. In Fig. 4.13, we see its strength. Over a long period of time, it is growing.

Inputs for Fig. 4.13

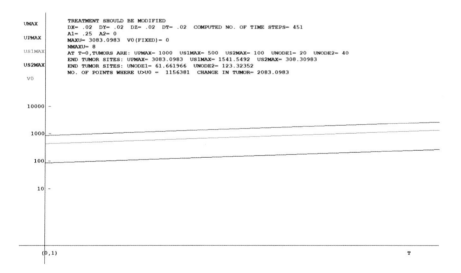

Fig. 4.13 Very slow growing tumor, with no medicine

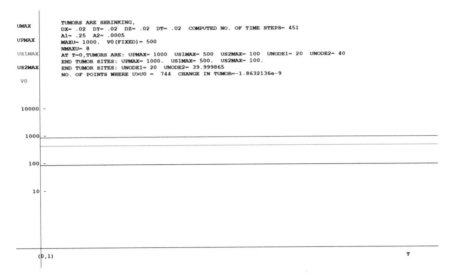

Fig. 4.14 Very slow growing tumor, against a weak immune response

```
!**********************************************************************************************
! CHAPTER 4, FIG 4.13
! THERE ARE THREE TUMORS , TWO LYMPH NODES ARE AFFECTED, S=0.4
! DU/DT - A1*U + (NU1)*DELSQ(U)                              ! ANTIGEN
! AT T=0, U0(OUTSIDE) IS - 20  AT THE PRIMARY SITE, U - 1000  SECOND TUMOR- 500   3RD - 100  V0-FIXED - 0
! DX- .02  DY- .02  DZ- .02  DT- .02  NX- 101  NY- 101  NZ - 101  NT- 450
! A1- .25
!
! NU1- 1.e-10
! GAMMA1- .015
! TWO LYMPH NODES ARE LOCATED AT: ( 3 , 3 , 3 ) AND  ( 35 , 35 , 35 ) BOTH ARE AFFECTED
! PRIMARY TUMOR IS LOCATED AT:   ( 13 , 13 , 13 ) UPRIMARY- 1000
! SECOND TUMOR IS LOCATED AT:    ( 26 , 26 , 26 ) US1- 500
! THIRD TUMOR IS LOCATED AT:     ( 47 , 47 , 47 ) US2- 100
! AT THE LYMPH NODE(L11,L12,L13) U0- 20  AT THE LYMPH NODE(L21,L22,L23) U0- 40
!**********************************************************************************************
```

Next assignment is to look into a minimal chemo treatment to keep it contained. That is shown in Fig. 4.14.

Inputs for Fig. 4.14

```
!**********************************************************************************************
! CHAPTER 4, FIG 4.14
! VERY SLOW GROWING THREE TUMORS , TWO LYMPH NODES ARE AFFECTED
! DU/DT - A1*U - A2*V0 + (NU1)*DELSQ(U)                      ! ANTIGEN
! AT T=0, U0(OUTSIDE) IS - 20  AT THE PRIMARY SITE, U - 1000  SECOND TUMOR- 500   3RD - 100  V0-FIXED - 500
! DX- .02  DY- .02  DZ- .02  DT- .02  NX- 101  NY- 101  NZ - 101  NT- 450
! A1- .25   A2- .0005
! NU1- 1.e-10
! GAMMA1- .015
! TWO LYMPH NODES ARE LOCATED AT: ( 3 , 3 , 3 ) AND  ( 35 , 35 , 35 ) BOTH ARE AFFECTED
! PRIMARY TUMOR IS LOCATED AT:   ( 13 , 13 , 13 ) UPRIMARY- 1000
! SECOND TUMOR IS LOCATED AT:    ( 26 , 26 , 26 ) US1- 500
! THIRD TUMOR IS LOCATED AT:     ( 47 , 47 , 47 ) US2- 100
! AT THE LYMPH NODE(L11,L12,L13) U0- 20  AT THE LYMPH NODE(L21,L22,L23) U0- 40
!**********************************************************************************************
```

Here UPMAX (max value of u, accumulation of cancer cells at the primary site) and UMAX (max value of u in the entire field) are the same.

4.10 Intratumoral Cancer Treatment

"Patients who are undergoing intravenous immunotherapy may consider intratumoral immunotherapy to enhance their treatments or add this to their radiation treatments or chemotherapy. According to experts, combination treatments are actually helpful, cost-effective, and safer for cancer patients since the therapy is more targeted. Furthermore, the effects of immunotherapy and ablation, when combined, are synergistic." https://williamscancerinstitute.com/advantages-of-intratumoral-immunotherapy-in-cancer-treatment.

In order to apply intratumoral immunotherapy, we proceeded as follows: We have considered cubic computational field with $(NX + 4) \times (NY + 4) \times (NZ + 4)$ number of points where $NX = NY = NZ = 101$.

$U_{ijk}^0 =$ initial value of malignant cells at the point $(i, j, k) =$ an accumulation of extra mass, distributed at each and every point in the field, except at the points where there are tumors or affected lymph nodes.

At each point of a tumor, U_{ijk}^0 has accumulated a large amount of mass. For an intratumoral therapy, a drug has been injected into each tumor at each t_n such that

$$U_{ijk}^n = a_4(n) \times U_{ijk}^n \qquad (4.40)$$

where $0 < a_4(n) < 1$.

In our code, $a_4(n)$ is a constant $\forall n$

The model is

$$\partial u / \partial t = (a_1 - a_2 V_0 - a_3 W_0)u + \kappa(\partial^2 u/\partial x^2 + \partial^2 u/\partial y^2 + \partial^2 u/\partial z^2) \qquad (4.41)$$

where V_0 and W_0 are two immunotherapeutic drugs to reduce u, the mass of the cancer cells.

At the tumor sites, $U_{ijk}^n = a_4 \times U_{ijk}^n$. (If $0 < a_4 < 1$, the medication is activated).

We have considered three cases: (1) $a_4 = 1$. This means no intratumoral therapy has been done. The results have been presented in Fig. 4.15; (2) $a_4 = 0.01$ with the other inputs remaining the same. Here the intratumoral shot is reducing the tumor, (3) $a_4 = 0.1$.

Here a_4 is fixed. Later we have made it a variable changing with time also the constant of dispersion $= \kappa$ has been increased by 100. We have assumed that when the intratumoral drug has been injected in the tumor, some malignant cells start dispersing fast.

"Despite reducing the size of primary tumors, chemotherapy changes the tumor microenvironment, resulting in an increased escape of cancer cells into the bloodstream. Furthermore, chemotherapy changes the tissue microenvironment at the distant sites, making it more hospitable to cancer cells upon their arrival" [16]. So during intratumoral immunotherapy, it is very likely that cancer cells will migrate faster from the tumor sites that means faster dispersion.

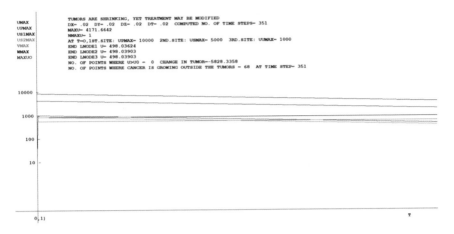

Fig. 4.15 No intratumoral shot. Immunotherapy only

With these ideas in view, we considered the first case where no intratumoral medication was used. Only strong immunotherapy.

Inputs for Fig. 4.15.

```
*********************************************************************************
! TREATMENT WITH IMMUNOTHERAPY AND   MONOCLONAL DRUG. NO INTRATUMORAL SHOTS. FIG.4.15
! DU/DT = A1*U-A2*U*VO - A3*U*WO + NU1*DELSQ(U) ! ANTIGEN
!    DX= .02   DY= .02   DZ= .02   DT= .02   NX= 101  NY= 101   NZ = 101   NT= 350
!    A1= 7.5   A2= .005  A3= .0045 A4= 1   VO= 650   WO= 1000   U0 IN THE FIELD = 25
!    B1= 0   B2= 0   B3= 0
!    C1= 0   C2= 0   C3= 0
! NU1= .0000001
! GAMMA1= .015   U AT THREE LYMPH NODES ARE 1200 EACH
! TWO LYMPH NODES ARE LOCATED AT: ( 3 , 3 , 3 ) AND    ( 35 , 35 , 35 ) ( 42 , 42 , 42 )
! THREE TUMORS LOCATED AT:    ( 13 , 13 , 13 ) AND ( 26 , 26 , 26 )
!( 47 , 47 , 47 ) THE VALUES OF U ARE: 10000   5000   1000   RESPECTIVELY.
!
*********************************************************************************
```

From the value of $a_1 = 7.5$, it is a very fast growing tumor.

Inputs for Fig. 4.16.

```
*********************************************************************************
! TREATMENT WITH INTRATUMORAL IMMUNOTHERAPY AND   MONOCLONAL DRUG. FIG.4.16
! DU/DT = A1*U-A2*U*VO - A3*U*WO + NU1*DELSQ(U) ! ANTIGEN
!    DX= .02   DY= .02   DZ= .02   DT= .02   NX= 101  NY= 101   NZ = 101   NT= 350
!    A1= 7.5   A2= .005  A3= .0045 A4= .01   VO= 650   WO= 1000   U0 IN THE FIELD = 25
!    B1= 0   B2= 0   B3= 0
!    C1= 0   C2= 0   C3= 0
! NU1= .0000001
! GAMMA1= .015   GAMMA2= 0   GAMMA3= 0   U AT THREE LYMPH NODES ARE 1200 EACH
! THREE LYMPH NODES ARE LOCATED AT: ( 3 , 3 , 3 ) AND   ( 35 , 35 , 35 ) ( 42 , 42 , 42 )
! AT EACH LYMP NODE U=1200
! THREE TUMORS LOCATED AT:    ( 13 , 13 , 13 ) AND ( 26 , 26 , 26 )
!( 47 , 47 , 47 ) THE VALUES OF U ARE: 10000   5000   1000   RESPECTIVELY.
!
*********************************************************************************
```

While, $V_0 = fixed$, and $W_0 = fixed$ were applied intravenously. The monoclonal drug is serving another purpose. It is getting attached to all the cancer cells in

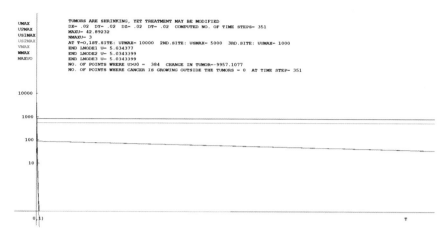

Fig. 4.16 Intratumoral therapy is added to immunotherapy

the tumors and inhibits their rate of growth. For that we have introduced the parameter a_4.

So, inside the tumors, $u_{ijk}^n = a_4 u_{ijk}^n$. This is an intratumoral treatment (applicable only to tumors).

With these inputs, tumors started shrinking drastically. However, it should be noted that the runaways are not getting smaller that fast (the redline) although the slope is negative. The graphs for $VMAX = V_0$ and $WMAX = W_0$ (because they remained constants), shown by the green line and the blue line, respectively, remained parallel to the horizontal line for time. The runaway cancer cells are declining very very slowly.

We took this graph for $NT = 351$.

Drastically, tumors started shrinking, which is a good news, but what runaways will do later is of some concern. This we call only a local solution.

To check what is happening after $NT = 51$, we took another run and found the result given in the second graph of Fig. 4.16. We called it Fig. 4.16, because it is a part of the same computer run. This shows that the redline representing the accumulations of cancer cells outside the tumors is hardly changing. This is of great concern!!!

Such information could be misleading to the doctors. The point is, regardless how strong the treatment is, the results may not be noticed in a short time. Also, intratumoral treatment may not be very effective. Because it only reduces the tumors and practically does nothing to the runaways. So eventually they are likely to form metastatic tumors elsewhere.

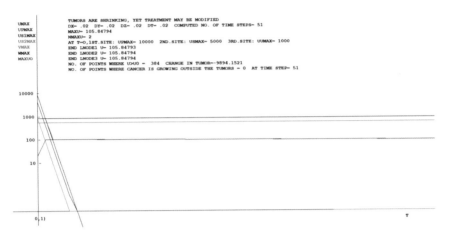

Fig. 4.17 Intratumoral therapy is added to immunotherapy

Inputs for Fig. 4.17.

```
********************************************************************************!
!TREATMENT WITH A WEAKER INTRATUMORAL DRUG FOR IMMUNOTHERAPY AND A MONOCLONAL DRUG. FIG.4.18
! DU/DT = A1*U-A2*U*VO - A3*U*WO + NU1*DELSQ(U) ! ANTIGEN
!    DX= .02  DY= .02  DZ= .02  DT= .02  NX= 101  NY= 101  NZ = 101  NT= 120
!    A1= 7.5  A2= .005  A3= .0045 A4= .1  VO= 650  WO= 1000  UO IN THE FIELD = 25
!    B1= 0  B2= 0  B3= 0
!    C1= 0  C2= 0  C3= 0
! NU1= .00001
! GAMMA1= .015  GAMMA2= 0  GAMMA3= 0  U AT THREE LYMPH NODES ARE 1200 EACH
! THREE LYMPH NODES ARE LOCATED AT: ( 3 , 3 , 3 ) AND   ( 35 , 35 , 35 ) ( 42 , 42 , 42 )
! AT EACH LYMP NODE U=1200
! THREE TUMORS LOCATED AT:     ( 13 , 13 , 13 ) AND ( 26 , 26 , 26 )
!( 47 , 47 , 47 ) THE VALUES OF U ARE: 10000  5000  1000  RESPECTIVELY.
!
********************************************************************************!
```

In the inputs, we notice at the tumors $U_{ijk}^n = a_4 U_{ijk}^n$ at each of them where $a_4 = 0.1$. We have also considered a faster dispersion $NU1 = 0.00001$. In the inputs for Fig. 4.16, it was $NU1 = 0.0000001$. Apparently, the treatment worked well. All tumors started shrinking. Yet runaway cancer cells are of great concern.

It is very strange that they did not totally vanish. Instead they became negligibly small which in the log scale is a negative number (meaning they lie between 0 and 1). Cancer is practically gone, but their remnants remained alive minutely!!! (Fig. 4.18). This is a sign of danger for the future.

Some aspects of this danger are presented in Fig. 4.19. The redline shows that the runaways are definitely going to be serious threats to life as discussed below.

Now we will consider a case where only intratumoral shots were used and no other medication for immunotherapy was administered. We have also considered the rate of growth of cancer $a_1 = 4.5$.

Figure 4.19 shows that all the three tumors did shrink quite fast as expected, whereas cancer in the field started growing drastically. So the treatment failed drastically.

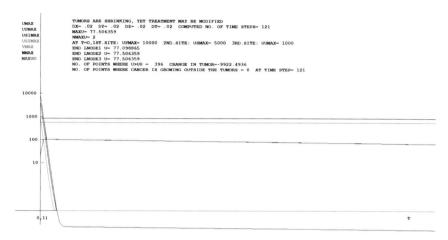

Fig. 4.18 Intratumoral therapy is added to immunotherapy with faster dispersion

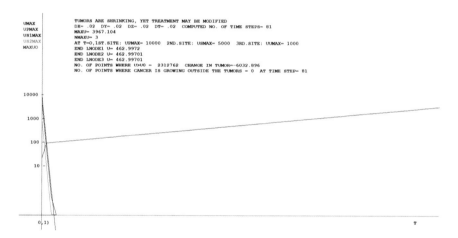

Fig. 4.19 Intratumoral therapy only. Outside the tumor, cancer cells (the red line) are increasing

Inputs for Fig. 4.19.

```
*****************************************************************************
!TREATMENT WITH NO IMMUNOTHERAPY OR MONOCLONAL DRUG.ONLY INTRATUMORAL FIG.4.19
! DU/DT = A1*U + NU1*DELSQ(U) ! ANTIGEN
! A1= 4.5 A4= .01   U0 IN THE FIELD = 25
! THERE ARE THREE TUMORS AND AT EACH TUMOR U = U*A4, DONE BY INTRATUMORAL SHOTS
! DX= .02  DY= .02  DZ= .02  DT= .02  NX= 101  NY= 101  NZ = 101  NT= 80
! NU1= .0000001 GAMMA1= .015
! U AT THREE LYMPH NODES IS 1200 EACH
! THREE LYMPH NODES ARE LOCATED AT: ( 3 , 3 , 3 ) AND  ( 35 , 35 , 35 ) ( 42 , 42 , 42 )
! AT EACH LYMP NODE U=1200
! THREE TUMORS LOCATED AT:   ( 13 , 13 , 13 ) AND ( 26 , 26 , 26 )
! ( 47 , 47 , 47 ) THE VALUES OF U ARE: 10000   5000   1000   RESPECTIVELY.
!
*****************************************************************************
```

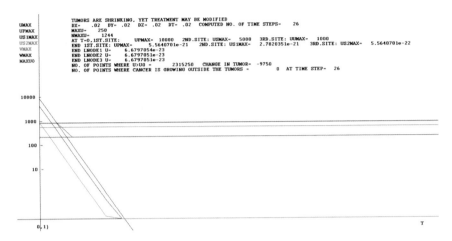

Fig. 4.20 Intravenous therapy only. Medicines not attacking boundaries (hideouts)

Here $u^n_{ijk} = a_4 u^n_{ijk}$, $a_4 = 0.01$. Tumors are gone. No trace. However, the runaways are increasing very rapidly.

These results validate what authors in the Ref. [16] have stated. Here all tumors are definitely shrinking drastically, however, outside the locations of the tumors they are spreading in full swing. The redline displays that grim tragedy.

Numerically/computationally, Fig. 4.19 demonstrates how stiff the system is.

We will compare this with similar intravenous treatment (Fig. 4.20). However, since the same amount of medication will be at each grid point, total amount in this case will be much more than that used in Fig. 4.19. Since intravenous infusion also means the medications will penetrate the tumors, so in the input we have mentioned that aspect of treatment. So intravenous also includes intratumoral in our models.

Inputs for Fig. 4.20.

```
!TREATMENT WITH NOIMMUNOTHERAPY OR MONOCLONAL DRUG.ONLY INTRATUMORAL FIG.4.20, INTRAVENOUS
!MEDICATIONS NOT ATTACKING POINTS ON THE BOUNDARY LINES
!  DU/DT = A1*U-A2*U*V0 - A3*U*W0 + NU1*DELSQ(U) ! ANTIGEN
!    DX-    .02  DY-   .02  DZ-   .02  DT-  .02  NX-  101  NY- 101   NZ - 101   NT- 25
!    A1-   7.5  A2-  .0005  A3-  .00045 A4-  .1  V0- 650  W0- 1000  U0 IN THE FIEL =   250
!    B1-    0  B2-  0  B3-  0
!    C1-    0  C2-  0  C3-  0
!  NU1-   .0000001
!  GAMMA1-   .015   GAMMA2-   0  GAMMA3-   0  U AT THREE LYMPH NODES ARE 1200 EACH
!  THREE LYMPH NODES ARE LOCATED AT:         ( 3 , 3 , 3 ) AND  ( 35 , 35 , 35 ) ( 42 , 42 , 42 )
!  AT EACH LYMP NODE U-1200
!  THREE TUMORS LOCATED AT:          ( 13 , 13 , 13 ) AND ( 26 , 26 , 26 )
!  ( 47 , 47 , 47 ) THE VALUES OF U ARE:  10000   5000   1000   RESPECTIVELY.
!
```

The results are as follows:

All tumors are being destroyed, and cancer cells went to their hideouts on 1244 ($=105 \times 12$–16) points on the border lines of the cubical field of computation on each side of which there are 105 grid points. So they should be tracked and attacked at every point in the computational field. In Fig. 4.20, MAXU = max value of $u = 250$, which is the value of u on the boundary. At the very beginning, this was the value at every point outside the cancer sites. Now all cancer sites have been eliminated inside the field as shown by the fast decreasing graphs; however, in the border lines of the cube, cancer survived as shown by the redline parallel to the t $axis$. On this line, MAXU = 250. NMAXU = number of points where MAXU is present is 1244.

In the next section, medications will penetrate the boundary lines. Let us discuss that.

4.10.1   The Entire Field Including the Boundaries is a War Zone

For most cases we have extrapolated the values of the variables on the external boundaries. In a few cases, this procedure has been changed. We transformed the entire computational field into a war zone.

Everywhere cancer cells move, and the defense forces are fighting them off right there. No more hideouts for cancer cells. Let us briefly discuss how this was done.

Here we have considered very similar inputs as follows:

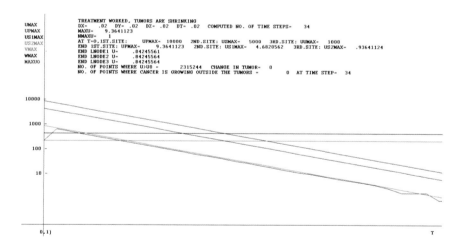

Fig. 4.21 Intravenous therapy only. Medicines attacking boundaries too

Inputs for Fig. 4.21

```
•••••••••••••••••••••••••••••••••••••••••••••••••••••••••••••••••••••••••••••••••••••••••
|TREATMENT WITH NO IMMUNOTHERAPY OR MONOCLONAL DRUG.ONLY INTRATUMORAL FIG.4.21, INTRAVENOUS
|MEDICATIONS ATTACKING POINTS ON THE BOUNDARY LINES TOO
| DU/DT = A1•U-A2•U•V0 - A3•U•W0 + NU1•DELSQ(U)  | ANTIGEN
|   DX=    .02   DY= .02   DZ= .02   DT= .02   NX= 101   NY= 101   NZ = 101   NT= 50
|   A1=   7.5   A2= .00005   A3= .000045  A4= .75   V0= 250   W0= 500   U0 IN THE FIELD =    250
|   B1=     0   B2= 0   B3= 0
|   C1=     0   C2= 0   C3= 0
| NU1=    .0000001
| GAMMA1=   .015   GAMMA2=   0   GAMMA3=   0   U AT THREE LYMPH NODES ARE 1200 EACH
| THREE LYMPH NODES ARE LOCATED AT:       ( 3 , 3 , 3 ) AND   ( 35 , 35 , 35 ) ( 42 , 42 , 42 )
| AT EACH LYMP NODE U=1200
| THREE TUMORS LOCATED AT:           ( 13 , 13 , 13 ) AND ( 26 , 26 , 26 )
| ( 47 , 47 , 47 ) THE VALUES OF U ARE:   10000   5000   1000   RESPECTIVELY.
|
•••••••••••••••••••••••••••••••••••••••••••••••••••••••••••••••••••••••••••••••••••••••••
```

Here there are no hideouts for cancer cells. Medications worked like guided missiles against all cancer cells all over the body carrying this foreign antigen. From Fig. 4.21, the largest size of the tumor has MAXU = 9.3641, fast shrinking tumors. (There are three tumors in the computational field).

Observing that the entire defense forces must follow the principle of "chemotaxis," meaning an assemble in large numbers wherever the enemies are, we assumed that on all external borders, wherever cancer cells are present, defense must exhibit its strong presence and in fact they should demonstrate overwhelming presence surrounding the enemies. So, it could be logical to assume that there is no dispersion on the border. Cancer cells cannot escape neither they can get in. Even if some cancer cells manage to get out, they will be attacked right there. Thus, over the boundaries, we should solve:

$$\partial u/\partial t = (a_1 - a_2 v - a_3 w)u + \kappa_1(\partial^2 u/\partial x^2 + \partial^2 u/\partial y^2 + \partial^2 u/\partial z^2)$$

$$\partial v/\partial t = b_1 u - (b_2 u + b_3 w)v + \kappa_2(\partial^2 v/\partial x^2 + \partial^2 v/\partial y^2 + \partial^2 v/\partial z^2)$$

$$\partial w/\partial t = c_1 u - (c_2 u + c_3 v)w + \kappa_3(\partial^2 w/\partial x^2 + \partial^2 w/\partial y^2 + \partial^2 w/\partial z^2)$$

Solving this model is our next task. CDey-Simpson method was applied.

4.10.2 Conclusion

It is very important to understand that the equations which we often call the reaction–diffusion equation are reaction–dispersion, because neither the cancer cells nor any other medications are really diffusing. Only if all the boundary conditions are zeros, the variables will be diffused in due time.

In a regular heat equation, heat finally settles down with the boundary conditions. Still, we could think that heat gets diffused in the atmosphere ultimately. Here in our models, although most cancer cells should die, many will not and will take shelter in some parts of the body. They could stay there for years till they find some suitable conditions to appear as metastatic tumors. So cancer treatments in some form often go on for years.

In the heat conduction equation, the coefficient of thermal conductivity is a constant caused by temperature differences. Here ν_1 is the coefficient of dispersion, meaning how cancer cells are moving inside the body to spread. This is the coefficient of viscosity in flow models. Malignant cells have the ability to move toward organs which may have already hosted many more cancer cells. During such travels, they use flow of blood, flow of lymph, and interstitial fluid. And fortunately not all, but some find new hosts and settle there. In the neighborhoods of these hosts, mathematically, they settle down to be dispersed in due time. So our equations are reaction–dispersion equations, not reaction–diffusion equations. The value of ν_1 could be sometimes significantly larger than the value of coefficient of viscosity of blood.

Please Note

We have often interchanged ν_1, ν_2, and ν_3 with κ_1, κ_2, and κ_3, respectively. Because ν and $v(v)$ look very much alike in this font.

Various mathematical models reveal that intratumoral treatment does not work very well. Of course, tumors start shrinking, but the cancer cells start growing at all other locations.

All our discussions on cancer treatments will now lead us to model how immunotherapy could be modeled mathematically. Here we will consider a larger model and advance our ideas on this aspect of cancer treatment mathematically.

References

1. Gyton, A.C., Hall, J.E: Text Book on Medical Physiology. W.B. Saunders (2000)
2. Britton, N.F: Reaction-Diffusion Equations and Their Applications to Biology. Academic, New York (1986)
3. Hattaf, K., Yousfi, N.: Global Stability for reaction-diffusion equations in biology. Comput. Math. Appl. **66**(8) (2013)
4. American Cancer Society. https://www.cancer.org/
5. Dreher, D., Junod, A.F.: Role of oxygen free radicals in cancer development. Eur. J. Cancer **32A**(1) (1996)
6. Valko, M., et al: Free radicals, metals, and antioxidants in oxidative stress-induced cancer. Chem. Bio. Interact. **160**(1) (2006)
7. Dey, S.: Large-scale parallel computations by perturbed functional iterations. Parallel Algorithms Appl. **3**(3–4) (1994)
8. Komen, S.G.: https://ww5.komen.org/
9. Dey, S., Dey, C.: An Explicit Predictor-Corrector Solver with Applications to Burgers' Equation. NASA Technical Memorandum 84402 (1983)
10. https://www.texasoncology.com/cancer-treatment/chemotherapy
11. Dey, S.K.: Computational modeling of breast cancer treatment by immunotherapy, radiation and estrogen inhibition. Sci. Math. Jpn. **58**(2) (2003)
12. Basu, A.K.: DNA damage, mutagenesis and cancer. Int. J. Mol. Sci. **19**(4) (2018)
13. Phadke, S., Bharadwaj, D., Dey, S.K.: An explicit predictor-corrector solver with application to seismic wave modelling. Comput. Geosci. **26**(9) (2000)

14. Osman, H., Wood, A.S.: School of Mathematics, University of Bradford. Stability of Charlie's Method on Linear Heat Conduction Equation. Mathematika.Jilid.15 (2001)
15. Steinar, E., Waldeland, J.O.: How tumor cells can make use of interstitial fluid flow in a strategy for metastasis. Cellul. Molecul. Bioengin. 12(3) (2019)
16. Middleton, J.D., Stover, D.J., Hai, T.: Chemotherapy-exacerbated breast cancer metastasis: a paradox explainable by dysregulated adaptive-response. Int. J. Molecul. Sci. (2018)
17. Redig, A.J., McAllister, S.S.: Breast cancer as a systemic disease: a view of metastasis. J. Int. Med. 274(2)
18. Alameddine, A.K, Conlin, F., Binnall, B.: An introduction to the mathematical modeling in the study of cancer systems biology. Cancer Inform. 17, 1–10 (2018)
19. Dey, C., Dey, S.K.: Explicit finite difference predictor and convex corrector with applications to hyperbolic partial differential equations. Comput. Math. Appl. 9(3) (1983)
20. Dey, C.: Simulation of geometrical curves on a microcomputer. Simulation 38 (1982)

Chapter 5
Modeling Advanced Immunotherapy with Monoclonal Drugs

Abstract In this chapter, applications of monoclonal drugs to control and contain cancer have been studied mathematically. We have used several computer outputs in the form of tables. The primary purpose is to show the actual numerical results that explain the success/failure of a method of treatment. These reveal what we expect the immune system and the medications used for inhibition of malignant cells to do and validate that they have made their best attempts to do their jobs. For instance, they should be at the height of their activities near the tumors and affected lymph nodes, they should get stronger as malignancy gets worse, and they should not increase their strengths if malignancy starts getting weaker. Computational results show that they worked as expected.

5.1 Rationale

Our objective is now to look into a more comprehensive picture of cancer treatment as being performed at the present time. We will look into this through mathematical modeling and computational studies.

In the introduction, we have discussed some aspects of our immune system. Immunotherapy means strengthening the immune system to kill pathogens (or sometimes, it means immunosuppression to reduce immunoactivities during organ transplants). We generally think of strengthening the leukocytes. But strong leukocytes may not give a strong fight. There are many checkpoints. Some epigenetic changes may make the immune system completely unresponsive when cancer cells flourish. In the previous chapter, we have discussed, theoretically, several forms of immunotherapy applicable to practically all forms of cancer. In the practical world, there are some restrictions. We will now discuss some of them and look into corresponding mathematical modeling. Unfortunately, immunotherapy is not yet very much in practice for breast cancer treatment. However, what we have done could be usable in the future.

Regarding breast cancer, immunotherapy generally means strengthening the immune system to fight against foreign antigens and annihilate every one of them

© Springer Nature Singapore Pte Ltd. 2021
S. Dey and C. Dey, *Mathematical and Computational Studies on Progress, Prognosis, Prevention & Panacea of Breast Cancer*, Forum for Interdisciplinary Mathematics,
https://doi.org/10.1007/978-981-16-6077-1_5

whenever and wherever they could be detected in the body. Conceptually, mathematical studies on this topic were undertaken with regard to all forms of immunosurveillance and immune activities to cure breast cancer at NASA Ames by the first author during his tenure in 2001. The models deal with strengthening the entire immune system. Later, both authors worked jointly on a NASA grant, presented their works at various universities and research centers around the world, and received both praise and positive inputs to improve their works.

Let us briefly discuss some of the immune activities, so that we may follow how and why we have constructed our models. Some of these have been discussed earlier. However, we felt that such repetitions are necessary, so that readers could understand our discussions more easily.

We have an innate immune system [2], our first line of defense, which attacks all pathogens indiscriminately. These are natural killer cells (NK-cells), neutrophils, eosinophils, and macrophages (phagocytic cells). Our second line of defense is the adaptive immune system. They are the T-cells and the B-cells. Both originate in the bone marrow. T-cells mature in the thymus which secretes the hormone thymosin. They are some of the most relentless fighters. They mostly move through the lymphatic vessels. Lymph nodes, spleen, and tonsils are their castles. If and when they recognize foreign antigens on the surface of a pathogen, they immediately launch a very antigen-specific attack to destroy them.

In general, our immune system attacks all foreign attackers in a very organized way. These attackers could be biomolecules or proteins identified by their corresponding antigens which have unique surface features or epitopes. The immune system releases cytokines like interferon, interleukin, and growth factors, for cell-to-cell communications. These are cell signaling molecules. They do cell-to-cell communication to awaken immune responses and stimulate them to move ahead to promote chemotaxis. Chemokines are secreted by infected cells to trigger an immune attack. B-cells release antibodies, which are varieties of proteins known as immunoglobulins. Both B-cells and cytotoxic T-cells or CTL, a lymphocyte (also known as CD8 + T-cell), release hundreds of very antigen-specific cytokines to destroy foreign antigens. Having assistance from helper T-cells, B-cells differentiate into plasma cells producing antibodies against a specific foreign antigen. It causes destruction or dissolution of that specific antigen. Debris are being cleared by phagocytic macrophages. This is humoral immunity or antibody-mediated immunity.

Cellular immunity is mediated by T lymphocytes. Recognizing the MHC–antigen (major histocompatibility complex) on an infected cell, helper T-cells release cytokines to wake up cytotoxic T-cells, and the war starts against the invading antigens. Then, the infected cells undergo lysis as T-cells win.

During these mediations, B- and T-cells create memory cells. These are exact duplicates of the B- and T-cells which did the fighting. They remain in the blood. So, if and when the body is threatened by the same antigen, they fight just the way their mothers did in the past. This concept is being used to manufacture vaccines.

In this respect, the role of natural killer (NK) cells, which are also lymphocytes, is remarkable. They are neither antigen specific nor do they depend on recognition of the MHC on the infected cells. "While on patrol NK-cells constantly contact other cells, whether or not the NK-cell kills these cells depends on a balance of signals from activating receptors and inhibitory receptors on the NK-cell surface. Activating receptors recognize molecules that are expressed on the surface of cancer cells and infected cells, and 'switch on' the NK-cells. Inhibitory receptors act as a check on NK-cell killing. Most normal healthy cells express MHC I receptors which mark these cells as 'self.' Inhibitory receptors on the surface of the NK-cell recognize cognate MHC I, and this 'switches off' the NK-cells, preventing them from killing. Cancer cells and infected cells often lose their MHC I, leaving them vulnerable to NK-cell killing. Once the decision is made to kill, the NK-cell releases cytotoxic granules containing perfin and granzymes, which leads to lysis of the target cell" [12]. NK-cells also activate the dendritic cells and macrophages to detect cancer cells. In [1], while modeling immunotherapy, we have modeled this mathematically (to be discussed in Chap. 6). In reality, this has not yet been done. However, researchers are working on this project. Recently, in September 2018, in a report, Dr. Roulet from Cancer Research Institute (CRI, New York) has stated: "In the cancer immunotherapy field there has been a singular focus on mobilizing anti-tumor T cells .... We believe that NK cells have an important place at the table. Checkpoint therapy combined with other NK-directed immunotherapies may enable us to target many types of tumors that are currently non-responsive to available therapies."

Regarding immunotherapy with B-cells, not much work has yet been done. B-cells are best known for their role in humoral response by producing antibodies. An antibody is a protein that sticks to a specific antigen (another protein). They travel all over the body, get attached to cancer cells, and trigger attacks by the immune system. That means they flag the cancer cells so that T-cells recognize these enemies. Recently, it has been found that they also play a role as antigen presenters. They also secrete cytokines and chemokines to regulate appropriate immune responses [9]. Antibodies are antigen (foreign)-specific proteins that circulate throughout the body until they find and attach to the foreign antigen. Once attached, they recruit other partners of the immune system to destroy the cells containing that foreign antigen. Researchers can design antibodies that specifically target certain foreign antigens, such as the one found on cancer cells. Then, they can make many copies of that antibody in the laboratory. These are known as *monoclonal antibodies* (mAbs). Naked mAbs are antibodies that work by themselves. There is no drug or radioactive material attached to them. These are the most common types of mAbs used to treat cancer. Most naked mAbs attach to antigens on cancer cells, but some work by binding to antigens on other, non-cancerous cells or even free-floating proteins.

Next are the dendritic cells and macrophages. They both are antigen-presenting cells. However, macrophages phagocytize them too. Immunotherapy should promote stronger activities to recognize foreign elements carrying unique antigens. Because, if they do not aggressively search for them, cytotoxic T-cells will not be able to do their jobs effectively. They badly need the assistance of the antigen-presenting cells.

Unfortunately, at the present time not all breast cancer patients are candidates for immunotherapy. It is not applied in general for many cancer patients, because of the common side effects like fever, nausea, skin rashes, etc. For triple-negative breast cancer patients who are PD-L1 positive, FDA approved the drug atezolizumab to be used along with a standard chemotherapy. We have attempted in this chapter to model these applications and mathematical analysis of the effects of using this form of immunotherapy.

The protein programmed death-ligand 1 (PD-L1) is expressed by many normal cells. T-cells express a protein programmed cell death protein 1 (PD-1). When PD-L1 binds with PD-1, T-cells recognize that this cell is harmless and do not attack it. That is why these proteins are called the checkpoint proteins. Unfortunately, several cancer cells have a lot of PD-L1 which they express and avoid an attack by the T-cells. So, an immune checkpoint inhibitor that stops PD-1 from binding with PD-L1 will allow T-cells to attack the cancer cells. This is a form of checkpoint immunotherapy. Patients, whose cancer cells have a high amount of PD-L1, could benefit from this form of immunotherapy.

As of 2019, for locally advanced triple-negative breast cancer, the drug atezolizumab is approved, if the cancer cells express PD-L1 protein. Researchers found that adding immunotherapy to the chemotherapy drug may be beneficial for a subset of patients, although it could increase harmful side effects.

This scenario and many more have been mathematically modeled and computationally solved in this chapter. All codes have been written in such a way that computations stop when at a particular time step $\max(U_{ijk}^n) < \xi \times max(U_{ijk}^0)$, where $\xi = 0.01$ or 0.001 according to what procedure has been undertaken. Here, $max(U_{ijk}^0)$ is the max value of the foreign antigen at $t = 0$, and $\max(U_{ijk}^n)$ is the max value of the antigen at t_n. ($U_{ijk}^n = U(x_i, y_j, z_k, t_n)$). Also depending upon the treatment plan, if V means strength of the immune system, then in the code we put a restriction on V. If $\max(V)$ at a time step exceed $5 \times \max(V^0)$, where V^0 is the initial value of V, then it is reduced by 5%, and if $\max(V) < 75\%$ of $\max(V^0)$, then V_{ijk}^n is increased by a factor of η, $where\ 1 < \eta < 1.25$. This has been done consistent with the information that we have collected from medical practitioners. These concepts need to be emphasized to study modeling the war on breast cancer through mathematical equations.

As mentioned before, malignant cells often hide their antigens and pretend that they are not foreign or use other techniques to hide their identities to avoid being attractants of leukocytes. Here, we have assumed that their identities have been exposed, chemotaxis has taken place, and the body's defensive forces have started fighting. This is valid in all through the monograph.

5.2 Three Reaction–Dispersion (Diffusion) Equations: The Model

Let us briefly repeat more properties of monoclonal antibodies. They are laboratory-produced molecules engineered to serve as substitute antibodies that can restore, enhance, or mimic the immune system's attack on cancer cells. They are designed to bind to antigens that are generally more numerous on the surface of cancer cells than healthy cells. They can block the rate of growth of cancer cells by preventing angiogenesis. They can block all checkpoint inhibitors of the immune response. So, the immune system becomes more effective against cancer. They can also directly attack cancer cells. Because a monoclonal antibody has the ability to flag cancer cells, it can be engineered as a delivery vehicle for radiation treatment. When a monoclonal antibody is attached to a small radioactive particle, it transports the radiation directly to cancer cells. That certainly could minimize the effect of radiation on healthy cells. Sometimes, two monoclonal particles are used. One is attached to the cancer cell and the other to an immune cell and that makes immunotherapy more effective. Such immunotherapies have been used for breast cancer treatment.

At the present time, monoclonal drugs have been designed to target cancer cells and destroy them. They can do several different jobs to keep cancer under strict control.

The activities of monoclonal drugs are summed up by the scientists and doctors at Mayo Clinic as follows:

"The role of the drug in helping the immune system may include the following:

- Flagging cancer cells. Some immune system cells depend on antibodies to locate the target of an attack. Cancer cells that are coated in monoclonal antibodies may be more easily detected and targeted for destruction.
- Triggering cell membrane destruction. Some monoclonal antibodies can trigger an immune system response that can destroy the outer wall (membrane) of a cancer cell.
- Blocking cell growth. Some monoclonal antibodies block the connection between a cancer cell and proteins that promote cell growth—an activity that is necessary for tumor growth and survival.
- Preventing blood vessel growth. In order for a cancerous tumor to grow and survive, it needs a blood supply. Some monoclonal antibody drugs block protein–cell interactions necessary for the development of new blood vessels.
- Blocking immune system inhibitors. Certain proteins that bind to immune system cells are regulators that prevent overactivity of the system. Monoclonal antibodies that bind to these immune system cells give the cancer-fighting cells an opportunity to work with less inhibition.
- Directly attacking cancer cells. Certain monoclonal antibodies may attack the cell more directly, even though they were designed for another purpose. When some of these antibodies attach to a cell, a series of events inside the cell may cause it to self-destruct.
- Delivering radiation treatment. Because of a monoclonal antibody's ability to connect with a cancer cell, the antibody can be engineered as a delivery vehicle for other treatments. When a monoclonal antibody is attached to a small radioactive particle, it transports the radiation treatment directly to cancer cells and may minimize the effect of radiation on healthy cells. This variation of standard radiation therapy for cancer is called radioimmunotherapy.

- Delivering chemotherapy. Similarly, some monoclonal antibodies are attached to a chemotherapeutic drug in order to deliver the treatment directly to the cancer cells while avoiding healthy cells.
- Binding cancer and immune cells. Some drugs combine two monoclonal antibodies, one that attaches to a cancer cell and one that attaches to a specific immune system cell. This connection may promote immune system attacks on the cancer cells."

The reference is: [https://www.mayoclinic.org/diseases-conditions/cancer/in-depth/monoclonal-antibody/art-20047808].

Our research shows that the most significant monoclonal drug is the one which (1) continuously reduces the rate of growth of cancer at every point in the body where they could be found, (2) slows down the rate of dispersion, the means for metastasization, and (3) strengthens the immune system, explicitly. We will do a short study on this and elaborately later.

Our model is:

Here, u = cancer cells (foreign antigen), v = the immune system, and w = the monoclonal drug.

$$\partial u / \partial t = a_1 u - a_2 uv - a_3 uw + \kappa_1 (\partial^2 u / \partial x^2 + \partial^2 u / \partial y^2 + \partial^2 u / \partial z^2) \qquad (5.1)$$

$$\partial v / \partial t = b_1 u - b_2 vu + b_3 vw + \kappa_2 (\partial^2 v / \partial x^2 + \partial^2 v / \partial y^2 + \partial^2 v / \partial z^2) \qquad (5.2)$$

$$\partial w / \partial t = c_1 u - c_2 wu - c_3 wv + \kappa_3 (\partial^2 w / \partial x^2 + \partial^2 w / \partial y^2 + \partial^2 w / \partial z^2) \qquad (5.3)$$

In the previous chapters, we have discussed about the coefficients a_1, a_2, a_3; b_1, b_2, b_3; and c_1, c_2, c_3.

These are the rate constants when two chemicals interact among themselves. Here again, we have considered variable dosages of drugs applied at each unit of time depending upon the growth and/or decay of cancer cells. From (5.2), we note that the coefficient of (vw) is positive. It is because w is a helper for v, but in doing so, w must lose some of its potency. So, the coefficient of (wv) in Eq. (5.3) is negative.

We have considered boundaries where values of u, v, w have been extrapolated (as shown in Chap. 4), and cancer is infiltrating all parts of the body. Initial/boundary conditions are:

At $t = 0$, $u(x, y, z, 0) = u_0$; $u(L_1, y, z, 0) = P_1(L_1, y, z, 0)$; $u(L_2, y, z, 0) = P_2(L_2, y, z, 0)$. $u(x, M_1, z, 0) = Q_1(x, M_1, z, 0)$; $u(x, M_2, z, 0) = Q_2(x, M_2, z, 0)$;

$$u(x, y, N_1, 0) = R_1(x, y, N_1, 0); u(x, y, N_2, 0) = R_2(x, y, N_2, 0).$$

$$0 \le L_1 < L_2; 0 \le M_1 < M_2; 0 \le N_1 < N_2 \qquad (5.4)$$

Similar boundary conditions were set up for v and w at $t = 0$. They change as $x, y, z,$ and t change.

It should be noted that when computations begin boundary conditions must remain variable. So, we must extrapolate values on the boundaries. It is also possible to set up our computations such that the war against cancer could be done on the boundary. This is also done here (and discussed in Chap. 4).

Our computational field is a cube. So, $L_1 = L_2 = M_1 = M_2 = N_1 = N_2$. Inside this cubical region, we consider several cubical tumors at $t = 0$, given by: $u\left(x_p^*, y_q^*, z_r^*, 0\right) = u_{pqr}^*.$

where each p, q, r has its own range. For computational notations, this condition could be expressed as:

$$u_{ijk}^{0,*} = U^{0,*}(estimated); 0 \le i_p^* \le i_q^* \le i_r^*; 0 \le j_p^* \le j_q^* \le j_r^*; 0 \le k_p^* \le k_q^* \le k_r^*.$$

where i_p^* denotes the x-coordinate x_p for several values of p, j_p^* denotes corresponding values of y_p, k_p^* represents corresponding values of z_p, etc.

From Eq. (5.1), a necessary condition that cancer will be damped out is: coefficient of $u < 0$.

That is:

$$\sigma = (a_1 - a_2 v - a_3 w) < 0 \qquad (5.5)$$

It is extremely important to check this condition at the very beginning of any treatment, by using initial conditions. For example, if V_0 and W_0 are the initial dosages of v and w, then V_0 and W_0 must satisfy (5.5), meaning $\sigma = (a_1 - a_2 V_0 - a_3 W_0) < 0$.

Here, v does not represent only the immune response of the body. It is also the immune response conducted by immunotherapy, whereas w is a monoclonal drug that also boosts the immune system and weakens the strength of cancer while fighting against these notorious cells.

5.2.1 Numerical Solution by the CDey-Simpson

For numerical solutions, the CDey-Simpson algorithm (as given in detail in Appendix A and in Chap. 4) has been applied. Following the studies on anatomy and pathogenesis of cancer, it has been assumed that when cancer strikes the body the malignant cells do not just settle at a few points. Using lymphatic and blood vessels, they start moving toward other organs. Fortunately, during these motions most of them die. However, a few survive and they settle wherever they find a suitable place to survive and thrive. Finally, many take shelter under the skin. Although this was primarily an assumption in this monograph, researchers from the American Osteopathic College reported that "Metastatic skin cancer may originate from melanoma, breast, lung, colon, and various other types of cancer. The cancer cells travel through the lymphatics or blood to reach the skin or cancer may spread directly to the skin through a surgical scar" (Ref.: www.aocd.org).

Let our computational field be 105 × 105 x 105. So it is a cube. There are 24 lines surrounding this cube. But only 12 of them are unique (non-repetitive). On each line, if there are 105 numbers of grid points, then there are 1260 points. However, there are 8 vertices and each vertex is repeated 3 times (because of 3 intersecting planes). They should be added only once. So we get $1260 - 2 \times 8 = 1244$.

Boundary conditions are not fixed. They are evaluated by (i) extrapolation, (ii) extension of the same values from the neighboring cells, or (iii) computations without normal dispersions. It has been assumed that these lines around the boundaries simulate the skin. (If we consider a human body inside out and shape it as a cube, then the skin will be the border lines of the cube.) Since we have considered that all boundary conditions are not fixed, the immune system, drugs, and cancer cells all could change on the boundaries. However, at 1244 points medications may not reach in the codes when just simple extrapolation techniques have been used to compute values of U_{ijk}^n, V_{ijk}^n, W_{ijk}^n on the boundaries.

Cancer is indeed a whole-body issue. Like a global not local problem of the body. When cancer starts, within a short time, with practically no apparent sign, the malignant cells start moving all over the body. It can spread all around the body and right under the skin too. So, these are the points in the computational field where medicines may not reach directly. These are the hideouts of malignant cells in our computations. At these points, the malignant cells do not necessarily form any tumor. So, computationally, if we assume that initially, the number of U_{ijk}^0 outside the tumors is N_0, then if after n number of time steps, this number is N_n and $N_n > N_0$, that implies that the number of cancer cells is growing outside the tumors. This is the mechanism of pathogenesis of metastasis of cancer. From these discussions, we can state that if the cube with 105 points on each side is our computational field, then $N_0 = 1244$. So, if at any time t_n, $N_n > N_0$, that implies that the cancer is spreading. This will not be true, if the immune system and drugs could reach those points and destroy the malignant cells.

5.2.2   A Special Note

In this chapter for all computer runs, we have assumed that: The primary tumor has metastasized at two different locations while still growing at the primary site. At $t = 0$, UPMAX = value of u at the primary site = 10,000 (the max value of u) and at the two secondary sites US1MAX = 5000 and US2MAX = 1000, respectively, at each point of the tumor. So in all, there are three tumors (This is multicentric breast cancer). However, they are housing different amounts of malignant cells. The reason this is a strong possibility is that inside a tumor there could be pockets of necrotic cells. These dead cells could be as dangerous as the living cells. "Researchers from Harvard Medical School, Boston, and the Institute for Systems Biology, Seattle, have discovered that the remains of tumor cells killed by chemotherapy or other cancer treatments can actually stimulate tumor growth by inducing an inflammatory reaction" [11]. These 3 tumors occupied 64 grid points each. So, the total number of points occupied by cancer cells in the computational field is equal to 3 x 64=192 grid points. Also, there are three lymph nodes at three grid points which are affected. At each such node, there are 1200 cancer cells. So in all, there are $192 + 3 = 195$ grid points where cancer cells are prominently present. Since by the time metastatic tumors are found, cancer cells are generally invisibly microstatically present all over the body, and our models took these into account. If during computations the max values $U_{ijk}^n > U_{ijk}^{n-1}$ at more than 195 points, that will imply that cancer cells are growing and spreading at other locations where they were not present before. This is an indication that cancer is possibly metastasizing elsewhere. All over the field $u = 25$ at the beginning. In the code. MAXOUT represents the largest value of the antigen outside the metastatic tumors. As before, in the computer code U, V, W are the net functions corresponding to u, v, w.

5.3 Computational Studies

Before using any monoclonal drug, let us study how immunotherapy could be done by using medications that will reduce the rate of growth of cancer all over the computational field representing the human body and at the same time destroy cancer cells. In this regard, we will consider two cases, one with a very poor immune system.

$\partial u/\partial t = a_1(t)u - a_2uV_0 + \kappa_1(\partial^2 u/\partial x^2 + \partial^2 u/\partial y^2 + \partial^2 u/\partial z^2)$, $V_0 = $ fixed for all t. = the initial value of V (*is kept unchanged so that the same amount of drug always stays in the blood.*)

We solved this equation numerically. Regarding the values of u on the boundaries, in most cases we have extrapolated these values from the cells near to the boundaries. We have also assumed sometimes that on the boundary $u_{ijk}^n = u_{i-1,j-1,k-1}^n$ $\forall n$. There is a reason for that. If cancer cells move individually, they are generally destroyed by our defense mechanism. So very cleverly, they move in a group forming thereby a joint fight against T-cells and B-cells. So within one mesh size possibly they change very little to nothing. However, this may not be happening in every case.

We have considered a case where cancer is growing at the rate of 450%. Medications for immunotherapy are given intravenously, and a fixed dosage is maintained constantly in the blood. The inputs are given in the inputs for Fig. 5.1. $V_0 = $ the defending cells $= 100$, which is indeed very poor in comparison with the number of cancer cells in the primary site which is 10000 at each grid point and all over the body $u = 1000$ at each grid point all over the computational field. Also $a_2 = 0.0001$, the rate at which cancer cells are being eliminated by each element of V_0. This is weak in this setup.

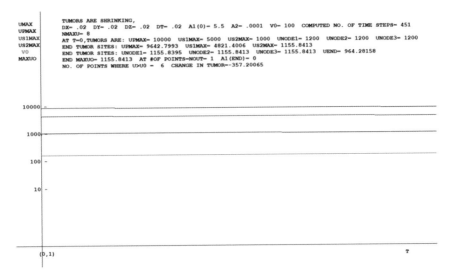

Fig. 5.1 Advanced immunotherapy, body's immune response is weak

Input for Fig. 5.1.

```
*****************************************************************************************
! CHAPTER 5, ADVANCED IMMUNOTHERAPY
! THERE ARE THREE TUMORS  & THREE LYMPH NODES ARE AFFECTED
! DU/DT = A1(N)*U-A2*U*VO  + (NU1)*DELSQ(U)                          ! ANTIGEN
! U0 AT T=0, U0 IS = 10000  AT THE PRIMARY SITE, U = 10000  SECOND TUMOR= 5000  3RD = 1000  VO=FIXED = 100
! DX= .02  DY= .02  DZ= .02  DT= .02  NX= 101  NY= 101  NZ = 101  NT= 450
! A1(0)= 5.5  A2= .0001    A1(N) = A1(N-1)*EXP(-DT*VO) BEING CONTINUOUSLY REDUCED.
! NU1= .0000001  GAMMA1= .015
! THREE LYMPH NODES ARE LOCATED AT: ( 3 , 3 , 3 )
! AND  ( 35 , 35 , 35 ) ARE AFFECTED
! PRIMARY TUMOR IS LOCATED AT: ( 13 , 13 , 13 ) UPRIMARY= 10000
! SECOND TUMOR IS LOCATED AT:  ( 26 , 26 , 26 ) US1= 5000
! THIRD TUMOR IS LOCATED AT:   ( 47 , 47 , 47 ) US2= 1000
! AT(L11,L12,L13) U0= 1200  AT(L21,L22,L23) U0= 1200  AT(L31,L32,L33) U0= 1200
! AT THE BOUNDARIES, NEIGHBORING VALUES HAVE BEEN USED.
*****************************************************************************************
```

We took the first computer run for $NT = 450$ time steps with $\Delta t = 0.02$. Due to immunotherapy, the rate of growth of cancer is getting smaller and smaller, and long before $NT = 451$, it has gone to zero. The formula for that is: $da(t_n)/dt = -\lambda V_0 a(t_n)$. Integrating from t_{n-1} to t_n and choosing $\lambda = 1$, we get the formula

$$a_n = a_{n-1} exp(-\Delta t V_0)$$

Figure 5.1 shows that cancer is very slowly receding everywhere in the field. From $U_0 = 10000$, the largest tumor at the primary site, it has gone down to $U_{451} = 9643$, a small drop at the time step $t_n = 451$. However, all tumors are uniformly shrinking and because we have assumed that at the boundaries, values of U are the same as the values next to the boundaries, that means we have assumed that the immunotherapies have attacked the hiding pockets of these malignant cells. So, there is no safe place in the body for cancer cells to hide. Such a patient is possibly an older person with a weak immune system.

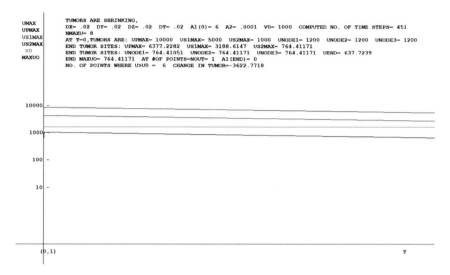

Fig. 5.2 Advanced immunotherapy, body's immune response is better

In the next case, V_0 is 10000, ten times better than in the previous case and $a(t = 0) = 6$, and cancer was increasing 500%. Figure 5.2 shows that cancer is slowing down at a better speed than before.

Input for Fig. 5.2.

```
.....................................................................................................
! CHAPTER 5, ADVANCED IMMUNOTHERAPY
! THERE ARE THREE TUMORS  & THREE LYMPH NODES ARE AFFECTED
! DU/DT = A1(N)*U-A2*U*V0  + (NU1)*DELSQ(U)                          ! ANTIGEN
! U0 AT T=0, U0 IS = 10000  AT THE PRIMARY SITE, U = 10000  SECOND TUMOR= 5000  3RD = 1000  V0-FIXED = 1000
! DX= .02  DY= .02  DZ= .02  DT= .02  NX= 101  NY= 101  NZ = 101  NT= 450
! A1(0)= 6  A2= .0001    A1(N) = A1(N-1)*EXP(-DT*V0) BEING CONTINUOUSLY REDUCED.
! NU1= .0000001  GAMMA1= .015
! THREE LYMPH NODES ARE LOCATED AT: ( 3 , 3 , 3 )
! AND   ( 35 , 35 , 35 ) ARE AFFECTED
! PRIMARY TUMOR IS LOCATED AT:  ( 13 , 13 , 13 ) UPRIMARY= 10000
! SECOND TUMOR IS LOCATED AT:  ( 26 , 26 , 26 ) US1= 5000
! THIRD TUMOR IS LOCATED AT:  ( 47 , 47 , 47 ) US2= 1000
! AT(L11,L12,L13) U0= 1200  AT(L21,L22,L23) U0= 1200  AT(L31,L32,L33) U0= 1200
! AT THE BOUNDARIES, NEIGHBORING VALUES HAVE BEEN USED.
*****************************************************************************************************
```

Here, the medications worked much better, although the growth of cancer is 500%. All tumors shrunk. Even at the affected lymph nodes, the values of U have been reduced. So, $V_0 = 100$ was a poor choice in Fig. 5.1 and a better choice in Fig. 5.2.

From Fig. 5.2, it is clear that just immunotherapy may not be enough. A second medication may be necessary.

Finally, at the third stage, a monoclonal drug W with a fixed amount, 500 units, was injected intravenously. It destroys cancer cells at the rate $a_3 = 0.001$, and that worked far better as we can see from Fig. 5.3.

Here, cancer cells are replicating at the speed of 650% very aggressively. The destructive power of W_0 is 10 times stronger than V_0. Here, the monoclonal drug helped V_0 to reduce the rate of growth of cancer. The simple formula that we have used is:

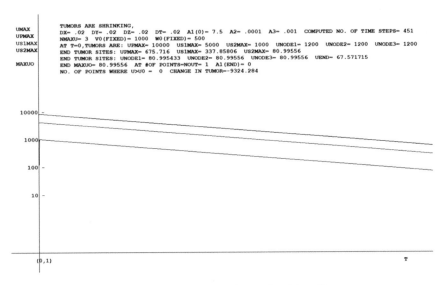

Fig. 5.3 Advanced immunotherapy, introduction of monoclonal drug W

$$a_1(n) = a_1(n-1)exp(-\Delta t \times V_0)$$

Input for Fig. 5.3.

```
**********************************************************************************************
! CHAPTER 5, ADVANCED IMMUNOTHERAPY
! THERE ARE THREE TUMORS  & THREE LYMPH NODES ARE AFFECTED
! DU/DT = A1(N)*U-A2*U*V0-A3*U*W0 + (NU1)*DELSQ(U)                          ! ANTIGEN
! U0 AT T=0, U0 IS = 10000   AT THE PRIMARY SITE, U = 10000   SECOND TUMOR= 5000   3RD = 1000
! DX= .02  DY= .02   DZ= .02   DT= .02   NX= 101   NY= 101   NZ = 101   NT= 450
! A1(0)= 7.5   A2= .0001   A3= .001    A1(N) = A1(N-1)*EXP(-DT*V0) BEING CONTINUOUSLY REDUCED.
! V0(FIXED)= 1000   W0(FIXED)= 500
! NU1= .0000001   GAMMA1= .015
! THREE LYMPH NODES ARE LOCATED AT: ( 3 , 3 , 3 )
! AND   ( 35 , 35 , 35 ) ARE AFFECTED
! PRIMARY TUMOR IS LOCATED AT:   ( 13 , 13 , 13 ) UPRIMARY= 10000
! SECOND TUMOR IS LOCATED AT:   ( 26 , 26 , 26 ) US1= 5000
! THIRD TUMOR IS LOCATED AT:    ( 47 , 47 , 47 ) US2= 1000
! AT(L11,L12,L13) U0= 1200   AT(L21,L22,L23) U0= 1200   AT(L31,L32,L33) U0= 1200
! AT THE BOUNDARIES, NEIGHBORING VALUES HAVE BEEN USED.
**********************************************************************************************
```

The warzone is the entire body. There are no hideouts for cancer cells. Both V_0 and W_0 are attacking these villains at every point of the entire field including the boundaries.

From Table 5.1, Figure 5.3, it is clear that from the very beginning, cancer cells were being destroyed. At $t_n = 2$, growth and metastasis were strictly controlled Table 5.1. Even at the end point $(NX + 4, NY + 4, NZ + 4)$, U became 556.09477, reduced from 1000. Although this is a mathematical finding, the rate constants applied here are realistic. However, all depend on the state of health of the patient. It is noticeable that from $t_n = 2$, at no point in the field $U_{max}^n > U_{max}^{n-1}$. No metastasis is going on from the very outset.

Table 5.1 Figure 5.3. All tumors are shrinking. (Observe the values of UMAX(N). They are decreasing). Treatment is working well. At no point cancer is winning, because nowhere U(n) > U(n-1)

```
*********************************************************************************************************
 SYSTEM TIME -09:54:35  MAX CONCENTRATION OVER THE FIELD AT TIME - 2  TIME STEP -  100  NT- 450
 A1(N)- 0  A2- .0001  U0- 10000  V0- 1000  W0- 500
 IN THE ENTIRE FIELD UMAX(N)- 5527.7482
 AT  15 , 14 ,  14  UMAX(N)- 5527.7482
 TOTAL NO.OF POINTS WHERE: U IS MAX-  4
 CHANGE OF: UMAX--33.199043
 AT THE TUMOR SITES: UPMAX- 5527.7482  US1MAX- 2763.3294  US2MAX- 663.16647
 U AT THE TUMORS:  UPRIMARY- 5524.0719  U SECONDARY1- 2762.2402  U SECONDARY2- 552.77486
 UNODE1- 663.16642  UNODE1-1- 552.77486  UNODE2- 663.16647  UNODE2+1- 552.77486  U(NX+4,NY+4,NZ+4)- 556.09477
 UNODE3- 663.16647
 NO.OF POINTS WHERE U(N)>U(N-1)-   0  MAX.CHANGE FROM THE START - UMAX - UOMAX--4472.2518
*********************************************************************************************************
 SYSTEM TIME -09:54:43  MAX CONCENTRATION OVER THE FIELD AT TIME - 2.02  TIME STEP -  101  NT- 450
 A1(N)- 0  A2- .0001  U0- 10000  V0- 1000  W0- 500
 IN THE ENTIRE FIELD UMAX(N)- 5494.7473
 AT  15 , 14 ,  14  UMAX(N)- 5494.7473
 TOTAL NO.OF POINTS WHERE: U IS MAX-  4
 CHANGE OF: UMAX--33.000843
 AT THE TUMOR SITES: UPMAX- 5494.7473  US1MAX- 2746.8267  US2MAX- 659.2057
 U AT THE TUMORS:  UPRIMARY- 5491.0561  U SECONDARY1- 2745.7332  U SECONDARY2- 549.47478
 UNODE1- 659.20565  UNODE1-1- 549.47478  UNODE2- 659.2057  UNODE2+1- 549.47478  U(NX+4,NY+4,NZ+4)- 552.77486
 UNODE3- 659.2057
 NO.OF POINTS WHERE U(N)>U(N-1)-   0  MAX.CHANGE FROM THE START - UMAX - UOMAX--4505.2527
*********************************************************************************************************
 SYSTEM TIME -09:54:51  MAX CONCENTRATION OVER THE FIELD AT TIME - 2.04  TIME STEP -  102  NT- 450
 A1(N)- 0  A2- .0001  U0- 10000  V0- 1000  W0- 500
 IN THE ENTIRE FIELD UMAX(N)- 5461.9435
 AT  15 , 14 ,  14  UMAX(N)- 5461.9435
 TOTAL NO.OF POINTS WHERE: U IS MAX-  4
 CHANGE OF: UMAX--32.803827
 AT THE TUMOR SITES: UPMAX- 5461.9435  US1MAX- 2730.4226  US2MAX- 655.2686
 U AT THE TUMORS:  UPRIMARY- 5458.2377  U SECONDARY1- 2729.3248  U SECONDARY2- 546.1944
 UNODE1- 655.26854  UNODE1-1- 546.1944  UNODE2- 655.2686  UNODE2+1- 546.1944  U(NX+4,NY+4,NZ+4)- 549.47478
 UNODE3- 655.2686
 NO.OF POINTS WHERE U(N)>U(N-1)-   0  MAX.CHANGE FROM THE START - UMAX - UOMAX--4538.0565
```

The values given in Table 5.1. Figure 5.3 with the fixed values of $V_0 = 1000$ and $W_0 = 500$ show how tumors at all the sites are shrinking. The rate of growth of U is zero; however, the fight is going on. (Rate of Growth of cancer cells $A(N) = 0$, does not mean that the war is over).

The tabular values of actual computations given in Table 5.1 describe how fast tumors are shrinking. For instance, at $t_n = 100$, the original largest tumor has been diminished by 4472.25, and at $t_n = 101$, it is the largest value has been diminished by 4505.35. Note: The point (15,14,14) is in the primary site. U is certainly decreasing slowly. The computed values clearly depict that at each tumor site, tumors are shrinking.

At the first node, at the TIME STEP=102, *UNODE1* is reduced from 1200 to 655.27. The same is true for the other nodes. Also, $UNODE1 - 1$ is the point next to $UNODE1$. Cancer is not increasing there. Also, $UNODE2 + 1$ is a point in front of the second affected node. Cancer is not increasing there either. $US1MAX =$ the max value of U at the first metastatic tumor (secondary tumor), and $US2MAX =$ the max value of U at the second metastatic tumor site. This also shows that cancer cells have failed to use lymph vessels to metastasize. So the treatment is a success.

Note that at the grid point (15, 14, 14), which is a point inside the tumor, U is a max at each time step. That is also the max value in the entire field. Throughout the computations, the values of V_0 and W_0 were kept fixed. So, there is no need to print their graphs.

All tumors steadily kept on decreasing. As the tumor shrunk, the cancer cells all over the field are dying.

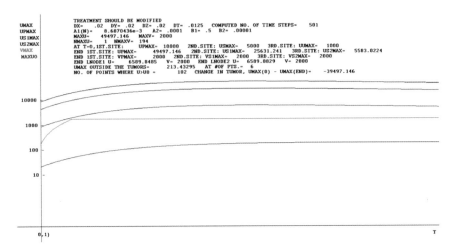

Fig. 5.4 Immunotherapy that reduces the rate of growth of cancer

These results reveal the truth that breast cancer treatment with immunotherapy along with monoclonal drugs could be very effective. Yet, we have decided to take another look at a very similar situation by minimizing the drug's inputs.

Next, we chose $a_1 = 4.25$ and used the formula $a_1(t_n) = 0.99 \times a_1(t_{n-1})$ for immunotherapy and a variable V without any monoclonal drug W. With the absence of monoclonal drugs, cancer cells find the places on the boundary where they can hide. One of these points is $(NX + 4, NY + 4, NZ + 4)$. Initially, $V_0 = 200$ and it attained a max value which is 2000. Still all tumors slowly started shrinking, and from the entire field, cancer cells started vanishing. The graphs show that the cancer is under control. There is no need to give any details of the initial conditions because all are clearly given in Fig. 5.4.

We notice that in Table 5.2. Figure 5.4, as U started increasing, V also started increasing. The values of U displayed a strong tendency to metastasize because of their sharp increase. However, it is worth noticing that at the points where U is very large, V is also large. This elucidates chemotaxis. This is what we expect to happen in any effective treatment. The entire computation consists of $105 \times 105 \times 105 - 1244 = 1156381$ number of points, out of which at 1,156,378 points $U_{max}^n > U_{max}^{n-1}$. So, this is of great concern. This means almost at every point in the field cancer is on the rise.

From Table 5.2 for Fig. 5.4, it is clear that U has significantly gone up and so did V. At the primary tumor site (14,14,14), U increased from 10,000 to 17,394 at $t_n = 24$ and at the same location and at the same time V moved up significantly from mere 200 to 1168. Also, being attacked, cancer cells spread all over the entire field of computation, because it went up at 1,156,378 points out of 1,156,381 points. This is the very nature of cancer cells. The moment they are attacked, they spread more. (Thus to prevent that, this metastatic mechanism must be destroyed.)

At $t_n = 26$, $UMAX(in\ the\ entire field) = U0MAX$(initial value of U) $+ 8106.62$, and it is on the rise. However, $UMAX$ did not keep on rising. As V attained

Table 5.2 For Fig. 5.4. Where tumor is the largest (at the grid point (14, 14, 14), primary site), V is the largest. From Table 5.2 around the tumors V is strongly present. This is chemotaxis, a built-in feature of the model. The same is true in other Tables

```
**************************************************************************************
SYSTEM TIME = 10:25:55   MAX CONCENTRATION OVER THE FIELD AT TIME =      .3   TIME STEP =    24   NT= 500   A1(N)= 3.3756146
AT  14 , 14 , 14   UMAX(N)=  17394.52    V=  1167.9937
AT  14 , 14 , 14   VMAX(N)=  1167.9937   U= 17394.52
TOTAL NO.OF POINTS WHERE: U IS MAX=     1 , V IS MAX=    1
CHANGE OF: UMAX=    353.113  , OF VMAX=   52.907553
AT THE PRIMARY SITE:   UMAX=  17394.52    VMAX=  1167.9937
AT THE SECONDARY SITE:   UMAX=  8726.0416   VMAX=  687.04924
AT THE UNDETECTED SITE:   UMAX=  1749.8604   VMAX=  297.90314
U AT THREE TUMORS:   USTAR1=  17392.63    USTAR2=  8725.0935   USTAR3 =  1749.6737
UNODE1=  2099.1023   UNODE1-1=  43.775019   UNODE2=  2099.1023   UNODE2+1=  43.775019   U(NX+4,NY+4,NZ+4)=    25
VNODE1=  317.42837   VNODE2=  317.42837   V(NX+4,NY+4,NZ+4)=    200
AT THE PRIMARY SITE: V=   1167.8868   AT THE SECONDARY SITE: V=   686.99533
MAX U OUTSIDE THE THREE TUMORS =   43.775093   AT NUMBER OF POINTS=    6
NO.OF POINTS WHERE U(N)>U(N-1)=   1156378   MAX.CHANGE FROM THE START = UOMAX - UMAX=   -7394.52
**************************************************************************************
SYSTEM TIME = 10:26:14   MAX CONCENTRATION OVER THE FIELD AT TIME =      .3125   TIME STEP =    25   NT= 500   A1(N)= 3.333682
AT  14 , 14 , 14   UMAX(N)=  17749.632    V=  1221.925
AT  14 , 14 , 14   VMAX(N)=  1221.925    U= 17749.632
TOTAL NO.OF POINTS WHERE: U IS MAX=     1 , V IS MAX=    1
CHANGE OF: UMAX=   355.11234  , OF VMAX=   53.931256
AT THE PRIMARY SITE:   UMAX=  17749.632    VMAX=  1221.925
AT THE SECONDARY SITE:   UMAX=  8907.0196   VMAX=  714.38984
AT THE UNDETECTED SITE:   UMAX=  1786.6128   VMAX=  303.43225
U AT THREE TUMORS:   USTAR1=  17747.621    USTAR2=  8906.01    USTAR3 =  1786.414
UNODE1=  2143.1422   UNODE1-1=  44.697255   UNODE2=  2143.1423   UNODE2+1=  44.697255   U(NX+4,NY+4,NZ+4)=    25
VNODE1=  324.05702   VNODE2=  324.05702   V(NX+4,NY+4,NZ+4)=    200
AT THE PRIMARY SITE: V=   1221.8074   AT THE SECONDARY SITE: V=   714.33051
MAX U OUTSIDE THE THREE TUMORS =   44.697336   AT NUMBER OF POINTS=    6
NO.OF POINTS WHERE U(N)>U(N-1)=   1156378   MAX.CHANGE FROM THE START = UOMAX - UMAX=   -7749.6323
**************************************************************************************
SYSTEM TIME = 10:26:33   MAX CONCENTRATION OVER THE FIELD AT TIME =      .325   TIME STEP =    26   NT= 500   A1(N)= 3.2922703
AT  14 , 14 , 14   UMAX(N)=  18106.619    V=  1276.8822
AT  14 , 14 , 14   VMAX(N)=  1276.8822   U= 18106.619
TOTAL NO.OF POINTS WHERE: U IS MAX=     1 , V IS MAX=    1
CHANGE OF: UMAX=   356.98655  , OF VMAX=   54.957196
AT THE PRIMARY SITE:   UMAX=  18106.619    VMAX=  1276.8822
AT THE SECONDARY SITE:   UMAX=  9089.2075   VMAX=  742.2754
AT THE UNDETECTED SITE:   UMAX=  1823.6524   VMAX=  309.07562
U AT THREE TUMORS:   USTAR1=  18104.483    USTAR2=  9088.1347   USTAR3 =  1823.4409
UNODE1=  2187.523   UNODE1-1=  45.626948   UNODE2=  2187.5231   UNODE2+1=  45.626948   U(NX+4,NY+4,NZ+4)=    25
VNODE1=  330.82231   VNODE2=  330.82231   V(NX+4,NY+4,NZ+4)=    200
AT THE PRIMARY SITE: V=   1276.7533   AT THE SECONDARY SITE: V=   742.21034
MAX U OUTSIDE THE THREE TUMORS =   45.627038   AT NUMBER OF POINTS=    6
NO.OF POINTS WHERE U(N)>U(N-1)=   1156378   MAX.CHANGE FROM THE START = UOMAX - UMAX=   -8106.6189
```

its max $V = 2000$, and practically caused $UMAX$ to move into a steady state. This is noticeable in Table 5.3 for Fig. 5.4. Also, the number of points at which $U_{max}^n > U_{max}^{n-1}$ is 102. So, cancer cells could not metastasize with a large army. The process of metastases has been controlled to a great extent. This shows that even though the rate of

Table 5.3 For Fig. 5.4. Cancer winning only at 102 number of points, given by U(N) > U(N-1) is true only at 102 points

```
**************************************************************************************
SYSTEM TIME = 13:04:57   MAX CONCENTRATION OVER THE FIELD AT TIME =     6.1375   TIME STEP =   491   NT= 500   A1(N)=   .00984371
AT  14 , 14 , 14   UMAX(N)=  50129.699    V=  2000
AT  50 , 50 , 50   VMAX(N)=  2000    U=  5638.1057
TOTAL NO.OF POINTS WHERE: U IS MAX=     1 , V IS MAX=   194
CHANGE OF: UMAX=   -63.522463  , OF VMAX=   0
AT THE PRIMARY SITE:   UMAX=  50129.699    VMAX=  2000
AT THE SECONDARY SITE:   UMAX=  25949.062   VMAX=  2000
AT THE UNDETECTED SITE:   UMAX=  5650.4693   VMAX=  2000
U AT THREE TUMORS:   USTAR1=  50017.592    USTAR2=  25890.687   USTAR3 =  5638.1057
UNODE1=  6670.1691   UNODE1-1=  213.78789   UNODE2=  6670.2024   UNODE2+1=  213.7883   U(NX+4,NY+4,NZ+4)=    25
VNODE1=  2000   VNODE2=  2000   V(NX+4,NY+4,NZ+4)=    200
AT THE PRIMARY SITE: V=   2000   AT THE SECONDARY SITE: V=   2000
MAX U OUTSIDE THE THREE TUMORS =   213.92889   AT NUMBER OF POINTS=    6
NO.OF POINTS WHERE U(N)>U(N-1)=   102   MAX.CHANGE FROM THE START = UOMAX - UMAX=   -40129.699
**************************************************************************************
SYSTEM TIME = 13:05:21   MAX CONCENTRATION OVER THE FIELD AT TIME =     6.15   TIME STEP =   492   NT= 500   A1(N)= 9.7214295e-3
AT  14 , 14 , 14   UMAX(N)=  50066.224    V=  2000
AT  50 , 50 , 50   VMAX(N)=  2000    U=  5631.3324
TOTAL NO.OF POINTS WHERE: U IS MAX=     1 , V IS MAX=   194
CHANGE OF: UMAX=   -63.475342  , OF VMAX=   0
AT THE PRIMARY SITE:   UMAX=  50066.224    VMAX=  2000
AT THE SECONDARY SITE:   UMAX=  25917.182   VMAX=  2000
AT THE UNDETECTED SITE:   UMAX=  5643.7064   VMAX=  2000
U AT THREE TUMORS:   USTAR1=  49954.035    USTAR2=  25858.762   USTAR3 =  5631.3324
UNODE1=  6662.1147   UNODE1-1=  213.73886   UNODE2=  6662.1481   UNODE2+1=  213.73927   U(NX+4,NY+4,NZ+4)=    25
VNODE1=  2000   VNODE2=  2000   V(NX+4,NY+4,NZ+4)=    200
AT THE PRIMARY SITE: V=   2000   AT THE SECONDARY SITE: V=   2000
MAX U OUTSIDE THE THREE TUMORS =   213.88034   AT NUMBER OF POINTS=    6
NO.OF POINTS WHERE U(N)>U(N-1)=   102   MAX.CHANGE FROM THE START = UOMAX - UMAX=   -40066.224
**************************************************************************************
SYSTEM TIME = 13:05:41   MAX CONCENTRATION OVER THE FIELD AT TIME =     6.1625   TIME STEP =   493   NT= 500   A1(N)= 9.600668e-3
AT  14 , 14 , 14   UMAX(N)=  50002.796    V=  2000
AT  50 , 50 , 50   VMAX(N)=  2000    U=  5624.563
TOTAL NO.OF POINTS WHERE: U IS MAX=     1 , V IS MAX=   194
CHANGE OF: UMAX=   -63.42725  , OF VMAX=   0
AT THE PRIMARY SITE:   UMAX=  50002.796    VMAX=  2000
AT THE SECONDARY SITE:   UMAX=  25885.324   VMAX=  2000
AT THE UNDETECTED SITE:   UMAX=  5636.9474   VMAX=  2000
U AT THREE TUMORS:   USTAR1=  49890.527    USTAR2=  25826.858   USTAR3 =  5624.563
UNODE1=  6654.065   UNODE1-1=  213.68959   UNODE2=  6654.0985   UNODE2+1=  213.69001   U(NX+4,NY+4,NZ+4)=    25
VNODE1=  2000   VNODE2=  2000   V(NX+4,NY+4,NZ+4)=    200
AT THE PRIMARY SITE: V=   2000   AT THE SECONDARY SITE: V=   2000
MAX U OUTSIDE THE THREE TUMORS =   213.83156   AT NUMBER OF POINTS=    6
NO.OF POINTS WHERE U(N)>U(N-1)=   102   MAX.CHANGE FROM THE START = UOMAX - UMAX=   -40002.796
```

growth is very insignificant at this time $a_1(t) = 0.0096$ and steadily decreasing, the fight is going in full swing.

A very stunning picture of the dreadful nature of this deadly disease is although the rate of growth of cancer is practically zero, it did grow. The question is how?

The only answer that we could get, must come from the last term. Cells are coming from their hideouts, because that term represents continuous accumulations of cells from outside the tumor!

So behind this macrostatic model, there must exist microstatic states which are entering into the tumor locations or necrotic cells inside the tumors generating new cells. At least one of these two statements has to be correct.

5.4   Applications of Immunotherapy with Variable Dosages of Monoclonal Drugs

In general as strategies of treatments change, use of drugs varies. So, it is very essential to look into how amounts of drugs should change as cancer treatments progress. In many cases only by enhancing the strengths of the immune systems, progress of cancer cannot be stopped. In such cases, monoclonal drugs are introduced.

The American Cancer Society has stated that "Trastuzumab (Herceptin, others): Trastuzumab can be used to treat both early-stage and advanced breast cancer. This drug is often given with chemo, but it might also be used alone (especially if chemo alone has already been tried)."

Monoclonal antibodies are man-made antibodies (immune system proteins). They form targeted therapy, attacking cancer cells only. So, collateral damages are minimized. In the case of breast cancers, they "attach to the HER2 protein on cancer cells, which can help stop the cells from growing."

There are various monoclonal drugs for targeted therapies.

In Fig. 5.5, W is a monoclonal drug. It is used just like a direct cancer killer. The inputs are: Here, $\Delta x = \Delta y = \Delta z = 0.02$ and $\Delta t = 0.02\ a_1 = 4.5$ (cancer is growing at the rate of 350%), $a_2 = 0.0005$, $a_3 = 0.0025$, $b_1 = c_1 = 1.5$, $b_2 = 0.00001$, $b_3 = 0.001$, $c_2 = c_3 = b_2$. Here, $a_2 = 50 \times b_2$ and $a_3 = 250 \times c_2$. The strengths of the immunotherapy drug V are given by a_2, its ability to destroy cancer cells, and the strength of the monoclonal drug w is a_3.

So, both V and W are relatively strong. At the primary site, $U^0 = 10000$. Mathematically, we know that if $\forall V$ and W, the coefficient of U in Eq. (5.1) should satisfy the necessary condition that the treatment should work which is: $(a_1 - a_2 V - a_3 W) < 0$, meaning that the rate of growth of cancer is negative. This is evident from Fig. 5.5.

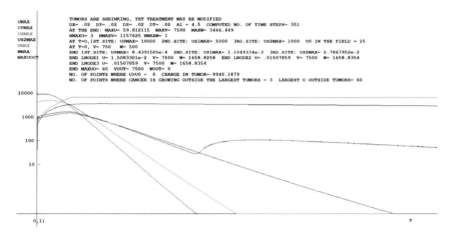

Fig. 5.5 Immunotherapy (Y) with a monoclonal drug(W)

Input for Fig. 5.5.

```
••••••••••••••••••••••••••••••••••••••••••••••••••••••••••••••••••••••••••••••••••••••••••••
!TREATMENT WITH IMMUNOTHERAPY AND  MONOCLONAL DRUG.HELPING V .
! DU/DT = A1*U-A2*U*V - A3*U*W + NU1*DELSQ(U) ! ANTIGEN
! DV/DT = B1*U-B2*V*U + B3*V*W + NU2*DELSQ(V) !IMMUNE RESPONSE
! DW/DT = C1*U -C2*W*U - C3*W*V+ NU3*DELSQ(W) ! MONOCLONAL DRUG HELPING V.
! DX= .02   DY= .02   DZ= .02   DT= .02   NX= 101   NY= 101   NZ = 101   NT= 350
! A1= 4.5   A2= .0005  A3= .0025
! B1= 1.5   B2= .00001  B3= .001   V0= 750
! C1= 1.5   C2= .00001  C3= .00001  W0= 500
! NU1= .0000001  NU2= .0000001  NU3= .0000001
! GAMMA1= .015  GAMMA2= .015  GAMMA3= .015
! THREE LYMPH NODES ARE LOCATED AT: ( 3 , 3 , 3 ) AND  ( 35 , 35 , 35 ) A THIRD LYMPH NODE IS AFFECTED
! THIRD AT   ( 42 , 42 , 42 )ULN10= 1200 ULN20= 1200 ULN30= 1200
! THREE TUMORS LOCATED AT:  ( 13 , 13 , 13 ) AND ( 26 , 26 , 26 )
!( 47 , 47 , 47 )
!
••••••••••••••••••••••••••••••••••••••••••••••••••••••••••••••••••••••••••••••••••••••••••••
```

The necessary condition shows an avenue so that a doctor could make estimations of how much drug should be injected intravenously to begin the treatment.

In many applications, we have made such estimation at the initial stage of computations, validating them by looking at the outputs.

From the point of view of numerical stability, the system is very stiff. Because whereas some variables are changing slowly, some are changing very fast.

From Fig. 5.5, we notice that around NT = 150 (an estimation), MAXOUT (the red curve), the largest value of U outside the tumors and affected lymph nodes, started rising suddenly. Then around the end, it started decreasing. At the beginning, $MAXU = 25$. At the end it is 60. But the slope is negative (the top redline which has a hump). So it is decreasing.

In the next run (Fig. 5.6), $a_1 = 1.5$ = rate of growth of U, $a_2 = 0.0005$ = rate of destruction of U by each unit of V per unit time, a_3 = rate of destruction of U by each unit of W per unit of time = 0.00005. $b_1 = 0.5 = c_1$, $b_2 = c_2 = 0.00001$, $b_3 = 0.005$ = enhancement of the strength of V by each unit of W per unit of time. $c_3 = 0.00001$. The initial dosages of V and W were 1000 and 1000, respectively.

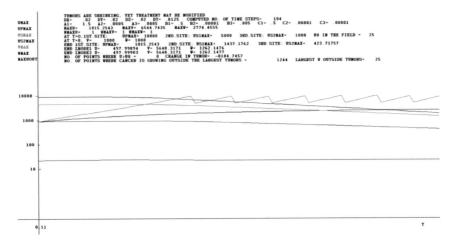

Fig. 5.6 Immunotherapy (V) with the monoclonal drug (W)

At the very outset, $\sigma = $ coefficient of $u = (1.5 - 0.0005(1000 + 1000)) = 0.5 >$ 0 (violation of the necessary condition for the decadence of u).

From here, we may estimate that initially cancer will be on the increase (on logarithmic scale it may not be detected properly) until the sum of the values of V *and* W exceeds 3000. At the end, $VMAX + WMAX = 6544.74 + 2774.46 >$ 3000, so $\sigma = 1.5 - 0.0005 \times 9319.2 = 1.5 - 4.6596 = -3.1596 < 0$. The necessary condition that treatment will succeed is satisfied.

In fact, if the least value of $VMAX + WMAX > 3000$, that indicates that treatment could be successful. This is seen in Fig. 5.6.

The graph for V is wavy. The reason is, to contain the antigen, the cancer, U, V must rise. However, the condition that it cannot exceed $10 \times V$ (at $t = 0$) brings it down. It is evident from the graph that cancer is under control. After 104 number of time steps, cancer cells are not spreading anymore because at no point, $U_{ijk}^{n} > U_{ijk}^{n-1}$. All the tumors are shrinking, and both V and W are going into a steady state. At 1244, the number of points on the boundary lines, U remained unchanged from the beginning. Those are the hideouts of cancer cells on the boundary as setup from the beginning. However, medications V and W are vigilant there such that cancer cannot move out to metastasize. That was not imposed on the boundary conditions in the code. The model is doing it on its own. The graphs have negative slopes indicating that all tumors are decreasing. The code is intentionally stopped here to check the state of the treatment. There is a mathematical reason for that. From Fig. 5.6, it is quite clear that the values of $VMAX$ and $WMAX$ and the max values at each t_n are approaching a steady pattern. So, we evaluated $(a_1 - a_2 v - a_3 w)$ using $VMAX$ and $WMAX$. Then, $(a_1 - a_2 VMAX - a_3 WMAX) = (1.5 - 0.0005(6544 + 2774)) = -3.159$. Since this is negative, it satisfies the necessary condition that U should keep on decreasing.

As we have noticed before, there are exactly 1244 grid points representing the skin which are the hideouts of the cancer cells. Later, we will study how this barrier could be overcome. "Breast cancer cells are notoriously nomadic and are known to travel and settle in the bones, brain, liver, and lungs. In some cases, the cancer cells may escape the primary tumor before the patient is diagnosed. Those cells may stay dormant or grow slowly in their hiding place, even as the patient goes through months of therapy and is then told doctors found no sign of cancer" (Ref. Cynthia Lynch, MD, Medical Oncologist, Cancer Treatment Center of America, Phoenix, Arizona). Also, oncologists from New York state: "Doctors may remove tumors with surgery, shrink them with radiation and chemotherapy or attack them by activating the body's immune system. But some cancers are wanderlust and migrate from their original tumors and hide from detection and traditional treatments. Then they may pop up months or years later, often more resistant to treatment than the original cancer" (Cancer treatment Centers of America, New York). Such provisions have been included in this model. That is what has been done and validated in the computer output.

If the strength of the medications is reduced from $a_2 = 0.0005$ to 0.0001 and $a_3 = 0.0005$ to 0.0001, then the output is just the opposite. A part of the output has been included below. It is noticeable that wherever U is a max (at the primary tumor according to the code) near that point both V and W are also max. Number of points where U is a max is 1. The same is true for V and W. And that point is the point $(15,15,15)$ in the primary site. That means both V and W are fighting that war very tactfully, so that U should not increase elsewhere. This is exactly what doctors expect to happen. At the two affected lymph nodes, both V and W were vigilant and did not allow cancer cells to move further. However, this hard fight did not become fruitful at the end. In fact, V raised from 1000 to 5000, yet it failed. Here, $a_1 = 1.5$ = rate of growth of u (a slow growing tumor, yet dangerous), $a_2 = 0.0001$ = rate of destruction of U by each unit of V per unit of time, a_3 = rate of destruction of U by each unit of W per unit of time = 0.0001. $b_1 = 0.5 = c_1$, $b_2 = c_2 = 0.00001$, $b_3 = 0.005$ = enhancement of the strength of V by each unit of W per unit of time. The initial dosages of V and W were 1000 and 1000, respectively. The primary tumor is from $(13,13,13)$ to $(16,16,16)$. In the numerical outputs, it can be noticed that while at $(15,15,15)$, the primary site, $U = UMAX = 12513.62$ (increased from 10,000), both V and W attended their max at the same site. This means the fighters are conducting a strategic fight. At both metastatic sites, the fight was equally strong. The reason cancer is gaining the upper hand is because the strengths of V and W, as given by a_2 and a_3, are not strong enough ($a_2 = 10 \times b_2$ and $a_3 = 10 \times c_2$). So, we go to the next computer run.

At t_{39}, fighting is mostly concentrated in the cancer sites. At the primary site $(15,15,15)$, $U = 12,513$ (Table 5.4 for Fig. 5.7). It was 10,000. So it has increased by 2513. V increased at the same point from 1000 to 5856.67, and W increased from 1000 to 2295.94. That indicates that both the immune system and the medication are surrounding U, which a doctor expects to happen. Numerical computations reveal that the values of U, V, W are continuously increasing and fighting is getting very intense and concentrated at the cancer site. Yet U is winning. At t_{41}, V and W

Table 5.4 For Fig. 5.7.

```
********************************************************************************
SYSTEM TIME =  11:07:52  MAX OVER THE FIELD AT TIME =     .4875    TIME STEP =    39   NT= 200
A1=   1.5   A2=   .0001   A3=    .0001   B1=  .5   B2=   .00001   B3=   .005   C1=  .5   C2=   .00001   C3=   .00001
AT   15 .  15 .  15   UMAX(N)=   12513.621      V=  5856.6724      W=  2295.9359
AT   14 .  14 .  14   VMAX(N)=   5856.6724      U=  12513.621      W=  2295.9359
AT   15 .  15 .  15   WMAX(N)=   2295.9359      U=  12513.621      V=  5856.6724
TOTAL NO.OF POINTS WHERE: U IS MAX=     1 ,  V IS MAX=     1 ,   W IS MAX=     1
CHANGE OF: UMAX=    54.305379 ,  OF VMAX=    446.9813 ,  OF WMAX=    36.467006
AT THE PRIMARY SITE:    UMAX=   12513.621     VMAX=   5856.6724    WMAX=   2295.9359
AT THE METASTATIC#1 SITE:   UMAX=   6465.7681    VMAX=   5589.2545    WMAX=   1658.2083
AT THE METASTATIC#2 SITE:   UMAX=   1322.2029    VMAX=   5376.9478    WMAX=   1129.662
UNODE1-  1584.5338    UNODE1-1=   33.222169    UNODE1+1=   33.222169    UNODE2-  1584.5338
UNODE2-1=   33.222169    UNODE2+1=   33.222169    U(NX+4,NY+4,NZ+4)=    25
VNODE1-  5387.4849    VNODE2=   5387.4849    V(NX+4,NY+4,NZ+4)=     5000
WNODE1-  1156.3982    WNODE2=   1156.3982    W(NX+4,NY+4,NZ+4)=     1000
NO. OF POINTS WHERE CANCER IS GROWING OUTSIDE THE LARGEST TUMORS =          6  LARGEST U OUTSIDE TUMORS=      33.222318
NO. OF POINTS WHERE U(N)>U(N-1)=       1156378    MAX.CHANGE FROM THE START = UMAX - UOMAX=      2513.6214
********************************************************************************
SYSTEM TIME =  11:08:30  MAX OVER THE FIELD AT TIME =      .5     TIME STEP =    40   NT= 200
A1=   1.5   A2=   .0001   A3=    .0001   B1=  .5   B2=   .00001   B3=   .005   C1=  .5   C2=   .00001   C3=   .00001
AT   15 .  15 .  15   UMAX(N)=   12564.102      V=  6344.8405      W=  2332.4452
AT   14 .  14 .  14   VMAX(N)=   6344.8405      U=  12564.102      W=  2332.4452
AT   15 .  15 .  15   WMAX(N)=   2332.4452      U=  12564.102      V=  6344.8405
TOTAL NO.OF POINTS WHERE: U IS MAX=     1 ,  V IS MAX=     1 ,   W IS MAX=     1
CHANGE OF: UMAX=    50.480217 ,  OF VMAX=    488.16812 ,  OF WMAX=    36.509307
AT THE PRIMARY SITE:    UMAX=   12564.102     VMAX=   6344.8405    WMAX=   2332.4452
AT THE METASTATIC#1 SITE:   UMAX=   6496.1058    VMAX=   5913.1448    WMAX=   1677.2037
AT THE METASTATIC#2 SITE:   UMAX=   1329.1956    VMAX=   5576.7182    WMAX=   1133.3262
UNODE1-  1592.7602    UNODE1-1=   33.400222    UNODE1+1=   33.400222    UNODE2-  1592.7602
UNODE2-1=   33.400222    UNODE2+1=   33.400222    U(NX+4,NY+4,NZ+4)=    25
VNODE1-  5593.2814    VNODE2=   5593.2814    V(NX+4,NY+4,NZ+4)=     5000
WNODE1-  1160.8519    WNODE2=   1160.8519    W(NX+4,NY+4,NZ+4)=     1000
NO. OF POINTS WHERE CANCER IS GROWING OUTSIDE THE LARGEST TUMORS =          6  LARGEST U OUTSIDE TUMORS=      33.40038
NO. OF POINTS WHERE U(N)>U(N-1)=       1156378    MAX.CHANGE FROM THE START = UMAX - UOMAX=      2564.1016
********************************************************************************
SYSTEM TIME =  11:09:10  MAX OVER THE FIELD AT TIME =     .5125    TIME STEP =    41   NT= 200
A1=   1.5   A2=   .0001   A3=    .0001   B1=  .5   B2=   .00001   B3=   .005   C1=  .5   C2=   .00001   C3=   .00001
AT   15 .  15 .  15   UMAX(N)=   12610.357      V=  6878.5366      W=  2368.9746
AT   14 .  14 .  14   VMAX(N)=   6878.5366      U=  12610.357      W=  2368.9746
AT   15 .  15 .  15   WMAX(N)=   2368.9746      U=  12610.357      V=  6878.5366
TOTAL NO.OF POINTS WHERE: U IS MAX=     1 ,  V IS MAX=     1 ,   W IS MAX=     1
CHANGE OF: UMAX=    46.255372 ,  OF VMAX=    533.69607 ,  OF WMAX=    36.529447
AT THE PRIMARY SITE:    UMAX=   12610.357     VMAX=   6878.5366    WMAX=   2368.9746
AT THE METASTATIC#1 SITE:   UMAX=   6525.113     VMAX=   6258.4972    WMAX=   1696.2367
AT THE METASTATIC#2 SITE:   UMAX=   1335.9685    VMAX=   5784.4491    WMAX=   1136.9949
UNODE1-  1600.8113    UNODE1-1=   33.575549    UNODE1+1=   33.575549    UNODE2-  1600.8113
UNODE2-1=   33.575549    UNODE2+1=   33.575549    U(NX+4,NY+4,NZ+4)=    25
VNODE1-  5807.5937    VNODE2=   5807.5937    V(NX+4,NY+4,NZ+4)=     5000
WNODE1-  1165.3127    WNODE2=   1165.3127    W(NX+4,NY+4,NZ+4)=     1000
NO. OF POINTS WHERE CANCER IS GROWING OUTSIDE THE LARGEST TUMORS =          6  LARGEST U OUTSIDE TUMORS=      33.575716
NO. OF POINTS WHERE U(N)>U(N-1)=       1156378    MAX.CHANGE FROM THE START = UMAX - UOMAX=      2610.357
```

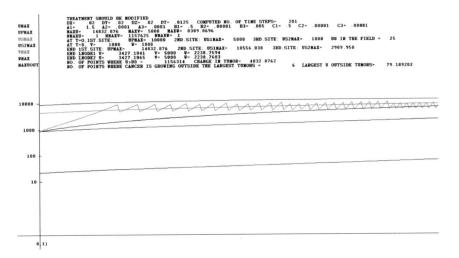

```
                TREATMENT SHOULD BE MODIFIED
UMAX            DX=    .02   DY=   .02   DZ=   .02   DT=   .0125   COMPUTED NO. OF TIME STEPS=      201
UPMAX           A1=    1.5   A2=   .0001   A3=   .0001   B1=  .5   B2=  .00001   B3=  .005   C1=  .5   C2=  .00001   C3=  .00001
USMAX           MAXU=   14832.876   MAXV=   5000   MAXW=   8389.0696
US2MAX          MAXU=       1   MAXV=   1157625   MAXV=      1
                AT T=0.1ST.SITE:  UPMAX=   10000   2ND.SITE: US1MAX=   5000   3RD.SITE: US2MAX=   1000   U0 IN THE FIELD =   25
VMAX            AT T=0,    U=   1000   V=   1000
WMAX            END 1ST.SITE: UPMAX=   3427.1041   V=   5000   W=   2238.7594   2ND.SITE: US1MAX=   10556.038   3RD.SITE: US2MAX=   2909.958
MAXUOUT         END 1NODE1 U=   3427.1065   V=   5000   W=   2238.7603
                NO. OF POINTS WHERE U>U0 =        1156314    CHANGE IN TUMOR=   4832.8762
                NO. OF POINTS WHERE CANCER IS GROWING OUTSIDE THE LARGEST TUMORS =          6  LARGEST U OUTSIDE TUMORS=      79.189202

10000 ─

1000 ─

100 ─

10 ─

0 | 1)
```

Fig. 5.7 Immunotherapy (V) with a monoclonal drug (W)

are steadily increasing exactly at the vicinities where U is growing and yet they have failed to contain U. At all sites, cancer is increasing very slowly, although apparently treatment seems to be effective. The same pattern is valid throughout the computations. At t_{176}, U has grown from 10000 at t_0 to 15,126.76 and $U_{ijk}^{176} > U_{ijk}^{175}$ at 1,156,378 points in the field, which indicates metastasization over a field of computation with 1,157,625 points. This is a terribly sad situation!!!

Table 5.5 For Fig. 5.7

```
*********************************************************************************
SYSTEM TIME =  12:26:04  MAX OVER THE FIELD AT TIME =    2.175  TIME STEP =    174  NT= 200
A1=   1.5   A2=  .0001   A3=  .0001   B1=   .5   B2=  .00001   B3=  .005   C1=  .5   C2=  .00001   C3=  .00001
AT   15 .  15 .  15  UMAX(N)=   15115.542   V=  10607.507   W=  7432.9976
AT   14 .  14 .  14  VMAX(N)=   10607.507   U=  15115.542   V=  10607.507
AT   15 .  15 .  15  WMAX(N)=   7432.9976   U=  15115.542   V=  10607.507
TOTAL NO.OF POINTS WHERE: U IS MAX=     1 . V IS MAX=     1 . W IS MAX=     1
CHANGE OF  UMAX=  -22.52283  . OF VMAX=  2348.7383  . OF WMAX=   35.635671
AT THE PRIMARY SITE:     UMAX=   15115.542   VMAX=  10607.507   WMAX=   7432.9976
AT THE METASTATIC#1 SITE:   UMAX=  10031.204   VMAX=  8047.7453   WMAX=   4765.236
AT THE METASTATIC#2 SITE:   UMAX=  2565.3352   VMAX=  5973.6929   WMAX=   1831.3827
UNODE1=  3034.7691   UNODE1-1=   68.303881   UNODE1+1=  68.303888   UNODE2=  3034.7707
UNODE2-1=  68.303888   UNODE2+1=   68.303888   U(NX+4,NY+4,NZ+4)=   25
VNODE1=   6074.4397   VNODE2=   6074.44   V(NX+4,NY+4,NZ+4)=   5000
WNODE1=  1998.0378   WNODE2=  1998.0383   W(NX+4,NY+4,NZ+4)=   1000
NO. OF POINTS WHERE CANCER IS GROWING OUTSIDE THE LARGEST TUMORS=     6   LARGEST U OUTSIDE TUMORS=   68.310159
NO.OF POINTS WHERE U(N)>U(N-1)=    1156314   MAX.CHANGE FROM THE START = UMAX - UOMAX=   5115.5416
*********************************************************************************
SYSTEM TIME =  12:26:38  MAX OVER THE FIELD AT TIME =    2.1875  TIME STEP =    175  NT= 200
A1=   1.5   A2=  .0001   A3=  .0001   B1=   .5   B2=  .00001   B3=  .005   C1=  .5   C2=  .00001   C3=  .00001
AT   15 .  15 .  15  UMAX(N)=   15129.61   V=  6452.5362   W=  7470.4088
AT   14 .  14 .  14  VMAX(N)=   6452.5362   U=  15129.61   V=  7470.4088
AT   15 .  15 .  15  WMAX(N)=   7470.4088   U=  15129.61   V=  6452.5362
TOTAL NO.OF POINTS WHERE: U IS MAX=     1 . V IS MAX=     1 . W IS MAX=     1
CHANGE OF  UMAX=   14.068808  . OF VMAX=  -4154.9703  . OF WMAX=   37.411135
AT THE PRIMARY SITE:     UMAX=   15129.61   VMAX=  6452.5362   WMAX=   7470.4088
AT THE METASTATIC#1 SITE:   UMAX=  10059.804   VMAX=  5875.1176   WMAX=   4791.9625
AT THE METASTATIC#2 SITE:   UMAX=  2578.0797   VMAX=  5308.4918   WMAX=   1838.5349
UNODE1=  3049.4524   UNODE1-1=   68.685578   UNODE1+1=  68.685585   UNODE2=  3049.454
UNODE2-1=  68.685585   UNODE2+1=   68.685585   U(NX+4,NY+4,NZ+4)=   25
VNODE1=   5338.7754   VNODE2=   5338.7755   V(NX+4,NY+4,NZ+4)=   5000
WNODE1=  2006.5085   WNODE2=  2006.5091   W(NX+4,NY+4,NZ+4)=   1000
NO. OF POINTS WHERE CANCER IS GROWING OUTSIDE THE LARGEST TUMORS=     6   LARGEST U OUTSIDE TUMORS=   68.691963
NO.OF POINTS WHERE U(N)>U(N-1)=    1156378   MAX.CHANGE FROM THE START = UMAX - UOMAX=   5129.6105
*********************************************************************************
SYSTEM TIME =  12:27:11  MAX OVER THE FIELD AT TIME =    2.2  TIME STEP =    176  NT= 200
A1=   1.5   A2=  .0001   A3=  .0001   B1=   .5   B2=  .00001   B3=  .005   C1=  .5   C2=  .00001   C3=  .00001
AT   15 .  15 .  15  UMAX(N)=   15126.762   V=  8321.0949   W=  7506.9546
AT   14 .  14 .  14  VMAX(N)=   8321.0949   U=  15126.762   V=  7506.9546
AT   15 .  15 .  15  WMAX(N)=   7506.9546   U=  15126.762   V=  8321.0949
TOTAL NO.OF POINTS WHERE: U IS MAX=     1 . V IS MAX=     1 . W IS MAX=     1
CHANGE OF  UMAX=   -2.848437  . OF VMAX=  1868.5587  . OF WMAX=   36.545852
AT THE PRIMARY SITE:     UMAX=   15126.762   VMAX=  8321.0949   WMAX=   7506.9546
AT THE METASTATIC#1 SITE:   UMAX=  10082.062   VMAX=  6903.435   WMAX=   4818.4321
AT THE METASTATIC#2 SITE:   UMAX=  2590.3491   VMAX=  5636.8355   WMAX=   1845.6834
UNODE1=  3063.5032   UNODE1-1=   69.062479   UNODE1+1=  69.062486   UNODE2=  3063.5048
UNODE2-1=  69.062486   UNODE2+1=   69.062486   U(NX+4,NY+4,NZ+4)=   25
VNODE1=   5701.4297   VNODE2=   5701.43   V(NX+4,NY+4,NZ+4)=   5000
WNODE1=  2014.9723   WNODE2=  2014.9729   W(NX+4,NY+4,NZ+4)=   1000
NO. OF POINTS WHERE CANCER IS GROWING OUTSIDE THE LARGEST TUMORS=     6   LARGEST U OUTSIDE TUMORS=   69.068972
NO.OF POINTS WHERE U(N)>U(N-1)=    1156314   MAX.CHANGE FROM THE START = UMAX - UOMAX=   5126.762
```

These facts are represented in Fig. 5.7.

We have already discussed the numerical outputs given in Table 5.5, Fig. 5.7. As computations continue, we may notice that at the time steps $t_n = 175$, and 176, respectively, the growths of $UMAX$, the largest value of the assembly of cancerous mass has been successfully halted. Table 5.5, Fig. 5.7 shows that the fighters V and W are successful in their efforts.

Figure 5.7 shows that speed of growth of U has gone down considerably and it is increasing slowly. However, it is also attempting to metastasize very slowly and possibly very secretly. At $t_n = 201$, $UMAX$ outside the original tumor sites has become 79.189202 (lower bottom redline).

Here $\sigma = 1.5 - 0.0001 \times 2000 = 1.3 > 0$, so the necessary condition that cancer must slow down is violated at the beginning and cancer is on the rise. At the end, $\sigma = 1.5 - 0.0001(5000 + 8389.87) = 0.161013 > 0$, again violation of the necessary condition. So cancer must keep on rising. But not rising as fast as it did initially because $0.161013 < 1.3$. The bottom line is a growing tumor (here it was growing at the rate of 50%) which could be as dangerous as a fast growing tumor. Moreover, it should be noted that this notoriously dangerous cell can change its growth rate at any time.

For the run in Fig. 5.8, we have used $a_1 = 2.5$, $a_2 = 0.005$, $a_3 = 0.0025$, $b_1 = 0.5$, $b_2 = 0.00001$, $b_3 = 0.005$, $c_1 = 0.5$, $c_2 = 0.00001$, $c_3 = 0.00001$. Initial values of V and W were only 75, and the max values are 750 for each. So, the strengths of V and W have been significantly increased with the efficiency of delivery of drugs. That is why we have chosen: $a_2 = 500 \times b_2$ and $a_3 = 250 \times c_2$. The wavy pattern of the values of V vanished, because V is uniformly annihilating

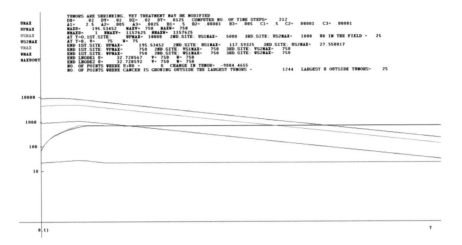

Fig. 5.8 Immunotherapy (V) with a monoclonal drug (W)

U. And as U decreases, V and W also decrease. The initial values of U at the primary and at the secondary sites were exactly the same as they were before. However, the rate of growth of U was increased to $a_1 = 2.5$ from $a_1 = 1.5$. The results appeared to be better. Some cancer cells took shelter on the planes of the boundary as we have discussed earlier. They are not metastasizing anywhere. Max value of U was reduced from 10,000 to 195.54. Thus, it becomes very clear that not only the amount of medications, but the intensities of their strengths matter most in the treatment plans. In this regard, some computational results (where final results are mentioned in Fig. 5.8) have been given: Let us repeat that: As before there are three cubical tumors at, respectively, $\{(13,13,13)$ to $(16,16,16)\}$ where $U_0(primary) = 10000$, at $\{(26,26,26)$ to $(29,29,29)\}$, where $U_0(Secondary\#1) = 5000$, and at $\{(47,47,47)$ to $(50,50,50)\}$ where $U_0(Secondary\#2) = 1000$. Three lymph nodes at $(3,3,3)$, $(35,35,35)$, and $(42,42,42)$ are also affected. At each node $U = 1200$, at $t = t_0$, respectively.

Up to first eight time steps (as noticed in the outputs), cancer started increasing, tumors getting bigger and bigger at the primary and the secondary sites. After that, they started decreasing. Also, it may be noticed that as U increases, V and W both increase accordingly, and as U started decreasing, both V and W started decreasing.

As before numerical results point out clearly that the highest concentrations of V and W are around the tumors (the primary cancer site is from $(13, 13, 13)$ to $(16, 16, 16)$ in the computational field), representing in anatomical terms the toughest fights between the antigens and the medications for the therapy are happening at the medically right locations.

At $t = t_7$ (Table 5.6 for Fig. 5.8), UMAX increased from UMAX(0) = 10,000 to UMAX$(t = t_7)$ 10,433.79, at $(15,15,15)$ which is a point inside the primary tumor site, and at that site, both VMAX(N) and WMAX(N) are the largest. In fact throughout the entire computations, we noticed that all our models reveal that the

Table 5.6 For Fig. 5.8

```
**************************************************************************************************
SYSTEM TIME =  18:59:39  MAX OVER THE FIELD AT TIME =      .0875  TIME STEP =    7  NT= 211  A1= 2.5
AT  15 . 15 . 15   UMAX(N)=  10433.789    V= 274.15494    W= 267.29824
AT  14 . 14 . 14   VMAX(N)=  274.15494    U= 10433.789    W= 267.29824
AT  15 . 15 . 15   WMAX(N)=  267.29824    U= 10433.789    V= 274.15494
TOTAL NO.OF POINTS WHERE: U IS MAX=      1 . V IS MAX=      1 . W IS MAX=      1
CHANGE OF: UMAX=    32.950251  . OF VMAX=    34.639172  . OF WMAX=    32.440934
AT THE PRIMARY SITE:      UMAX= 10433.789   VMAX=  274.15494   WMAX=  267.29824
AT THE METASTATIC#1 SITE:  UMAX=  5296.5564   VMAX=  175.08963   WMAX=  171.81232
AT THE METASTATIC#2 SITE:  UMAX=  1072.2794   VMAX=   95.865511  WMAX=   94.468367
UNODE1=  1285.8803   UNODE1-1=   26.88677   UNODE1+1=   26.88677   UNODE2=  1285.8803
UNODE2-1=   26.88677   UNODE2+1=   26.88677   U(NX+4,NY+4,NZ+4)=     25
VNODE1=   99.824723   VNODE2=   99.824723   V(NX+4,NY+4,NZ+4)=     75
WNODE1=   98.354546   WNODE2=   98.354546   W(NX+4,NY+4,NZ+4)=     75
NO. OF POINTS WHERE CANCER IS GROWING OUTSIDE THE LARGEST TUMORS =        6  LARGEST U OUTSIDE TUMORS=   26.886773
NO.OF POINTS WHERE U(N)>U(N-1)=      1156378   MAX.CHANGE FROM THE START =  UMAX - U0MAX=   433.78884
**************************************************************************************************
SYSTEM TIME =  19:00:09  MAX OVER THE FIELD AT TIME =      .1    TIME STEP =    8  NT= 211  A1= 2.5
AT  15 . 15 . 15   UMAX(N)=  10450.007    V= 309.42647    W= 299.77644
AT  14 . 14 . 14   VMAX(N)=  309.42647    U= 10450.007    W= 299.77644
AT  15 . 15 . 15   WMAX(N)=  299.77644    U= 10450.007    V= 309.42647
TOTAL NO.OF POINTS WHERE: U IS MAX=      1 . V IS MAX=      1 . W IS MAX=      1
CHANGE OF: UMAX=    16.217932  . OF VMAX=    35.271532  . OF WMAX=    32.478197
AT THE PRIMARY SITE:      UMAX= 10450.007   VMAX=  309.42647   WMAX=  299.77644
AT THE METASTATIC#1 SITE:  UMAX=  5332.7803   VMAX=  192.78049   WMAX=  188.4012
AT THE METASTATIC#2 SITE:  UMAX=  1084.1997   VMAX=   99.541456  WMAX=   97.843182
UNODE1=  1299.8869   UNODE1-1=   27.213883   UNODE1+1=   27.213883   UNODE2=  1299.8869
UNODE2-1=   27.213883   UNODE2+1=   27.213883   U(NX+4,NY+4,NZ+4)=     25
VNODE1=  104.19992   VNODE2=  104.19992   V(NX+4,NY+4,NZ+4)=     75
WNODE1=  102.40042   WNODE2=  102.40042   W(NX+4,NY+4,NZ+4)=     75
NO. OF POINTS WHERE CANCER IS GROWING OUTSIDE THE LARGEST TUMORS =        6  LARGEST U OUTSIDE TUMORS=   27.213887
NO.OF POINTS WHERE U(N)>U(N-1)=      1156378   MAX.CHANGE FROM THE START =  UMAX - U0MAX=   450.00677
**************************************************************************************************
SYSTEM TIME =  19:00:38  MAX OVER THE FIELD AT TIME =      .1125  TIME STEP =    9  NT= 211  A1= 2.5
AT  15 . 15 . 15   UMAX(N)=  10449.22     V= 345.35507    W= 332.23888
AT  14 . 14 . 14   VMAX(N)=  345.35507    U= 10449.22     W= 332.23888
AT  15 . 15 . 15   WMAX(N)=  332.23888    U= 10449.22     V= 345.35507
TOTAL NO.OF POINTS WHERE: U IS MAX=      1 . V IS MAX=      1 . W IS MAX=      1
CHANGE OF: UMAX=   - .7871758  . OF VMAX=    35.928602  . OF WMAX=    32.462442
AT THE PRIMARY SITE:      UMAX= 10449.22    VMAX=  345.35507   WMAX=  332.23888
AT THE METASTATIC#1 SITE:  UMAX=  5364.8242   VMAX=  210.7798   WMAX=  205.08625
AT THE METASTATIC#2 SITE:  UMAX=  1096.0653   VMAX=  103.27644   WMAX=  101.25477
UNODE1=  1313.7783   UNODE1-1=   27.544707   UNODE1+1=   27.544707   UNODE2=  1313.7783
UNODE2-1=   27.544707   UNODE2+1=   27.544707   U(NX+4,NY+4,NZ+4)=     25
VNODE1=  108.64596   VNODE2=  108.64596   V(NX+4,NY+4,NZ+4)=     75
WNODE1=  106.48927   WNODE2=  106.48927   W(NX+4,NY+4,NZ+4)=     75
NO. OF POINTS WHERE CANCER IS GROWING OUTSIDE THE LARGEST TUMORS =        6  LARGEST U OUTSIDE TUMORS=   27.544713
```

medications are mostly concentrated where cancer cells are mostly concentrated (which is obviously what doctors would like to see). Initially at two lymph nodes, $U(0) = 1200$. These increased by 85.88. V increased from 75 to almost 274, and W increased from 75 to 267 (Fig. 5.8, initial values are recorded). Outside the field, U increased from 25 to 26.88. The code is written in such a way that neither V nor W is able to attack U in its hideouts. Here, these hideouts are on the border. That is shown by the values of U, V, W at the end point (NX + 4, NY + 4, NZ + 4).

The entire Table 5.7 for Fig. 5.8 shows that both medications V and W have kept on strong vigilance surrounding the cancer cells. So, the fight is what we expect to see. Also, we notice that there are 1,156,378 out of a total of 1,157,625 points in the field where $U_{ijk}^n > U_{ijk}^{n-1}$ at t_7. That means that cancer is wildly increasing at almost all points in the computational field. This is abruptly changed at $t = t_{117}$. At no point $U_{ijk}^{117} > U_{ijk}^{116}$. It should be noted that when $U_{NX+4,NY+4,NZ+4}^n = 25$, that means U is not increasing any more at the boundary points (possibly at their hideouts).

It should be noted that at $t = t_{117}$, UMAX(N) = 1298.61 at the primary site and started decreasing at $t = t_{118}$ and $t = t_{119}$ ($U_{ijk}^n < U_{ijk}^{n-1}$). A patient may not like this scenario (because it took some time for the medications to be effective). But oncologists understand how much medications a patient may tolerate. More amounts may cause more harm than good. That is one of the primary reasons why oncologists do regular blood tests and complete blood counts (CBCs).

The values of V and W reveal that they are moving uniformly toward steady states. That means those are the required amounts of drugs that the patient shall need to live a normal life. Note that the values of $VMAX$ and $WMAX$ hardly changed from t_{141} to t_{143} (Table 5.8 for Fig. 5.8). The pattern stayed the same throughout the

Table 5.7 For Fig. 5.8 Cancer winning nowhere

```
*************************************************************************************************
SYSTEM TIME =  19:56:00  MAX OVER THE FIELD AT TIME =    1.4625   TIME STEP =   117  NT= 211  A1= 2.5
AT  15 . 15 . 15   UMAX(N)=  1298.6117   V=  772.09395   W=  753.97509
AT  15 . 15 . 15   VMAX(N)=  772.09395   U=  1298.6117   W=  753.97509
AT  15 . 15 . 15   WMAX(N)=  753.97509   U=  1298.6117   V=  772.09395
TOTAL NO.OF POINTS WHERE: U IS MAX=      1 . V IS MAX=      8 . W IS MAX=       7
CHANGE OF: UMAX=  -26.257708  . OF VMAX=  -.08436696  . OF WMAX=  -8.1105502e-2
AT THE PRIMARY SITE:    UMAX=  1298.6117   VMAX=  772.09395   WMAX=  753.97509
AT THE METASTATIC#1 SITE:    UMAX=  778.88799   VMAX=  770.42418   WMAX=  752.36979
AT THE METASTATIC#2 SITE:    UMAX=  181.91753   VMAX=  768.50651   WMAX=  750.526
UNODE1=  216.33669   UNODE1-1=   4.7305706   UNODE1+1=   4.7305708   UNODE2=  216.33675
UNODE2-1=   4.7305708   UNODE2+1=   4.7305708   U(NX+4,NY+4,NZ+4)=     25
VNODE1=  768.61706   VNODE2=  768.61706   V(NX+4,NY+4,NZ+4)=    750
WNODE1=  750.6323    WNODE2=  750.6323    W(NX+4,NY+4,NZ+4)=    750
NO. OF POINTS WHERE CANCER IS GROWING OUTSIDE THE LARGEST TUMORS=       1244   LARGEST U OUTSIDE TUMORS=    25
NO.OF POINTS WHERE U(N)>U(N-1)=        0  MAX.CHANGE FROM THE START = UMAX - UOMAX=      -8701.3883
*************************************************************************************************
SYSTEM TIME =  19:56:30  MAX OVER THE FIELD AT TIME =    1.475    TIME STEP =   118  NT= 211  A1= 2.5
AT  15 . 15 . 15   UMAX(N)=  1272.8784   V=  772.01127   W=  753.8956
AT  15 . 15 . 15   VMAX(N)=  772.01127   U=  1272.8784   W=  753.8956
AT  15 . 15 . 15   WMAX(N)=  753.8956    U=  1272.8784   V=  772.01127
TOTAL NO.OF POINTS WHERE: U IS MAX=      1 . V IS MAX=      8 . W IS MAX=       4
CHANGE OF: UMAX=  -25.733291  . OF VMAX= -2681428e-2  . OF WMAX=  -.07948547
AT THE PRIMARY SITE:    UMAX=  1272.8784   VMAX=  772.01127   WMAX=  753.8956
AT THE METASTATIC#1 SITE:    UMAX=  763.5012    VMAX=  770.37475   WMAX=  752.32227
AT THE METASTATIC#2 SITE:    UMAX=  178.33656   VMAX=  768.495    WMAX=  750.51494
UNODE1=  212.07541   UNODE1-1=   4.6375501   UNODE1+1=   4.6375502   UNODE2=  212.07546
UNODE2-1=   4.6375502   UNODE2+1=   4.6375502   U(NX+4,NY+4,NZ+4)=     25
VNODE1=  768.60337   VNODE2=  768.60337   V(NX+4,NY+4,NZ+4)=    750
WNODE1=  750.61914   WNODE2=  750.61914   W(NX+4,NY+4,NZ+4)=    750
NO. OF POINTS WHERE CANCER IS GROWING OUTSIDE THE LARGEST TUMORS=       1244   LARGEST U OUTSIDE TUMORS=    25
NO.OF POINTS WHERE U(N)>U(N-1)=        0  MAX.CHANGE FROM THE START = UMAX - UOMAX=      -8727.1216
*************************************************************************************************
SYSTEM TIME =  19:57:02  MAX OVER THE FIELD AT TIME =    1.4875   TIME STEP =   119  NT= 211  A1= 2.5
AT  15 . 15 . 15   UMAX(N)=  1247.6589   V=  771.93024   W=  753.8177
AT  15 . 15 . 15   VMAX(N)=  771.93024   U=  1247.6589   W=  753.8177
AT  15 . 15 . 15   WMAX(N)=  753.8177    U=  1247.6589   V=  771.93024
TOTAL NO.OF POINTS WHERE: U IS MAX=      1 . V IS MAX=      8 . W IS MAX=       7
CHANGE OF: UMAX=  -25.219504  . OF VMAX= -8.1030088e-2  . OF WMAX= -7.7898288e-2
AT THE PRIMARY SITE:    UMAX=  1247.6589   VMAX=  771.93024   WMAX=  753.8177
AT THE METASTATIC#1 SITE:    UMAX=  748.41975   VMAX=  770.3263    WMAX=  752.27569
AT THE METASTATIC#2 SITE:    UMAX=  174.82616   VMAX=  768.48373   WMAX=  750.5041
UNODE1=  207.89816   UNODE1-1=   4.5463588   UNODE1+1=   4.5463589   UNODE2=  207.89821
UNODE2-1=   4.5463589   UNODE2+1=   4.5463589   U(NX+4,NY+4,NZ+4)=     25
VNODE1=  768.58995   VNODE2=  768.58995   V(NX+4,NY+4,NZ+4)=    750
WNODE1=  750.60624   WNODE2=  750.60624   W(NX+4,NY+4,NZ+4)=    750
NO. OF POINTS WHERE CANCER IS GROWING OUTSIDE THE LARGEST TUMORS=       1244   LARGEST U OUTSIDE TUMORS=    25
NO.OF POINTS WHERE U(N)>U(N-1)=        0  MAX.CHANGE FROM THE START = UMAX - UOMAX=      -8752.3411
```

Table 5.8 For Fig. 5.8. Here, defense is concentrated most where the enemies are mostly present

```
*************************************************************************************************
SYSTEM TIME =  20:08:43  MAX OVER THE FIELD AT TIME =    1.7625   TIME STEP =   141  NT= 211  A1= 2.5
AT  15 . 15 . 15   UMAX(N)=  803.86556   V=  770.50442   W=  752.44694
AT  15 . 15 . 15   VMAX(N)=  770.50442   U=  803.86556   W=  752.44694
AT  15 . 15 . 15   WMAX(N)=  752.44694   U=  803.86556   V=  770.50442
TOTAL NO.OF POINTS WHERE: U IS MAX=      1 . V IS MAX=      8 . W IS MAX=       8
CHANGE OF: UMAX=  -16.204313  . OF VMAX= -5.2058266e-2  . OF WMAX= -5.0049919e-2
AT THE PRIMARY SITE:    UMAX=  803.86556   VMAX=  770.50442   WMAX=  752.44694
AT THE METASTATIC#1 SITE:    UMAX=  402.72578   VMAX=  769.47277   WMAX=  751.45506
AT THE METASTATIC#2 SITE:    UMAX=  112.90148   VMAX=  768.28482   WMAX=  750.31285
UNODE1=  134.22283   UNODE1-1=   2.9370864   UNODE1+1=   2.9370866   UNODE2=  134.22288
UNODE2-1=   2.9370866   UNODE2+1=   2.9370866   U(NX+4,NY+4,NZ+4)=     25
VNODE1=  768.3533    VNODE2=  768.3533    V(NX+4,NY+4,NZ+4)=    750
WNODE1=  750.37869   WNODE2=  750.37869   W(NX+4,NY+4,NZ+4)=    750
NO. OF POINTS WHERE CANCER IS GROWING OUTSIDE THE LARGEST TUMORS=       1244   LARGEST U OUTSIDE TUMORS=    25
NO.OF POINTS WHERE U(N)>U(N-1)=        0  MAX.CHANGE FROM THE START = UMAX - UOMAX=      -9196.1344
*************************************************************************************************
SYSTEM TIME =  20:09:14  MAX OVER THE FIELD AT TIME =    1.775    TIME STEP =   142  NT= 211  A1= 2.5
AT  15 . 15 . 15   UMAX(N)=  787.98297   V=  770.4534    W=  752.39788
AT  15 . 15 . 15   VMAX(N)=  770.4534    U=  787.98297   W=  752.39788
AT  15 . 15 . 15   WMAX(N)=  752.39788   U=  787.98297   V=  770.4534
TOTAL NO.OF POINTS WHERE: U IS MAX=      1 . V IS MAX=      8 . W IS MAX=       8
CHANGE OF: UMAX=  -15.882587  . OF VMAX= -5.1024473e-2  . OF WMAX= -4.9056138e-2
AT THE PRIMARY SITE:    UMAX=  787.98297   VMAX=  770.4534    WMAX=  752.39788
AT THE METASTATIC#1 SITE:    UMAX=  473.20644   VMAX=  769.44219   WMAX=  751.42566
AT THE METASTATIC#2 SITE:    UMAX=  110.67998   VMAX=  768.27769   WMAX=  750.30599
UNODE1=  131.58028   UNODE1-1=   2.8793331   UNODE1+1=   2.8793332   UNODE2=  131.58032
UNODE2-1=   2.8793332   UNODE2+1=   2.8793332   U(NX+4,NY+4,NZ+4)=     25
VNODE1=  768.34482   VNODE2=  768.34482   V(NX+4,NY+4,NZ+4)=    750
WNODE1=  750.37053   WNODE2=  750.37053   W(NX+4,NY+4,NZ+4)=    750
NO. OF POINTS WHERE CANCER IS GROWING OUTSIDE THE LARGEST TUMORS=       1244   LARGEST U OUTSIDE TUMORS=    25
NO.OF POINTS WHERE U(N)>U(N-1)=        0  MAX.CHANGE FROM THE START = UMAX - UOMAX=      -9212.017
*************************************************************************************************
SYSTEM TIME =  20:09:45  MAX OVER THE FIELD AT TIME =    1.7875   TIME STEP =   143  NT= 211  A1= 2.5
AT  15 . 15 . 15   UMAX(N)=  772.41566   V=  770.40338   W=  752.3498
AT  15 . 15 . 15   VMAX(N)=  770.40338   U=  772.41566   W=  752.3498
AT  15 . 15 . 15   WMAX(N)=  752.3498    U=  772.41566   V=  770.40338
TOTAL NO.OF POINTS WHERE: U IS MAX=      1 . V IS MAX=      8 . W IS MAX=       8
CHANGE OF: UMAX=  -15.56731   . OF VMAX= -5.0011406e-2  . OF WMAX= -4.8082276e-2
AT THE PRIMARY SITE:    UMAX=  772.41566   VMAX=  770.40338   WMAX=  752.3498
AT THE METASTATIC#1 SITE:    UMAX=  463.07536   VMAX=  769.41221   WMAX=  751.39684
AT THE METASTATIC#2 SITE:    UMAX=  109.50222   VMAX=  768.27069   WMAX=  750.29926
UNODE1=  128.98979   UNODE1-1=   2.8227153   UNODE1+1=   2.8227155   UNODE2=  128.98984
UNODE2-1=   2.8227155   UNODE2+1=   2.8227155   U(NX+4,NY+4,NZ+4)=     25
VNODE1=  768.3365    VNODE2=  768.3365    V(NX+4,NY+4,NZ+4)=    750
WNODE1=  750.36253   WNODE2=  750.36253   W(NX+4,NY+4,NZ+4)=    750
NO. OF POINTS WHERE CANCER IS GROWING OUTSIDE THE LARGEST TUMORS=       1244   LARGEST U OUTSIDE TUMORS=    25
NO.OF POINTS WHERE U(N)>U(N-1)=        0  MAX.CHANGE FROM THE START = UMAX - UOMAX=      -9227.5843
```

later computations. The fact of the matter is that a successful cancer treatment is sometimes painfully slow. At no point in the field $U(t_n) > U(t_{n-1})$.

However, what is very noticeable is that the medications are mostly present along the sites of the tumors and all around where malignancies are present and have reached almost a steady state. So, no more extra medications are necessary. V and

W are confident that the strengths that they have gained are good enough to destroy aggressiveness of the enemies.

More of such results are given in the computations the time steps t_{141} to t_{143}. In the entire field of computation, at no point $U_{ijk}^n > U_{ijk}^{n-1}$. The cancer sites along with the lymph nodes are surrounded by the medications (Table 5.8 of Fig. 5.8). But on the border, the hideouts of cancer, there are 1244 points and at those points medications are ineffective (as we have discussed earlier).

These treatments reveal a slow process of a determined victory. A fast cure of cancer is possible mathematically, but it could be hardly seen in reality. The best news at this point is that it is not growing anywhere.

At t_{141}, the largest value of U in the entire computational field $= U0MAX - 9196.13$, and at t_{142}, it is $U0MAX - 9212$. It is reduced by only 16 units. Most importantly, there is no sign of metastasis.

Because, at no point $U_{ijk}^n > U_{ijk}^{n-1}$. The values of V and W at the end point $(NX + 4, NY + 4, NZ + 4)$ show that they are strongly present. But since we are not solving the equations at these points, it means they are not active. That means this end point is a hideout point of cancer cells.

The most significant features are that the war is mostly concentrated at the tumors. At the infected nodes, cancer cells have been contained. They are unable to get out and infect other tissues. The primary reason that we have shown here the details of our computations is that the readers could see how the models are working and predict what the outcomes will be.

Figure 5.8 shows what we have discussed earlier.

The situation appeared to be much better (in Fig. 5.9) when with $a_1 = 2.0$, and all other inputs remain the same. Again from the computed results below, it could be noticed that at the location of the primary tumor site where U is the largest, medications V and W are mostly present there fighting U. Certainly, this is what

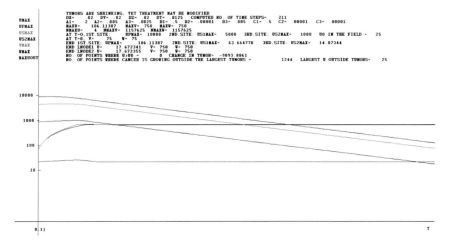

Fig. 5.9 Immunotherapy (V) with a monoclonal drug (W)

Table 5.9 For Fig. 5.9

```
*****************************************************************************************
SYSTEM TIME =  17:20:21  MAX OVER THE FIELD AT TIME =      .05  TIME STEP =   4  NT= 210
AT  15 . 15 . 15  UMAX(N)=  10190.004    V=  171.48869    W=  169.74992
AT  14 . 14 . 14  VMAX(N)=  171.48869    U=  10190.004    W=  169.74992
AT  15 . 15 . 15  WMAX(N)=  169.74992    U=  10190.004    V=  171.48869
TOTAL NO.OF POINTS WHERE: U IS MAX=      1 . V IS MAX=      1 . W IS MAX=     7
CHANGE OF: UMAX=     48.395229 . OF VMAX=    32.594576 . OF WMAX=    31.737684
AT THE PRIMARY SITE:     UMAX=  10190.004    VMAX=  171.48869    WMAX=  169.74992
AT THE METASTATIC#1 SITE:  UMAX=  5115.8704    VMAX=  123.51516    WMAX=  122.47769
AT THE METASTATIC#2 SITE:  UMAX=  1026.5294    VMAX=  85.127692    WMAX=  84.511272
UNODE1=  1231.5995    UNODE1-1=  25.683738    UNODE1+1=  25.683738    UNODE2=  1231.5995
UNODE2-1=  25.683738    UNODE2+1=  25.683738    U(NX+4,NY+4,NZ+4)=    25
VNODE1=  87.046863    VNODE2=  87.046863    V(NX+4,NY+4,NZ+4)=     75
WNODE1=  86.412358    WNODE2=  86.412358    W(NX+4,NY+4,NZ+4)=     75
NO.OF POINTS WHERE CANCER IS GROWING OUTSIDE THE LARGEST TUMORS =      6  LARGEST U OUTSIDE TUMORS=  25.683739
NO.OF POINTS WHERE U(N)>U(N-1)=    1156378    MAX.CHANGE FROM THE START = UMAX - UOMAX=  190.00396
*****************************************************************************************
SYSTEM TIME =  17:20:58  MAX OVER THE FIELD AT TIME =     .0625  TIME STEP =   5  NT= 210
AT  15 . 15 . 15  UMAX(N)=  10222.951    V=  204.54064    W=  201.57779
AT  14 . 14 . 14  VMAX(N)=  204.54064    U=  10222.951    W=  201.57779
AT  15 . 15 . 15  WMAX(N)=  201.57779    U=  10222.951    V=  204.54064
TOTAL NO.OF POINTS WHERE: U IS MAX=      1 . V IS MAX=      1 . W IS MAX=     7
CHANGE OF: UMAX=     32.946684 . OF VMAX=    33.051944 . OF WMAX=    31.82787
AT THE PRIMARY SITE:     UMAX=  10222.951    VMAX=  204.54064    WMAX=  201.57779
AT THE METASTATIC#1 SITE:  UMAX=  5147.1931    VMAX=  140.12947    WMAX=  138.50785
AT THE METASTATIC#2 SITE:  UMAX=  1035.2009    VMAX=  88.592421    WMAX=  87.735796
UNODE1=  1241.8483    UNODE1-1=  25.915296    UNODE1+1=  25.915296    UNODE2=  1241.8483
UNODE2-1=  25.915296    UNODE2+1=  25.915296    U(NX+4,NY+4,NZ+4)=    25
VNODE1=  91.168789    VNODE2=  91.168789    V(NX+4,NY+4,NZ+4)=     75
WNODE1=  90.280457    WNODE2=  90.280457    W(NX+4,NY+4,NZ+4)=     75
NO.OF POINTS WHERE CANCER IS GROWING OUTSIDE THE LARGEST TUMORS =      6  LARGEST U OUTSIDE TUMORS=  25.915297
NO.OF POINTS WHERE U(N)>U(N-1)=    1156378    MAX.CHANGE FROM THE START = UMAX - UOMAX=  222.95064
*****************************************************************************************
SYSTEM TIME =  17:21:35  MAX OVER THE FIELD AT TIME =     .075  TIME STEP =   6  NT= 210
AT  15 . 15 . 15  UMAX(N)=  10240.153    V=  238.06989    W=  233.44659
AT  14 . 14 . 14  VMAX(N)=  238.06989    U=  10240.153    W=  233.44659
AT  15 . 15 . 15  WMAX(N)=  233.44659    U=  10240.153    V=  238.06989
TOTAL NO.OF POINTS WHERE: U IS MAX=      1 . V IS MAX=      1 . W IS MAX=     1
CHANGE OF: UMAX=     17.202841 . OF VMAX=    33.529256 . OF WMAX=    31.868804
AT THE PRIMARY SITE:     UMAX=  10240.153    VMAX=  238.06989    WMAX=  233.44659
AT THE METASTATIC#1 SITE:  UMAX=  5174.6702    VMAX=  156.97501    WMAX=  154.62032
AT THE METASTATIC#2 SITE:  UMAX=  1043.7774    VMAX=  92.102561    WMAX=  90.986885
UNODE1=  1251.9418    UNODE1-1=  26.148694    UNODE1+1=  26.148694    UNODE2=  1251.9418
UNODE2-1=  26.148694    UNODE2+1=  26.148694    U(NX+4,NY+4,NZ+4)=    25
VNODE1=  95.345119    VNODE2=  95.345119    V(NX+4,NY+4,NZ+4)=     75
WNODE1=  94.179749    WNODE2=  94.179749    W(NX+4,NY+4,NZ+4)=     75
NO. OF POINTS WHERE CANCER IS GROWING OUTSIDE THE LARGEST TUMORS =      6  LARGEST U OUTSIDE TUMORS=  26.148696
NO.OF POINTS WHERE U(N)>U(N-1)=    1156378    MAX.CHANGE FROM THE START = UMAX - UOMAX=  240.15348
```

oncologists expect to see. Although $a_1 = 4 \times b_1$, which means that the cancer cells are growing 4 times faster than the rate of growth of V (dosage), cancer is defeated because of the strengths of the medications which are given by $a_2 = 500 \times b_2$ and $a_3 = 250 \times c_3$. This obviously means that mathematically, the strengths of medicines play major roles regarding their ability to cure/contain a disease, which is quite logical. It may also be seen that as cancer is dying, it is slowly leaving the lymph nodes (as may be noticed in the second set of computations from t_{53} to t_{55} in the Table 5.10). At no point in the field U_{ijk}^n is increasing. Both V and W are following the cancer cells very closely. Following previous discussions on numerical results, one may do the same for results given below. In Fig. 5.9, final results are recorded.

We notice in Table 5.9. Figure 5.9 that at the time steps 4, 5, 6, $UMAX$ is on the rise at the primary location. And exactly at that location, the medications are congregating most. Here, the code is written such that accumulation of runaway cancer cells on the boundary lines could be noticeable. So at the end boundary point $(NX + 4, NY + 4, NZ + 4)$, the number of cancer cells is slowly rising, whereas the values V and W remaining steady. However, in Table 5.10, Fig. 5.9, the situation changed near the boundary. At $(NX + 4, NY + 4, NZ + 4)$, values of V and W increased significantly and the value of U came back to its initial condition.

Comparing Tables 5.9 and 5.10 of Fig. 5.9, we notice that at time step $t_n = 6$, $U_{ijk}^n > U_{ijk}^{n-1}$ at 1,156,378 number of points, and at $t_n = 55$, at no points $U_{ijk}^n > U_{ijk}^{n-1}$. So metastatic tendency has been fully stopped.

The tables of actual computer outputs give all the details as shown in Fig. 5.9.

The results got much nicer when $a_1 = 1.5$, as expected, although the values of $V_0 = W_0 = 45$, much smaller than their previous values, and all other inputs remaining the same (Fig. 5.10).

Table 5.10 For Fig. 5.9. Cancer winning nowhere. $U_{ijk}^n > U_{ijk}^{n-1}$ at no point in the field

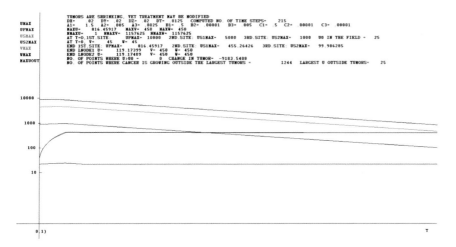

Fig. 5.10 Immunotherapy (V) with a monoclonal drug (W)

These graphs vividly display that as the rate of growth, meaning as aggressiveness of cancer cells becomes less and less, it becomes somewhat easier to treat. That means the prognosis gets better.

We need to remember that in the code there is no provision that a fight is being conducted at the boundary.

5.5 A Severe Case with Five Tumors

This is a special case with stage#4 states of cancer with four metastatic tumors. This simulates a breast cancer patient who has multiple tumors in her breast. This is known as multifocal breast cancer. Here, cancer is extremely aggressive, with $a_1 = 4.5$. The monoclonal drug W is doing three functions: (1) It is reducing the rate of growth a_1, (2) it is strengthening the immune system V, and (3) it is destroying some cancer cells as it moves through blood and lymphatic vessels. The inputs are mostly the same as before. The new assumption is (this is done by applying a monoclonal drug):

$$a_1(t_n) = a_1(t_{n-1})exp(-\Delta t). \qquad (5.4.1)$$

We have not yet associated this assumption with how it affects values of V and W, from $t = t_0$. Medical professionals make such assumptions after observing some special treatments and measuring the results statistically.

If $\Delta t = 0.0125$, $exp(-\Delta t) = 0.9875778$

.

The mathematical formula is: For each unit of W (the monoclonal drug), $da_1/dt = -a_1(t)W_0$, where $W_0 = 1$. If we integrate from t_{n-1} to t_n, we will get the formula (5.4.1).

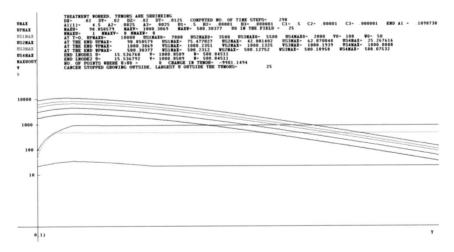

Fig. 5.11 Immunotherapy (V) with a monoclonal drug (W) that reduces A1, strenghens V and kills U

Input for Fig. 5.11.

```
!****************************************************************************************!
! TREATMENT WITH IMMUNOTHERAPY AND  MONOCLONAL DRUG.
! DU/DT = A1(T)*U-A2*U*V - A3*U*W + NU1*DELSQ(U) ! ANTIGEN    , A1(T) = A1(T)*EXP(-DT), DONE BY W
! DV/DT = B1*U-B2*V*U + B3*V*W + NU2*DELSQ(V) ! IMMUNE RESPONSE
! DW/DT = C1*U -C2*W*U - C3*W*V+ NU3*DELSQ(W) ! MONOCLONAL DRUG
! DX- .02  DY- .02  DZ- .02  DT- .0125  NX- 101  NY- 101  NZ - 101   NT- 300
! A1(1)= 4.5  A2- .0025  A3- .0025
! B1- .5  B2- .0001  B3- .000001  V0- 100
! C1- .5  C2- .0001  C3- .000001  W0- 50
! AT T=0: PRIMARY U= 10000  1ST.UMETA#1 - 7000  2ND.UMETA#2- 3500  3RD.UMETA#3- 5500  4RTH.UMETA#4- 2000
! NU1- .0000001  NU2- .0000001  NU3- .0000001
! GAMMA1= .015  GAMMA2= .015  GAMMA3- .015
! TWO LYMPH NODES ARE LOCATED AT: ( 3 , 3 , 3 ) AND   ( 35 , 35 , 35 ) SECOND LYMPH NODE IS AFFECTED
! FOUR TUMORS LOCATED AT:   ( 13 , 13 , 13 ) AND ( 26 , 26 , 26 )
!( 47 , 47 , 47 )
!( 67 , 67 , 67 );( 79 , 79 , 79 )
!
!****************************************************************************************!
```

At the outset, cancer spread all over the field. From Table 5.11. Figure 5.11 notice that where cancer is very strong, the presence of V and W is also gaining strength. As before war is going on, each side fighting very vigorously.

From the very beginning, fighting was concentrated near the cancer sites. The larger the tumor, the stronger the fight. As the monoclonal drug W attempts to reduce the rate of growth of cancer cells, and attacking them as they attempt to increase their strength all over the field and/or try to hide. So outside the cancer sites, $UMAX = 27.64$ which was 25.00 at the time step 5. Also at the primary site, they increased from 10,000 to 10,951.95. In fact at each tumor site, they are desperately trying to stimulate their presence to give a stronger fight (clear from the

Table 5.11 For Fig. 5.11. Defense is attacking most where enemies gathered most. Near the primary site

```
-------------------------------------------------------------------------------------------
SYSTEM TIME = 18:10:48  MAX OVER THE FIELD AT TIME =     .0375  TIME STEP =    3  NT= 303  A1(N)= 4.3888946
AT  15 .  15 .  15   UMAX(N)=   10494.464    V=  164.40294    V-  114.46714
AT  14 .  14 .  14   VMAX(N)=  164.40294    U-  10494.464    V-  114.46714
AT  14 .  14 .  14   WMAX(N)=  114.46714    U-  10494.464    V-  164.40294
TOTAL NO.OF POINTS WHERE: U IS MAX-        4 , V IS MAX-       1
CHANGE OF: UMAX-   243.13787    OF VMAX-   32.572818   . OF WMAX-   32.605239
AT THE PRIMARY SITE:      UMAX-  10494.464    VMAX- 164.40294    WMAX-  114.46714
AT THE METASTATIC#1 SITE:    UMAX-  7352.008    VMAX- 145.10686    WMAX-  95.151799
AT THE METASTATIC#2 SITE:    UMAX-  3679.4408    VMAX- 122.56794    WMAX-  72.590364
AT THE METASTATIC#3 SITE:    UMAX-  5778.8913    VMAX- 135.45087    WMAX-  85.486159
AT THE METASTATIC#4 SITE:    UMAX-  2103.3802    VMAX- 112.89954    WMAX-  62.912306
UNODE1-  1262.2743     UNODE1-1-   26.306131     UNODE2-  1262.2743
UNODE2-1-   26.306131     UNODE2+1-   26.306131   U(NX+4,NY+4,NZ+4)-    25
VNODE1-  107.7407     VNODE2-  107.7407   V(NX+4,NY+4,NZ+4)-    100
WNODE1-  61.748307   WNODE2-  57.748307   W(NX+4,NY+4,NZ+4)-    50
NO. OF POINTS WHERE CANCER IS GROWING OUTSIDE THE LARGEST TUMORS -      6  LARGEST U OUTSIDE TUMORS-   26.306131
NO.OF POINTS WHERE U(N)>U(N-1)-      1156378   MAX CHANGE FROM THE START - UMAX - UOMAX-       494.46411
*******************************************************************************************
SYSTEM TIME = 18:11:19  MAX OVER THE FIELD AT TIME =     .05  TIME STEP =    4  NT= 303  A1(N)= 4.3343749
AT  15 .  15 .  15   UMAX(N)=   10728.354    V-  197.68875    V-  147.78596
AT  14 .  14 .  14   VMAX(N)=  197.68875    U-  10728.354    V-  147.78596
AT  14 .  14 .  14   WMAX(N)=  147.78596    U-  10728.354    V-  197.68875
TOTAL NO.OF POINTS WHERE: U IS MAX-        4 , V IS MAX-       1  , U IS MAX-     1
CHANGE OF: UMAX-   233.88982    OF VMAX-   33.285812   . OF WMAX-   33.318823
AT THE PRIMARY SITE:      UMAX-  10728.354    VMAX- 197.68875    WMAX-  147.78596
AT THE METASTATIC#1 SITE:    UMAX-  7522.3957    VMAX- 168.45115    WMAX-  118.51924
AT THE METASTATIC#2 SITE:    UMAX-  3768.5403    VMAX- 134.26596    WMAX-  84.29997
AT THE METASTATIC#3 SITE:    UMAX-  5915.394    VMAX- 153.81021    WMAX-  103.86371
AT THE METASTATIC#4 SITE:    UMAX-  2155.2541    VMAX- 119.59048    WMAX-  69.609853
UNODE1-  1293.6937     UNODE1-1-   26.970392     UNODE2-  1293.6937
UNODE2-1-   26.970392     UNODE2+1-   26.970392   U(NX+4,NY+4,NZ+4)-    25
VNODE1-  111.7571     VNODE2-  111.7571   V(NX+4,NY+4,NZ+4)-    100
WNODE1-  61.768653   WNODE2-  61.768653   W(NX+4,NY+4,NZ+4)-    50
NO. OF POINTS WHERE CANCER IS GROWING OUTSIDE THE LARGEST TUMORS -      6  LARGEST U OUTSIDE TUMORS-   26.970393
NO.OF POINTS WHERE U(N)>U(N-1)-      1156378   MAX CHANGE FROM THE START - UMAX - UOMAX-       728.35393
*******************************************************************************************
SYSTEM TIME = 18:11:49  MAX OVER THE FIELD AT TIME =     .0625  TIME STEP =    5  NT= 303  A1(N)= 4.2805324
AT  15 .  15 .  15   UMAX(N)=   10951.95    V-  231.65463    V-  181.78538
AT  14 .  14 .  14   VMAX(N)=  231.65463    U-  10951.95    V-  181.78538
AT  14 .  14 .  14   WMAX(N)=  181.78538    U-  10951.95    V-  231.65463
TOTAL NO.OF POINTS WHERE: U IS MAX-        4 , V IS MAX-       1  , U IS MAX-     1
CHANGE OF: UMAX-   223.59562    OF VMAX-   33.96588   . OF WMAX-   33.999417
AT THE PRIMARY SITE:      UMAX-  10951.95    VMAX- 231.65463    WMAX-  181.78538
AT THE METASTATIC#1 SITE:    UMAX-  7688.3053    VMAX- 192.30406    WMAX-  142.39573
AT THE METASTATIC#2 SITE:    UMAX-  3857.0191    VMAX- 146.23746    WMAX-  96.283294
AT THE METASTATIC#3 SITE:    UMAX-  6049.4624    VMAX- 172.58207    WMAX-  122.65413
AT THE METASTATIC#4 SITE:    UMAX-  2207.1747    VMAX- 126.44242    WMAX-  76.468541
UNODE1-  1325.2697     UNODE1-1-   27.641902     UNODE1+1-   27.641902     UNODE2-  1325.2697
UNODE2-1-   27.641902     UNODE2+1-   27.641902   U(NX+4,NY+4,NZ+4)-    25
VNODE1-  115.87154     VNODE2-  115.87154   V(NX+4,NY+4,NZ+4)-    100
WNODE1-  65.887123   WNODE2-  65.887123   W(NX+4,NY+4,NZ+4)-    50
NO. OF POINTS WHERE CANCER IS GROWING OUTSIDE THE LARGEST TUMORS -      6  LARGEST U OUTSIDE TUMORS-   27.641903
NO. OF POINTS WHERE U(N)>U(N-1)-     1156378   MAX CHANGE FROM THE START - UMAX - UOMAX-       951.94955
```

Table 5.12 For Fig. 5.15. Defense is getting stronger and stronger

```
SYSTEM TIME =  17:08:02  MAX OVER THE FIELD AT TIME =     2.425   TIME STEP =     194   NT= 303   A1(N)=   .40315967
AT   14 :  14 :  14   UMAX(N)=   964.26373    V=  1002.9659    U=   502.98979
AT   14 :  14 :  14   VMAX(N)=  1002.9659     U=   964.26373    V=   502.98979
AT   15 :  14 :  14   WMAX(N)=   502.98979    V=   964.26373    V=   502.9659
TOTAL NO.OF POINTS WHERE: U IS MAX=       1 . V IS MAX=        1 . W IS MAX=       4
CHANGE OF: UMAX=    -20.312905  . OF VMAX=   -6.2399395e-2  . OF WMAX=  -6.3035126e-2
AT THE PRIMARY SITE:    UMAX=   964.26373    VMAX=  1002.9659    WMAX=   502.98979
AT THE METASTATIC#1 SITE:   UMAX=   735.69906    VMAX=  1002.2636    WMAX=  502.28036
AT THE METASTATIC#2 SITE:   UMAX=   409.73481    VMAX=  1001.262    WMAX=   501.26862
AT THE METASTATIC#3 SITE:   UMAX=   604.76038    VMAX=  1001.8613   WMAX=   501.87394
AT THE METASTATIC#4 SITE:   UMAX=   245.88905    VMAX=  1000.7586   WMAX=   500.76007
UNODE1=   151.28858    UNODE1-1=   3.2878216    UNODE1+1=   3.287822    UNODE2=   151.28868
UNODE2-1=   3.287822    UNODE2+1=   3.287822    U(NX+4,NY+4,NZ+4)=   25
VNODE1=  1000.468    VNODE2=  1000.468    V(NX+4,NY+4,NZ+4)=   1000
WNODE1=   500.46644    WNODE2=   500.46644    W(NX+4,NY+4,NZ+4)=    500
NO. OF POINTS WHERE CANCER IS GROWING OUTSIDE THE LARGEST TUMORS =      1244   LARGEST U OUTSIDE TUMORS=     25
NO.OF POINTS WHERE U(N)>U(N-1)=       0  MAX.CHANGE FROM THE START =  UMAX - UOMAX=   -9035.7363
**********************************************************************************************************************
SYSTEM TIME =  17:08:40  MAX OVER THE FIELD AT TIME =     2.4375   TIME STEP =     195   NT= 303   A1(N)=   .39815154
AT   14 :  14 :  14   UMAX(N)=   944.34222    V=  1002.9047    U=   944.34222    V=   502.92797
AT   15 :  15 :  15   VMAX(N)=  1002.9047     U=   944.34222    V=   502.92797
AT   14 :  14 :  14   WMAX(N)=   502.92797    U=   944.34222    V=  1002.9047
TOTAL NO.OF POINTS WHERE: U IS MAX=       1 . V IS MAX=        8 . W IS MAX=       4
CHANGE OF: UMAX=    -19.92151  . OF VMAX=   -6.1197255e-2  . OF WMAX=  -6.1820739e-2
AT THE PRIMARY SITE:    UMAX=   944.34222    VMAX=  1002.9047    WMAX=   502.92797
AT THE METASTATIC#1 SITE:   UMAX=   720.51276    VMAX=  1002.2169    WMAX=  502.23323
AT THE METASTATIC#2 SITE:   UMAX=   401.28747    VMAX=  1001.2361   WMAX=   501.2424
AT THE METASTATIC#3 SITE:   UMAX=   592.28308    VMAX=  1001.8229   WMAX=   501.83522
AT THE METASTATIC#4 SITE:   UMAX=   240.82279    VMAX=  1000.7431   WMAX=   500.74435
UNODE1=   148.17123    UNODE1-1=   3.2201423    UNODE1+1=   3.2201423    UNODE2=   148.17133
UNODE2-1=   3.2201423    UNODE2+1=   3.2201423    U(NX+4,NY+4,NZ+4)=   25
VNODE1=  1000.4584    VNODE2=  1000.4584    V(NX+4,NY+4,NZ+4)=   1000
WNODE1=   500.45677    WNODE2=   500.45677    W(NX+4,NY+4,NZ+4)=    500
NO. OF POINTS WHERE CANCER IS GROWING OUTSIDE THE LARGEST TUMORS =      1244   LARGEST U OUTSIDE TUMORS=     25
NO.OF POINTS WHERE U(N)>U(N-1)=       0  MAX.CHANGE FROM THE START =  UMAX - UOMAX=   -9055.6578
**********************************************************************************************************************
SYSTEM TIME =  17:09:15  MAX OVER THE FIELD AT TIME =     2.45   TIME STEP =     196   NT= 303   A1(N)=   .39320562
AT   14 :  14 :  14   UMAX(N)=   924.80558    V=  1002.8446    U=   924.80558    V=   502.86734
AT   15 :  15 :  14   VMAX(N)=  1002.8446     U=   924.80558    V=   502.86734
AT   15 :  15 :  14   WMAX(N)=   502.86734    V=   924.80558    V=  1002.8446
TOTAL NO.OF POINTS WHERE: U IS MAX=       1 . V IS MAX=        7 . W IS MAX=       7
CHANGE OF: UMAX=    -19.536644  . OF VMAX=   -6.0015396e-2  . OF WMAX=  -6.0626839e-2
AT THE PRIMARY SITE:    UMAX=   924.80558    VMAX=  1002.8446    WMAX=   502.86734
AT THE METASTATIC#1 SITE:   UMAX=   705.61928    VMAX=  1002.1712    WMAX=  502.18701
AT THE METASTATIC#2 SITE:   UMAX=   393.00257    VMAX=  1001.2106   WMAX=   501.21669
AT THE METASTATIC#3 SITE:   UMAX=   580.04611    VMAX=  1001.7853   WMAX=   501.79725
AT THE METASTATIC#4 SITE:   UMAX=   235.85382    VMAX=  1000.7278   WMAX=   500.72893
UNODE1=   145.11373    UNODE1-1=   3.1537598    UNODE1+1=   3.1537598    UNODE2=   145.11383
UNODE2-1=   3.1537598    UNODE2+1=   3.1537598    U(NX+4,NY+4,NZ+4)=   25
VNODE1=  1000.449    VNODE2=  1000.449    V(NX+4,NY+4,NZ+4)=   1000
WNODE1=   500.44728    WNODE2=   500.44728    W(NX+4,NY+4,NZ+4)=    500
NO. OF POINTS WHERE CANCER IS GROWING OUTSIDE THE LARGEST TUMORS =      1244   LARGEST U OUTSIDE TUMORS=     25
NO.OF POINTS WHERE U(N)>U(N-1)=       0  MAX.CHANGE FROM THE START =  UMAX - UOMAX=   -9075.1944
```

numerical computations). At each metastatic site, they are on the rise. This trend has been averted by the interactions of the drugs.

At $t_n = 194$, cancer is not spreading any more (Table 5.12, Fig. 5.11) and has taken a negative slope. It is continuously decreasing. However, even though the amounts of drugs are increasing at the boundaries, the number of cancer cells hiding at different sites along the boundaries is not diminishing very much.

This is a very intrinsic property of cancer cells. Drugs fail to destroy all of them. They manage to tolerate poisoning and burning. They smartly figure out how and where they could find a safe haven.

From these results, we think physicians may make some estimates regarding their treatments.

At this point, we wanted to look into what is happening as treatment proceeds. So, we have looked into an output with exactly the same inputs as before during the middle of our mathematical treatment. That is given in Fig. 5.12.

At the time step t_{104}, we notice the trend of treatment reveals that although everything is going in the right direction, one tumor at the second metastatic site (black line) is not showing any strong negative slope.

That indicates that during a treatment if results do not show much improvement that does not necessarily mean that treatment is not going to work.

However, we took another run with exactly the same inputs, simply turning the switch of reduction of growth of cancer cells off and at the same time we considered a slower rate of growth of cancer.

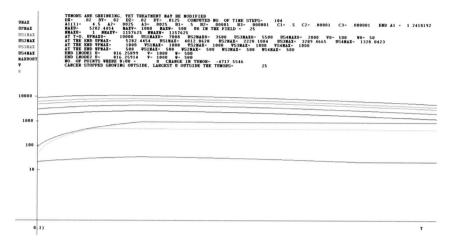

Fig. 5.12   Immunotherapy (V) with a monoclonal drug (W) that reduces A1, strenghens V and kills U

Before, in Fig. 5.12, it was $a_1 = 4.5$ (rate of growth is 350%), whereas in Fig. 5.13 it is only 2.5 (a 150% rate of growth). It failed. That indicates that the reduction of rate of growth of cancer should be a part of the overall treatment.

Let us assume that drugs V and W are administered in such a way that there are always in the blood V_0 and W_0 amounts of these medications. So, in order that treatments will be effective, coefficient of U must satisfy (a necessary condition): $(a_1 - a_2 V_0 - a_3 W_0) < 0$. So, if a_1 is large, larger values of a_2, a_3, V_0, W_0 (all or some according to the treatment procedure will be required).

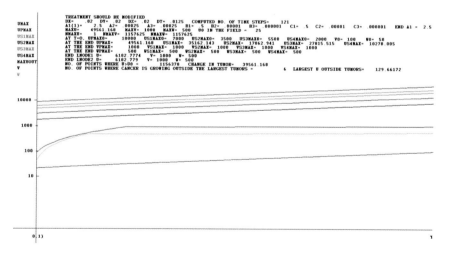

Fig. 5.13   Immunotherapy (V) with a monoclonal drug (W) that strengthens V and kills U

For the second set, (Fig. 5.11) cancer is highly aggressive ($a_1 = 4.5$). Monoclonal drug W is striking its very growth at every time step. $V_0 = 1000$, $W_0 = 500$. They hardly changed throughout the computations. The code was purposely stopped at t_{104}, because cancer must go as may be seen by the coefficient of U.

$$\sigma = (a_1 - a_2 V_0 - a_3 W_0) = 1.2418 - 0.0025 \times 1000 - 0.0025 \times 500 = -2.5082 < 0.$$

Cancer is losing.

In this case (Fig. 5.13), actually the medication aspect of reduction of the growth rate of cancer has been turned off.

Input for Fig. 5.13.

```
••••••••••••••••••••••••••••••••••••••••••••••••••••••••••••••••••••••••••••••••••••••••••••••••••
!TREATMENT WITH IMMUNOTHERAPY AND  MONOCLONAL DRUG
!
! DU/DT = A1(T)•U-A2•U•V - A3•U•W + NU1•DELSQ(U) ! ANTIGEN   . A1(T) = A1(T)•EXP(-DT). DONE BY W
!
! DV/DT = B1•U-B2•V•U + B3•V•W + NU2•DELSQ(V)  !IMMUNE RESPONSE
!
! DW/DT = C1•U -C2•V•U - C3•W•V+ NU3•DELSQ(W) ! MONOCLONAL DRUG
!
! DX=    .02  DY=  .02   DZ=  .02  DT=  .0125   NX= 101  NY= 101   NZ = 101   NT= 120
!
! A1(1)=   2.5   A2=  .00025   A3=  .00025
!
! B1=   .5  B2=  .00001   B3=  .000001   V0= 100
!
! C1=   .5  C2=  .00001   C3=  .000001   V0= 50
!
! AT T=0: PRIMARY U=    10000   1ST UMETA#1 =  7000   2ND UMETA#2=  3500   3RD UMETA#3=  5500   4RTH UMETA#4=  2000
!
! NU1=   .0000001   NU2=  .0000001   NU3=  .0000001
!
! GAMMA1=   .015   GAMMA2=  .015   GAMMA3=  .015
!
! TWO LYMPH NODES ARE LOCATED AT:       ( 3 . 3 . 3 ) AND  ( 35 . 35 . 35 ) SECOND LYMPH NODE IS AFFECTED
! FOUR TUMORS LOCATED AT:            ( 13 . 13 . 13 ) AND ( 26 . 26 . 26 )
!  ( 47 . 47 . 47 )
!  ( 67 . 67 . 67 ):( 79 . 79 . 79 )
••••••••••••••••••••••••••••••••••••••••••••••••••••••••••••••••••••••••••••••••••••••••••••••••••
```

In Fig. 5.13, coefficient of U, $(a_1 - a_2 V_0 - a_3 W_0) = 2.5 - 0.00025(1000 + 500) = 2.125 > 0$. The necessary condition for a successful treatment is violated, so U must increase with time.

Cancer is gaining more strength and is noticeable in Fig. 5.13.

Continuously, while V and W are going to steady state, all four tumors are getting larger. To stop this progress of cancer, if we could strengthen W (or V) by using more dosages of these medications, or using a stronger power of these medications to eliminate U that may serve the purpose, if and only if the oncologists agree that the patient should be able to tolerate that. This we can easily explain without running a new computer code.

Applying the necessary condition, we could compute what should be the most minimum dosages of medications such that cancer could be defeated and that makes

$$\sigma < 0$$

That means conditions for any treatment to work are:

$$a_2 MAXV + a_3 MAXW > a_1$$

Since this is a necessary condition, this is not a guarantee that the treatment shall work.

5.6   More Powerful Immunotherapy

Here, both reaction coefficients of v and w were increased using the following formulas:

$da_2/dt = \xi_1 a_2 v$, meaning that as time moves on, a_2 which is a measure of strength of the immune system that will be increased by immunotherapy. Integrating we get

$$a_2(n) = a_2(n-1)exp\left(\xi_1 \int_{t_{n-1}}^{t_n} vmax \, dt\right)$$

$$= a_2(n-1)exp(\xi_1(\Delta t/2)(vmax(t_n) + vmax(t_{n-1})).$$

We considered $\xi_1 = 0.005$. Also, $vmax(t_n) = max(v_{ijk})$ at $t = t_n$.

Similarly, the monoclonal drug will increase its strength, by the formula $da_3/dt = \xi_2 a_3 w$. We considered $\xi_1 = 0.001$.

It has to be noted that both the drug for immunotherapy and monoclonal drug have an upper limit of their respective strengths. So we have assumed: $a_2 \leq 0.005$ and $a_3 \leq 0.01$.

Here, the monoclonal drug is doing only one job; namely, it is increasing its fighting capabilities.

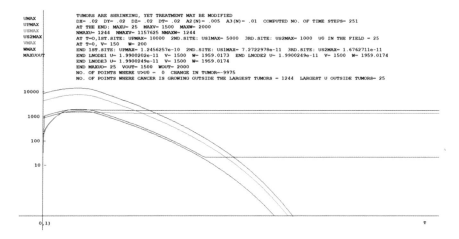

Fig. 5.14 Immunotherapy (V) increases A2(N) & monoclonal drug (W) increases A3(N)

Input for Fig. 5.14.

```
•••••••••••••••••••••••••••••••••••••••••••••••••••••••••••••••••••••••••••••••••••••••••••••••••••
!TREATMENT WITH IMMUNOTHERAPY AND  MONOCLONAL DRUG.HELPING V .
! DU/DT  = A1*U-A2(N)*U*V  - A3(N)*U*W + NU1*DELSQ(U) ! ANTIGEN
! DV/DT  = B1*U-B2*V*U + B3*V*W + NU2*DELSQ(V) ! IMMUNE RESPONSE
! DW/DT  = C1*U -C2*W*U - C3*W*V+ NU3*DELSQ(W) ! MONOCLONAL DRUG HELPING V.
! DX- .02   DY- .02   DZ- .02   NX- 101   NY- 101   NZ - 101   NT- 250
! A1- 4.5   A2(1)- .0005  A3(1)- .0005
! B1- 1    B2- .00001   B3- .001   V0- 150
! C1- 1    C2- .00001   C3- .00001   W0- 200            □
! NU1- .0000001  NU2- .0000001  NU3- .0000001
! GAMMA1- .015  GAMMA2- .015  GAMMA3- .015
! THREE LYMPH NODES ARE LOCATED AT: ( 3 , 3 , 3 ) AND ( 35 , 35 , 35 ) A THIRD LYMPH NODE IS AFFECTED
! THIRD AT  ( 42 , 42 , 42 )ULN10- 1200 ULN20- 1200 ULN30- 1200
! THREE TUMORS LOCATED AT: ( 13 , 13 , 13 ) AND ( 26 , 26 , 26 )
!( 47 , 47 , 47 )
! A2(N) =A2(N-1)*EXP(0.005*(DT/2)*(VMAX(N)+VMAX(N-1))), IF  A2(N) > 0.005 THEN A2(N) - 0.005
! A3(N)- A3(N-1)*EXP(0.001*(DT/2)*(WMAX(N)+WMAX(N-1))), IF A3(N) > 0.01 THEN A3(N) - 0.01
!
•••••••••••••••••••••••••••••••••••••••••••••••••••••••••••••••••••••••••••••••••••••••••••••••••••
```

The formula that we have used is very forceful as we may see in the graph (Fig. 5.14). Certainly, we do not expect the biochemist to follow these mathematical formulas literally. But we expect them to simulate these formulas in the laboratories which should simulate the concepts used in these mathematical formulas.

We are now considering a significantly more serious case (input for Fig. 5.15). At the outset $a_1^0 = 10$. The rate of growth of cancer is 900%. Here practically all methods used before failed drastically. Here, we have considered three properties that a monoclonal drug may use simultaneously:

1. It reduces the rate of growth of cancer cells.
2. It strengthens the abilities of the immune system to destroy cancer cells. One way is by removing the checkpoint brakes.
3. It strengthens its own ability to annihilate cancer cells.

Input for Fig. 5.15, Cancer cells hiding in the boundaries and possibly launching attacks from there.

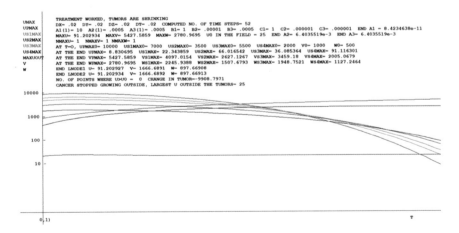

Fig. 5.15 Monoclonal, Drug (W) that strengthens V & ITSELF destroying cancer cells

```
***********************************************************************************************
!TREATMENT WITH IMMUNOTHERAPY WITH A MONOCLONAL DRUG.
! DU/DT = A1(N)*U-A2(N)*U*V - A3(N)*U*W + NU1*DELSQ(U)  ! ANTIGEN   , A1(N) = A1(N)*EXP(-DT*W0), DONE BY W
! A1(N)  = A1(N-1)*EXP(-0.025*DT*W0); A2(N) = A2(N-1)*EXP(0.0025*DT*W0);A3(N) = A3(N-1)*EXP(0.0025*DT*W0)
! DV/DT = B1*U-B2*V*W + B3*V*W + NU2*DELSQ(V)  !IMMUNE RESPONSE
! DW/DT = C1*U -C2*W*U - C3*W*V+ NU3*DELSQ(W)  ! MONOCLONAL DRUG
! DX= .02  DY= .02  DZ= .02  DT= .02  NX= 101  NY= 101  NZ = 101  NT= 300
! A1(1)= 10  A2(1)= .0005  A3(1)= .0005
! B1= 1  B2= .00001  B3= .0005  V0= 1000
! C1= 1  C2= .000001  C3= .0005  W0= 500
! AT T=0: PRIMARY U= 10000  1ST.UMETA#1 = 7000  2ND.UMETA#2= 3500  3RD.UMETA#3= 5500  4RTH.UMETA#4= 2000
! NU1= .0000001  NU2= .0000001  NU3= .0000001
! GAMMA1= .015  GAMMA2= .015  GAMMA3= .015
! TWO LYMPH NODES ARE LOCATED AT: ( 3 , 3 , 3 ) AND  ( 35 , 35 , 35 ) SECOND LYMPH NODE IS AFFECTED
! FIVE TUMORS LOCATED AT: ( 13 , 13 , 13 ) AND ( 26 , 26 , 26 )
!( 47 , 47 , 47 )
!( 67 , 67 , 67 );( 79 , 79 , 79 )
***********************************************************************************************
```

The term b_3 is positive, because the monoclonal drug W is strengthening the defense. It started working almost instantly. At the $t_n = 8$, there is not a single point in the field where $U_{max}^n > U_{max}^{n-1}$.

Here, we have assumed that there are some tissues where cancer cells can hide. So at the very beginning, some cells did hide at these locations and treatment did not cause them any harm. This is indicated by $U(NX + 4, NY + 4, NZ + 4) = 25$. Here, both medications V and W are strongly present. However, they are practically inactive!!!

In the output, we notice that wherever cancer is a max, amounts of drugs are also attaining the max. That means the treatment is going in the right direction: Table 5.13.

Figure 5.15 is self-explanatory. All tumors are dying. Cancer is being defeated. And the drugs V and W are going into a steady state. However, the safe havens of cancer cells (under the skin, at various tissues where no drug is going into) kept on protecting them. So their value, 25, remained unchanged. This is revealed by the bottom redline.

Table 5.13 For Fig. 5.15. Monoclonal drugs are more efficient. Cancer winning nowhere

```
***********************************************************************************************
SYSTEM TIME -18:54:08 TIME STEP =  8  NT= 300  A1(N)= .30197383  A2(N)= 7.0953377e-4  A3(N)= 7.0953377e-4
AT  15 , 15 , 15  UMAX(N)= 10496.997  V= 1788.5677  W= 1251.2547
AT  14 , 14 , 14  VMAX(N)= 1788.5677  U= 10496.997  W= 1251.2547
AT  15 , 15 , 15  WMAX(N)= 1251.2547  U= 1788.5677
TOTAL NO.OF POINTS WHERE: U IS MAX=  1 , V IS MAX=  1 , W IS MAX=  1
CHANGE OF: UMAX=-192.74976 , OF VMAX= 114.43356 , OF WMAX= 105.21213
AT THE PRIMARY SITE: UMAX= 10496.997  VMAX= 1788.5677  WMAX= 1251.2547
AT THE METASTATIC#1 SITE: UMAX= 7430.2324  VMAX= 1558.4467  WMAX= 1028.0395
AT THE METASTATIC#2 SITE: UMAX= 3763.9768  VMAX= 1288.738   WMAX= 765.28247
AT THE METASTATIC#3 SITE: UMAX= 5870.7717  VMAX= 1443.0217  WMAX= 915.7401
AT THE METASTATIC#4 SITE: UMAX= 2162.9693  VMAX= 1172.7339  WMAX= 651.89098
UNODE1= 1301.5513  UNODE1-1= 27.238537  UNODE1+1= 27.238537  UNODE2= 1301.5513
UNODE2-1= 27.238537  UNODE2+1= 27.238537  U(NX+4,NY+4,NZ+4)= 25
VNODE1= 1110.7513  VNODE2= 1110.7513  V(NX+4,NY+4,NZ+4)= 1000
WNODE1= 591.21169  WNODE2= 591.21169  W(NX+4,NY+4,NZ+4)= 500
NO. OF POINTS WHERE CANCER IS GROWING OUTSIDE THE LARGEST TUMORS = 6  LARGEST U OUTSIDE TUMORS= 27.238547
NO.OF POINTS WHERE U(N)>U(N-1)=  0  MAX.CHANGE FROM THE START = UMAX - UOMAX= 496.99654
***********************************************************************************************
SYSTEM TIME -18:54:39 TIME STEP =  9  NT= 300  A1(N)= .18315639  A2(N)= 7.4591235e-4  A3(N)= 7.4591235e-4
AT  15 , 15 , 15  UMAX(N)= 10268.084  V= 1902.3669  W= 1354.2202
AT  14 , 14 , 14  VMAX(N)= 1902.3669  U= 10268.084  W= 1354.2202
AT  15 , 15 , 15  WMAX(N)= 1354.2202  U= 10268.084  V= 1902.3669
TOTAL NO.OF POINTS WHERE: U IS MAX=  1 , V IS MAX=  1 , W IS MAX=  1
CHANGE OF: UMAX=-228.91246 , OF VMAX= 113.79925 , OF WMAX= 102.96545
AT THE PRIMARY SITE: UMAX= 10268.084  VMAX= 1902.3669  WMAX= 1354.2202
AT THE METASTATIC#1 SITE: UMAX= 7295.3584  VMAX= 1639.3318  WMAX= 1101.154
AT THE METASTATIC#2 SITE: UMAX= 3711.9394  VMAX= 1330.7331  WMAX= 802.45668
AT THE METASTATIC#3 SITE: UMAX= 5775.0431  VMAX= 1507.3059  WMAX= 973.60066
AT THE METASTATIC#4 SITE: UMAX= 2137.1224  VMAX= 1197.8935  WMAX= 673.28536
UNODE1= 1287.2881  UNODE1-1= 26.980833  UNODE1+1= 26.980833  UNODE2= 1287.2881
UNODE2-1= 26.980833  UNODE2+1= 26.980833  U(NX+4,NY+4,NZ+4)= 25
VNODE1= 1126.8863  VNODE2= 1126.8863  V(NX+4,NY+4,NZ+4)= 1000
WNODE1= 604.09299  WNODE2= 604.09299  W(NX+4,NY+4,NZ+4)= 500
NO. OF POINTS WHERE CANCER IS GROWING OUTSIDE THE LARGEST TUMORS = 6  LARGEST U OUTSIDE TUMORS= 26.980846
NO.OF POINTS WHERE U(N)>U(N-1)=  0  MAX.CHANGE FROM THE START = UMAX - UOMAX= 268.08408
```

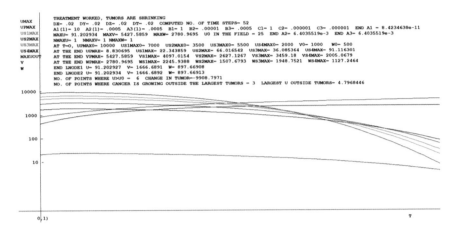

Fig. 5.16 Monoclonal, drug (W) that strengthens V & destroying cancer cells

The bottom redline shows cancer cells are unharmed all through.

In Fig. 5.16, that protection has been removed. No more cancer can hide. Then all over the field, including at the boundaries, cancer cells are dying. The largest accumulation of cancer cells was 25 in Fig. 5.15; now it is merely 3. If we do more computations, that will be 0.

Obviously, these results are attempts to see how mathematically this could be rendered possible.

Input for Fig. 5.16 (Same as in Fig. 5.14)

```
***********************************************************************************
!TREATMENT WITH IMMUNOTHERAPY WITH A MONOCLONAL DRUG.
! DU/DT - A1(N)*U-A2(N)*U*V - A3(N)*U*W + NU1*DELSQ(U) ! ANTIGEN   , A1(N) - A1(N)*EXP(-DT*W0), DONE BY W
! A1(N) - A1(N-1)*EXP(-0.025*DT*W0); A2(N)  - A2(N-1)*EXP(0.0025*DT*W0);A3(N) - A3(N-1)*EXP(0.0025*DT*W0)
! DV/DT - B1*U-B2*V*U + B3*V*W + NU2*DELSQ(V) !IMMUNE RESPONSE
! DW/DT - C1*U -C2*W*U - C3*W*V+ NU3*DELSQ(W) ! MONOCLONAL DRUG, IT ATTACKS CANCER EVEN AT THE BOUNDARY
! DX- .02  DY- .02  DZ- .02  DT- .02  NX- 101  NY- 101  NZ - 101  NT- 300
! A1(1)- 10  A2(1)- .0005  A3(1)- .0005
! B1- 1  B2- .00001  B3- .0005  V0- 1000
! C1- 1  C2- .000001  C3- .000001  W0- 500
! AT T=0: PRIMARY U= 10000  1ST.UMETA#1 - 7000   2ND.UMETA#2- 3500  3RD.UMETA#3- 5500   4RTH.UMETA#4- 2000
! NU1- .0000001  NU2- .0000001  NU3- .0000001
! GAMMA1- .015  GAMMA2- .015  GAMMA3- .015
! TWO LYMPH NODES ARE LOCATED AT: ( 3 , 3 , 3 ) AND ( 35 , 35 , 35 ) SECOND LYMPH NODE IS AFFECTED
! FIVE TUMORS LOCATED AT:   ( 13 , 13 , 13 ) AND ( 26 , 26 , 26 )
!( 47 , 47 , 47 )
!( 67 , 67 , 67 );( 79 , 79 , 79 )
***********************************************************************************
```

Here, we have used a rather simple technique to reach the boundaries through computations. That is, we have used projection of the values nearest to the boundaries on the boundaries. So, we have made an assumption that the defense and the drugs do reach the boundaries when they are closest to the boundaries. Hideouts of cancer cells have been attacked. These can be attained during targeted immune treatment. Just like cancer cells can fool the T-cells and B-cells by releasing various proteins, similarly man-made antibodies could fool the cancer cells by releasing antibodies that attract them and destroy them. So hideouts and tricks of the antigens will be eradicated. Here that scheme of fight has been meticulously adopted mathematically. The bottom redline demonstrates a negative slope.

Now war is at every nook and corner. Boundaries are no longer safe shelters of the malignant cells.

After this, there is no need to discuss any further on the applications of monoclonal drugs. Yet we need to go a bit further. The fact of the matter is the stronger the drugs, the stronger are the side effects in general.

So, we should discuss a few more applications of monoclonal drugs.

First, let us repeat the concept that safe havens of antigens could be destroyed.

5.7   Boundaries Are No Longer Safe Havens for Malignant Cells

For all previous cases, boundaries in the computational fields were kept as safe havens for cancer cells to simulate the actual facts that cancer cells may safely hide in adipose tissues and under the skin of the patients. Now that game is over. The fight is going on in the computational field at every point including the boundaries. There are several ways to do that computationally. We should look into this because it is indeed the very nature of breast cancer cells to metastasize.

"It is clear that the manner in which the body responds to certain breast malignancies creates a systemic environment that is amenable to the outgrowth of microscopic disseminated tumors that would otherwise have remained indolent" [19]. In [20], we can see the same concept.

The Duke researchers say some of these metastatic cells seek out the molecule E-selectin, which is often found in bone marrow. The cancer cells bind to E-selectin, allowing them to enter the marrow. There, another protein called CXCR4 allows the cells to nestle into the marrow, where they may hide for years. An E-selectin inhibitor called GMI-1271 may prevent the metastatic cells from entering the bone marrow. Researchers also believe the drug plerixafor (Mozobil®), which may be used to treat multiple myeloma and non-Hodgkin lymphoma, may act as a CXCR4 blocker and flush out hidden cells, exposing them to treatment. "We are hopeful that by understanding how these breast cancer cells migrate through the body and what their life cycle is, we can discover ways to make them more vulnerable and treatable," says Dr. Sipkins. The reference is given below: [https://www.cancercenter.com/community/blog/2016/12/how-does-cancer-do-that-attacking-cancer-cells-where-they-hide].

Researchers at MIT found that some cancer cells escape to the thymus and find safe havens. "Now, in a study of mice with lymphoma, MIT biologists have discovered that a few cancer cells escape chemotherapy by hiding out in the thymus, an organ where immune cells mature. In the thymus, the cancer cells are bathed in growth factors that protect them from the drugs' effects" [Ref: https://www.technologyreview.com/2010/12/21/198298/a-hideout-for-cancer-cells/.].

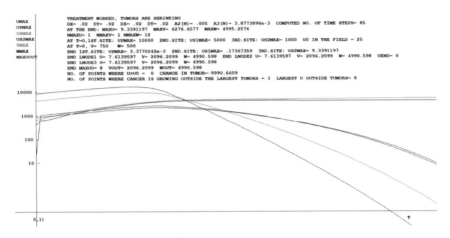

Fig. 5.17 Entire field is a warzone v increases A2 (N) & monoclonal w increases A3(N)

We need to look into this computationally. So, we have simulated this by making small changes in the code where drugs go everywhere and on the extreme boundaries there is no flux.

The inputs are given in the INPUT for Fig. 5.17. We have considered $a_1 = 4.5$.

In this case, medications for immunotherapy increase $a_2(n)$, as time changes. It is the strength of v, to kill cancer cells. Also $a_3(n)$ the cancer destroying capability of w, the monoclonal drug is being increased with time. The formulas for $a_2(n)$ are given by:

$$da_2/dt = \xi_1 \max v(n)a_2 \quad \text{and} \quad da_3/dt = \xi_2 \max w(n)a_3,$$ where $\max v(n) = \max(v_{ijk})$ over the entire field at time $t = t_n$. The same is true for $\max w(n)$, and both ξ_1 and ξ_2 are positive. We have chosen $\xi_1 = \xi_2 = 0.005$. Integrating them from t_{n-1} to t_n, we get:

$$a_2(n) = a_2(n-1)exp(\int_{t_{n-1}}^{t_n} \xi_1 \max v(n)dt) \quad \text{and}$$

$$a_3(n) = a_3(n-1)exp(\int_{t_{n-1}}^{t_n} \xi_2 \max w(n)dt).$$ We have integrated them by trapezoidal rule and used them in the code.

They are not fixed amounts and are being slowly injected as the patient gets better and better.

Input for Fig. 5.17. Defense is strengthened by Monoclonal Drugs.

```
!•••••••••••••••••••••••••••••••••••••••••••••••••••••••••••••••••••••••••••••••••••••••
!ENTIRE FIELD IS A WAR ZONE TREATMENT WITH IMMUNOTHERAPY AND  MONOCLONAL DRUG.HELPING V .
! CHARLIE-SIMPSON USED INSIDE THE FIELD AND ONLY CHARLIE WAS USED FOR THE FIGHT ON THE BOUNDARIES
! DU/DT = A1*U-A2(N)*U*V - A3(N)*U*W + NU1*DELSQ(U) ! ANTIGEN
! DV/DT = B1*U-B2*V*U + B3*V*W + NU2*DELSQ(V) !IMMUNE RESPONSE
! DW/DT = C1*U -C2*W*U - C3*W*V+ NU3*DELSQ(W) ! MONOCLONAL DRUG HELPING V.
! DX=   .02   DY= .02  DZ=  .02   DT= .02  NX= 101  NY= 101  NZ = 101  NT= 350
! A1=   4.5   A2(1)= .00012   A3(1)= .00012
! B1=   1  B2= .00001   B3= .00005   V0= 750
! C1=   1  C2= .00001   C3= .00001   W0= 500
! NU1=  .0000001  NU2= .0000001  NU3= .0000001
! GAMMA1=  .015  GAMMA2= .015  GAMMA3= .015
! THREE LYMPH NODES ARE LOCATED AT:   ( 3 , 3 , 3 ) AND  ( 35 , 35 , 35 ) A THIRD LYMPH NODE IS AFFECTED
! THIRD AT   ( 42 , 42 , 42 )ULN10= 1200 ULN20= 1200 ULN30= 1200
! THREE TUMORS LOCATED AT:   ( 13 , 13 , 13 ) AND ( 26 , 26 , 26 )
! ( 47 , 47 , 47 )
! A2(N)=A2(N-1)*EXP(0.005*(DT/2)*(VMAX(N)+VMAX(N-1))). IF A2(N) > 0.005 THEN A2(N) = 0.005
! A3(N)= A3(N-1)*EXP(0.001*(DT/2)*(WMAX(N)+WMAX(N-1))). IF A3(N) > 0.01 THEN A3(N) = 0.01
!•••••••••••••••••••••••••••••••••••••••••••••••••••••••••••••••••••••••••••••••••••••••••
```

Here, the necessary and sufficient condition that the treatment will work is:

$$\sigma = (a_1 - a_2(n)V(n) - a_3(n)W(n)) < 0 \text{ as } n \to \infty.$$

Figure 5.17 reveals that cancer is conquered. This is true that it is just a mathematical model. But such a model may someday be materialized in reality. However, we have to keep in mind that even when such medications are manufactured, they may not be applicable to all patients because of health restrictions.

5.8 More Applications of Monoclonal Drugs

In this section, one more forward step has been undertaken. Both the medications V and W were used to reduce the coefficient of dispersion. Cancer cells do not just move through the flow of blood and lymph. As they penetrate the tissues, they aggressively cut the barriers between the cells, destroying the law of homeostasis, and move through interstitial fluid to expand their bases. Mathematically, this is dispersion, and if it could be slowed down, cancer will lose its ability to spread. Both V and W are applied to do this job. Also they will be moving toward the tumor sites which are heavily attacked by malignant cells.

Input for Fig. 5.18 Defense strengthened by Monoclonal Drugs.

```
!•••••••••••••••••••••••••••••••••••••••••••••••••••••••••••••••••••••••••••••••••••••••••
! ENTIRE FIELD IS A WARZONE. TREATMENT WITH IMMUNOTHERAPY AND  MONOCLONAL DRUG.HELPING V .
! IMMUNOTHERAPY IMPROVES THE PERFORMANCE OF V, MONOCLONAL INCREASES V AND STRENGTHENS ITS CAPACITY TO FIGHT CANCER.
! THEY BOTH WEAKEN CANCER'S ABILITY TO SPREAD.
! DU/DT = A1*U-A2(N)*U*V - A3(N)*U*W + NU1*DELSQ(U) ! ANTIGEN
! DV/DT = B1*U-B2*V*U + B3*V*W + NU2*DELSQ(V) !IMMUNE RESPONSE
! DW/DT = C1*U -C2*W*U - C3*W*V+ NU3*DELSQ(W) ! MONOCLONAL DRUG HELPING V.
! DX= .02  DY= .02  DZ= .02  DT= .02  NX= 101  NY= 101  NZ = 101  NT= 350
! A1= 4.5  A2(1)= .0001  A3(1)= .0001
! B1= .5  B2= .00001  B3= .001  V0= 750
! C1= .5  C2= .00001  C3= .00001  W0= 500
! NU1= .0000001  NU2= .0000001  NU3= .0000001
! GAMMA1= .015  GAMMA2= .015  GAMMA3= .015
! THREE LYMPH NODES ARE LOCATED AT: ( 3 , 3 , 3 ) AND  ( 35 , 35 , 35 ) A THIRD LYMPH NODE IS AFFECTED
! THIRD AT ( 42 , 42 , 42 )ULN10= 1200 ULN20= 1200 ULN30= 1200
! THREE TUMORS LOCATED AT: ( 13 , 13 , 13 ) AND ( 26 , 26 , 26 )
!( 47 , 47 , 47 )
! A2(N)=A2(N-1)*EXP(0.005*(DT/2)*(VMAX(N)+VMAX(N-1))), IF  A2(N) > 0.005 THEN A2(N) = 0.005
! A3(N)= A3(N-1)*EXP(0.001*(DT/2)*(WMAX(N)+WMAX(N-1))), IF A3(N) > 0.01 THEN A3(N) = 0.01
! IF (VMAX(N) > WMAX(N) THEN NU1=NU1*EXP(-(WMAX(N)/VMAX(N)))
! IF VMAX(N) < WMAX(N) THEN NU1 = NU1*EXP(-(VMAX(N)/WMAX(N)))
!
!•••••••••••••••••••••••••••••••••••••••••••••••••••••••••••••••••••••••••••••••••••••••••
```

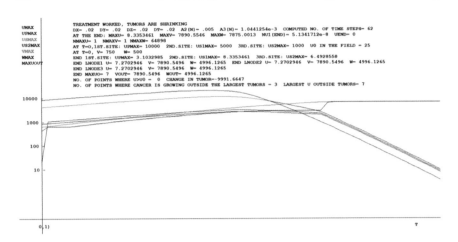

Fig. 5.18 The entire field including the boundaries is a war zone V increases A2(N) & W increases A3(N)

But all these should have medical/physiological limitations. Those are also stated in the inputs in INPUTS for Fig. 5.11. At the beginning, $\sigma = 4.5 - 0.0001(750 + 500) = 4.375$. So, cancer cells must increase. And they did.

Monoclonal drugs are not reducing a_1. They are simply increasing the strengths of a_2 and a_3 *slowly as seen in the TABLE*. This should be precalculated by the oncologists: Table 5.14.

Table 5.14 For Fig. 5.18. Defense strengthened. Defense, given by the values of VMAX(N) and WMAX(N), is still the strongest at the primary site (14,14,14)

```
*************************************************************************************************
SYSTEM TIME -18:26:36 TIME = .12 A2(N)= 1.2894378e-4 A3(N)= 1.0412167e-4 NU1= 5.1341712e-8 TIME STEP = 6 NT= 350
AT 42 , 42 , 42 MAXUO(N)= 1496 V= 803.70963 W= 533.48199 NOUT= 3
AT 14 , 14 , 14 UMAX(N)= 12450.904 V= 1058.5351 W= 779.46028
AT 14 , 14 , 14 VMAX(N)= 1058.5351 U= 12450.904 W= 779.46028
AT 14 , 14 , 14 WMAX(N)= 779.46028 U= 12450.904 V= 1058.5351
TOTAL NO.OF POINTS WHERE: U IS MAX= 1 , V IS MAX= 1 , W IS MAX= 1
CHANGE OF: UMAX= 529.28784 , OF VMAX= 68.498947 , OF WMAX= 13.811663
AT THE PRIMARY SITE: UMAX= 12450.904 VMAX= 1058.5351 WMAX= 779.46028
AT THE METASTATIC#1 SITE: UMAX= 6230.9648 VMAX= 913.66879 WMAX= 639.90979
AT THE METASTATIC#2 SITE: UMAX= 1247.0766 VMAX= 797.92799 WMAX= 527.87393
UNODE1= 1496.3809 UNODE1-1= 31.182304 UNODE1+1= 31.182304 UNODE2= 1496.3809
UNODE2-1= 31.182304 UNODE2+1= 31.182304 U(NX+4,NY+4,NZ+4)= 30.964511
VNODE1= 803.70963 VNODE2= 803.70963 V(NX+4,NY+4,NZ+4)= 769.6273
WNODE1= 533.48199 WNODE2= 533.48199 W(NX+4,NY+4,NZ+4)= 769.6273
UNODE3= 1496.3809 VNODE3= 803.70963 WNODE3= 533.48199
NO. OF POINTS WHERE CANCER IS GROWING OUTSIDE THE LARGEST TUMORS = 3 LARGEST U OUTSIDE TUMORS= 1496
NO.OF POINTS WHERE U(N)>U(N-1)= 1157625 MAX.CHANGE FROM THE START = UMAX - UOMAX= 2450.9039
*************************************************************************************************
SYSTEM TIME -18:27:06 TIME = .14 A2(N)= 1.3595219e-4 A3(N)= 1.0493643e-4 NU1= 5.1341712e-8 TIME STEP = 7 NT= 350
AT 42 , 42 , 42 MAXUO(N)= 1564 V= 815.72724 W= 541.10802 NOUT= 3
AT 14 , 14 , 14 UMAX(N)= 13000.453 V= 1130.8661 W= 842.78289
AT 14 , 14 , 14 VMAX(N)= 1130.8661 U= 13000.453 W= 842.78289
AT 14 , 14 , 14 WMAX(N)= 842.78289 U= 13000.453 V= 1130.8661
TOTAL NO.OF POINTS WHERE: U IS MAX= 1 , V IS MAX= 1 , W IS MAX= 1
CHANGE OF: UMAX= 549.54914 , OF VMAX= 72.331071 , OF WMAX= 63.322611
AT THE PRIMARY SITE: UMAX= 13000.453 VMAX= 1130.8661 WMAX= 842.78289
AT THE METASTATIC#1 SITE: UMAX= 6508.6986 VMAX= 951.62838 WMAX= 671.69874
AT THE METASTATIC#2 SITE: UMAX= 1303.0981 VMAX= 808.58563 WMAX= 534.22447
UNODE1= 1563.5635 UNODE1-1= 32.585735 UNODE1+1= 32.585739 UNODE2= 1563.5635
UNODE2-1= 32.585739 UNODE2+1= 32.585739 U(NX+4,NY+4,NZ+4)= 32.318706
VNODE1= 815.72724 VNODE2= 815.72724 V(NX+4,NY+4,NZ+4)= 773.63351
WNODE1= 541.10802 WNODE2= 541.10802 W(NX+4,NY+4,NZ+4)= 773.63351
UNODE3= 1563.5635 VNODE3= 815.72724 WNODE3= 541.10802
NO. OF POINTS WHERE CANCER IS GROWING OUTSIDE THE LARGEST TUMORS = 3 LARGEST U OUTSIDE TUMORS= 1564
NO.OF POINTS WHERE U(N)>U(N-1)= 1157625 MAX.CHANGE FROM THE START = UMAX - UOMAX= 3000.453
```

Table 5.15 For Fig. 5.18. Defense destroying tumors, yet maintaining a strong presence at those sites. Cancer is winning nowhere now (nowhere U(N) > U(N-1)). Cancer declining

```
**************************************************************************************************
  SYSTEM TIME =18:53:49 TIME = 1.2 A2(N)= .005 A3(N)= 8.9197208e-4 NU1= 5.1341712e-8  TIME STEP =  60  NT= 350
  AT  42 , 42 , 42  MAXUO(N)= 14  V= 7890.581  W= 4996.1556     NOUT= 3
  AT  29 , 29 , 29  UMAX(N)= 15.937714  V= 7890.5906  W= 4996.1645
  AT  29 , 29 , 29  VMAX(N)= 7890.5906  U= 15.937714  W= 4996.1645
  AT 105 , 105 , 105  WMAX(N)= 7875.0026  U= 0  V= 7875.0026
  TOTAL NO.OF POINTS WHERE: U IS MAX=  1 , V IS MAX=  1 , W IS MAX=  64898
  CHANGE OF: UMAX=-6.0298488 , OF VMAX=-.02858233 , OF WMAX=-9.5956885e-4
  AT THE PRIMARY SITE: UMAX= 5.9336866  VMAX= 7890.5434  WMAX= 4996.1208
  AT THE METASTATIC#1 SITE: UMAX= 15.937714  VMAX= 7890.5906  WMAX= 4996.1645
  AT THE METASTATIC#2 SITE: UMAX= 12.41469  VMAX= 7890.5739  WMAX= 4996.1491
  UNODE1= 13.901318  UNODE1-1= .43606726  UNODE1+1= .4360971  UNODE2= 13.901318
  UNODE2-1= .4360971  UNODE2+1= .4360971  U(NX+4,NY+4,NZ+4)=-.2251514
  VNODE1= 7890.581  VNODE2= 7890.581  V(NX+4,NY+4,NZ+4)= 7875.0026
  WNODE1= 4996.1556  WNODE2= 4996.1556  W(NX+4,NY+4,NZ+4)= 7875.0026
  UNODE3= 13.901318  VNODE3= 7890.581  WNODE3= 4996.1556
  NO. OF POINTS WHERE CANCER IS GROWING OUTSIDE THE LARGEST TUMORS = 3  LARGEST U OUTSIDE TUMORS= 14
  NO.OF POINTS WHERE U(N)>U(N-1)=  0  MAX.CHANGE FROM THE START = UMAX - UOMAX=-9984.0623
  **************************************************************************************************
  SYSTEM TIME =18:54:19 TIME = 1.22 A2(N)= .005 A3(N)= 9.6505477e-4 NU1= 5.1341712e-8  TIME STEP =  61  NT= 350
  AT  42 , 42 , 42  MAXUO(N)= 10  V= 7890.5628  W= 4996.1388     NOUT= 3
  AT  29 , 29 , 29  UMAX(N)= 11.538821  V= 7890.5697  W= 4996.1452
  AT  29 , 29 , 29  VMAX(N)= 7890.5697  U= 11.538821  W= 4996.1452
  AT 105 , 105 , 105  WMAX(N)= 7875.0019  U= 0  V= 7875.0019
  TOTAL NO.OF POINTS WHERE: U IS MAX=  1 , V IS MAX=  1 , W IS MAX=  64898
  CHANGE OF: UMAX=-4.3988935 , OF VMAX=-2.0833475e-2 , OF WMAX=-7.0116425e-4
  AT THE PRIMARY SITE: UMAX= 4.2959653  VMAX= 7890.5356  WMAX= 4996.1136
  AT THE METASTATIC#1 SITE: UMAX= 11.538821  VMAX= 7890.5697  WMAX= 4996.1452
  AT THE METASTATIC#2 SITE: UMAX= 8.9881942  VMAX= 7890.5577  WMAX= 4996.1341
  UNODE1= 10.064463  UNODE1-1= .31571069  UNODE1+1= .31573349  UNODE2= 10.064463
  UNODE2-1= .31573349  UNODE2+1= .31573349  U(NX+4,NY+4,NZ+4)=-.16494364
  VNODE1= 7890.5628  VNODE2= 7890.5628  V(NX+4,NY+4,NZ+4)= 7875.0019
  WNODE1= 4996.1388  WNODE2= 4996.1388  W(NX+4,NY+4,NZ+4)= 7875.0019
  UNODE3= 10.064463  VNODE3= 7890.5628  WNODE3= 4996.1388
  NO. OF POINTS WHERE CANCER IS GROWING OUTSIDE THE LARGEST TUMORS = 3  LARGEST U OUTSIDE TUMORS= 10
  NO.OF POINTS WHERE U(N)>U(N-1)=  0  MAX.CHANGE FROM THE START = UMAX - UOMAX=-9988.4612
```

At the time step $t_n = 61, \sigma = 4.5 - 0.005 \times 7890.56 - 9.6505477 \times 10^{-4} \times 4996 = -39.164 < 0$. So, cancer starts decreasing. Here, we also notice that as cancer is increasing (runaway malignant cells) at $(NX + 4, NY + 4, NZ + 4)$, the values of V and W are also increasing there: Table 5.14.

At the point where U is a max, at (29,29,29), second tumor, exactly at that point both V and W attained their max. The kinematic coefficient of viscosity is reduced slightly from $\kappa_1 = 10^{-7}$ to 5.1×10^{-8}. Here, interpretation is entirely different. Here, it means that both immunotherapy and monoclonal drugs are preventing flux, thereby keeping cancer contained. Both V and W are dominating the entire field including the boundaries. They have destroyed the hideouts of cancer cells.

Figure 5.18 shows all tumors and $MAXUOUT$, giving: Max value of the cancer cells outside all the sites of cancer is taking a drastic dive down with large negative slopes. Here, the warzone includes the boundaries. So, cancer cells have no hideouts.

Input For Fig. 5.19. Defense strengthened.

```
************************************************************************************************
! ENTIRE FIELD IS A WARZONE. TREATMENT WITH IMMUNOTHERAPY AND MONOCLONAL DRUG.HELPING V .
! IMMUNOTHERAPY IMPROVES THE PERFORMANCE OF V, MONOCLONAL INCREASES V AND STRENGTHENS ITS CAPACITY TO FIGHT CANCER.
! THEY BOTH WEAKEN CANCER'S ABILITY TO SPREAD.
! DU/DT = A1*U-A2(N)*U*V - A3(N)*U*W + NU1*DELSQ(U)  ! ANTIGEN
! DV/DT = B1*U-B2*V*U + B3*V*W + NU2*DELSQ(V)  !IMMUNE RESPONSE
! DW/DT = C1*U -C2*W*U - C3*W*V+ NU3*DELSQ(W)  ! MONOCLONAL DRUG HELPING V.
! DX= .02  DY= .02  DZ= .02  DT= .02  NX= 101  NY= 101  NZ = 101  NT= 350
! A1= 4.5  A2(1)= .0001  A3(1)= .0001
! B1= .5  B2= .00001  B3= .001  V0= 750
! C1= .5  C2= .00001  C3= .00001  W0= 500
! NU1(1)= .00001  NU2= .0000001  NU3= .0000001
! GAMMA1= .015  GAMMA2= .015  GAMMA3= .015
! THREE LYMPH NODES ARE LOCATED AT: ( 3 , 3 , 3 ) AND  ( 35 , 35 , 35 ) A THIRD LYMPH NODE IS AFFECTED
! THIRD AT  ( 42 , 42 , 42 )ULN10= 1200 ULN20= 1200 ULN30= 1200
! THREE TUMORS LOCATED AT: ( 13 , 13 , 13 ) AND ( 26 , 26 , 26 )
!( 47 , 47 , 47 )
! A2(N)=A2(N-1)*EXP(0.005*(DT/2)*(VMAX(N)+VMAX(N-1))),  IF  A2(N) > 0.005 THEN A2(N) = 0.005
! A3(N)= A3(N-1)*EXP(0.001*(DT/2)*(WMAX(N)+WMAX(N-1)), IF A3(N) > 0.01 THEN A3(N) = 0.01
! IF VMAX(N) > WMAX(N) THEN NU1=NU1*EXP(-0.1*(WMAX(N)/VMAX(N)))
! IF VMAX(N) < WMAX(N) THEN  NU1 = NU1*EXP(-0.1*(VMAX(N)/WMAX(N)))
!
************************************************************************************************
```

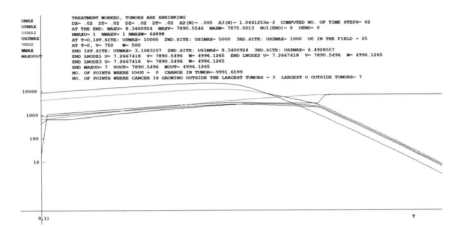

Fig. 5.19 The entire field including the boundaries is a war zone V increases A2(N) & W increases A3(N)

Here, V is a form of immunotherapy which strengthens the immune system by increasing its strength. This is represented by the formula:
$a_2^n = a_2^{n-1} exp(0.005(\Delta t/2)) \times (VMAX(n) + VMAX(n-1))$. However, it has a strict limitation. If $a_2^n > 0.005$, then
a_2^n is set equal to 0.005. That means a_2^n must not exceed 0.005.

A similar restriction has been imposed on a_3^n as given in the INPUTS for Fig. 5.19. Also, values of κ_1 (in the code it is NU1) are connected with the max values of $VMAX$ and $WMAX$.

The results are very similar as before that may be observed in Fig. 5.18.

In comparison with Fig. 5.18, we find that the results given by Fig. 5.19 are well expected.

5.9 A Very Severe Case

Here $a_1 = 10$. We have already considered this strength of cancer in Fig. 5.16. We will study another approach where a monoclonal drug is preventing cancer cells from escaping from any site where it is present. Cancer is increasing 900%. What we have found here is that the sooner the dispersion of cancer cells is stopped, treatment could get much better. We will justify this by considering the necessary condition that the treatment should work: $\sigma = $ coefficient of $u_{ijk}^n = \left(a_1 - a_2(n)v_{ijk}^n - a_3(n)w_{ijk}^n \right) < 0$, where $a_2(n)$ is controlled by a different immunotherapy, $a_3(n)$ is controlled by the monoclonal drug w, and both control the flux of cancer which causes metastasization. At the time step N = 57 (Fig. 5.20), $a_1 = 10, a_2(n) = 0.005, maxv(n) = 7890.56, a_3(n) = 0.0018283, maxw(n) = 7875$. That gives $\sigma = -43.85 < 0$.

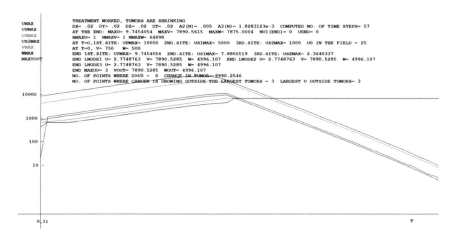

Fig. 5.20 The entire field & the boundaries = war zone V increases A2(N) & W increases A3(N), they reduce NU1

So the treatment did work. Again, this is an extreme case. The computer outputs showed that at the time step $N = 25$, at the primary site $UMAX = 66263.71$, which means it has gone up by 56,263.71 from its initial value which was 10,000 and outside in the field, $UMAX = 12030$, which was initially only 25.

So, the battle was tough and the victory was ceremonious. The inputs are given in the INPUTS for Fig. 5.20.

Input for Fig. 5.20. Very aggressive cancer. Defense appropriately strengthened.

```
****************************************************************************************************
! ENTIRE FIELD IS A WARZONE. TREATMENT WITH IMMUNOTHERAPY AND  MONOCLONAL DRUG.HELPING V .
! IMMUNOTHERAPY IMPROVES THE PERFORMANCE OF V, MONOCLONAL INCREASES V AND STRENGTHENS ITS CAPACITY TO FIGHT CANCER.
! THEY BOTH WEAKEN CANCER'S ABILITY TO SPREAD.
! DU/DT = A1*U-A2(N)*U*V - A3(N)*U*W + NU1*DELSQ(U) ! ANTIGEN
! DV/DT = B1*U-B2*V*U + B3*V*W + NU2*DELSQ(V) !IMMUNE RESPONSE
! DW/DT = C1*U -C2*W*U - C3*W*V+ NU3*DELSQ(W) ! MONOCLONAL DRUG HELPING V.
! DX= .02  DY= .02  DZ= .02  DT= .02  NX= 101  NY= 101  NZ = 101  NT= 350
! A1= 10  A2(1)= .0001  A3(1)= .0001
! B1= .5  B2= .00001  B3= .001  V0= 750
! C1= .5  C2= .00001  C3= .00001  W0= 500
! NU1(1)= .00001  NU2= .0000001  NU3= .0000001
! GAMMA1= .015  GAMMA2= .015  GAMMA3= .015
! THREE LYMPH NODES ARE LOCATED AT: ( 3 , 3 , 3 ) AND  ( 35 , 35 , 35 ) A THIRD LYMPH NODE IS AFFECTED
! THIRD AT  ( 42 , 42 , 42 )ULN10= 1200 ULN20= 1200 ULN30= 1200
! THREE TUMORS LOCATED AT:  ( 13 , 13 , 13 ) AND ( 26 , 26 , 26 )
!( 47 , 47 , 47 )
! A2(N)=A2(N-1)*EXP(0.005*(DT/2)*(VMAX(N)+VMAX(N-1))),  IF  A2(N) > 0.005 THEN A2(N) = 0.005
! A3(N)= A3(N-1)*EXP(0.001*(DT/2)*(WMAX(N)+WMAX(N-1))), IF A3(N) > 0.01 THEN A3(N) = 0.01
! IF VMAX(N) > WMAX(N) THEN NU1=NU1*EXP(-0.1*(WMAX(N)/VMAX(N)))
! IF VMAX(N) < WMAX(N) THEN NU1 = NU1*EXP(-0.1*(VMAX(N)/WMAX(N)))
!
****************************************************************************************************
```

5.10   Conclusion

All mathematical possibilities could be rendered practical by biochemists expertizing in oncology.

There are several very interesting results that are predominantly present in all these computer runs.

1. Wherever cancer cells are growing most, at the same points immune defense and monoclonal drugs are at their highest concentrations.
2. The moment cancer is attacked, it tries to move to different locations very fast.
3. Cancer cells look for more hideouts as treatment begins.
4. Immunotherapeutic drugs and monoclonal drugs are always dominating the entire field of computation.
5. Among the uses of V and W (the monoclonal antibodies), W appears to be more effective than V, even though they have the same dosages and almost the same strengths.
6. All the graphs reveal that CDey-Simpson explicit finite difference scheme is a powerful explicit finite difference scheme to solve stiff equations.

A few more information about monoclonal drugs are: "It has been more than three decades since the first monoclonal antibody was approved by the United States Food and Drug Administration (US FDA) in 1986, and during this time, antibody engineering has dramatically evolved. Current antibody drugs have increasingly fewer adverse effects due to their high specificity. As a result, therapeutic antibodies have become the predominant class of new drugs developed in recent years" [16].

What we have done in this chapter regarding the multiple tasks that monoclonal drugs are doing in the models are: (1) They (coat cancer antigens and) flag the immune system to attack these cancer cells, (2) block the rate of growth of cancer cells, (3) directly attack cancer cells, (4) make radiation therapy more effective, (5) deliver chemo to cancer cells, (6) block angiogenesis, and (7) can inhibit functions of proteins that could block activities of the immune system.

Mayo Clinic also stated that "Monoclonal antibodies are designed to function in different ways. A particular drug may actually function by more than one means." We have attempted to model this concept in this chapter. Several monoclonal drugs are designed by researchers at different laboratories around the world. This is an ongoing research. "Conjugated monoclonal antibodies carry a radioactive substance, drug, or toxin that kills cancer cells recognized by the antibody. Antibody–drug conjugates, antibodies that have chemotherapy drugs attached to them, target the surface of cancer cells and deliver the toxic substance to that specific area. This approach can eliminate some of the side effects of chemotherapy, which can damage healthy cells when used as a single agent.

Radioimmunotherapy is a treatment that uses monoclonal antibodies in combination with radiation. By attaching a monoclonal antibody to a radioactive molecule, this technique can deliver a dose of radiation therapy directly to tumor cells." Reference is given below: https://blog.dana-farber.org/insight/2018/03/monoclonal-antibody-therapy-cancer/.

Radioimmunotherapy can eliminate most, if not all side effects of radiation therapy. This concept has been applied in this chapter and in Chap. 7.

Depending upon the state of health of a patient, all drugs are applied accordingly.

References

1. Dey, S.K.: Computational modeling of the breast cancer treatment by immunotherapy, radiation and estrogen inhibition. Scientiac Mathematicac Japonicac 58(2) (2003)
2. Clancy, J.: Basic Concepts in Immunology. The McGraw-Hill Companies, New York (1998)
3. Chen, D.S., Mellman, I.: Oncology meets immunology: the cancer-immunity cycle. Immunity **39**, 1–10 (2013)
4. Chen, D.S., Irving, B.A., Hodi, F.S.: Molecular pathways: next-generation immunotherapy—inhibiting programmed death-ligand 1 and programmed death-1. Clin. Cancer Res. **18**, 6580–6587 (2012)
5. Henry, N.L.He. et.al.: Immunotherapy for breast cancer treatment: is it an option? cancer. Net. Revised (2019)
6. Intlekofer, A.M., Thompson, C.B.: At the bench: preclinical rationale for CTLA-4 and PD-1 blockade as cancer immunotherapy. J. Leukoc Biol. **94**(1), 25–39. 2 (2013)
7. Patrick, A.: Ott.immune checkpoint blockade in cancer: inhibiting CTLA-4 and PD-1/PD-L1 with monoclonal antibodies. Contem Oncol (2014)
8. Zhang, H. et al.: HIF-1 regulates CD47 expression in breast cancer cells to promote evasion of phagocytosis and maintenance of cancer stem cells. PNAS First Published October 28, 2015
9. Bates, J.P. et al.: Mechanisms of immune evasion in breast cancer. BMC Cancer Center, Article Number: 556 (2018)
10. Wennhold, K.A. et al.: B cell-based cancer immunotherapy. Transfus. Med. Hemotherapy. **46** (2019)
11. Wolters Kluwer Health, Inc.: What Happens to Tumor Cells After They Are Killed? Oncology Times **39**(24), 46–47 (2017)
12. British Society for Immunology. https://www.immunology.org/
13. Celià-Terrassa, T.C.-T., Kang, Y.K.: Distinctive properties of metastasis-initiating cells. Genes Dev. **30**(8) (2016)
14. Aznar, M.A. et al.: Intratumoral delivery of immunotherapy—act locally, think globally. J. Immunol. **198** (1) (2017)
15. Gynecol Oncol Rep. 2018 Aug; 25: 98–101.Nivolumab use for *BRCA* gene mutation carriers with recurrent epithelial ovarian cancer: A case seriesKoji Matsuo,a,b, *,1 Samantha E. Spragg,a,1 Marcia A. Ciccone,a Erin A. Blake,a Charité Ricker,c Huyen Q. Pham,a,b and Lynda D. Romana,b .;VPMCID: PMC6038829 .PMID: 29998185ournal List. Gynecol Oncol Rep V.25; 2018 Aug
16. Wu, L. et al.: Development of therapeutic antibodies for the treatment of diseases. J. Biomed. Sci. **27**(1) (2020). www.cancer.org/treatment/treatments-and-side-effects/treatment-types/immunotherapy/monoclonal-antibodies, https://www.cancerresearchuk.org/about-cancer/cancer-in-general/treatment/targeted-cancer-drugs/types/monoclonal-antibodies
17. Bayer,Virginia BSN, RN, OCN®An Overview of Monoclonal Antibodies. https://doi.org/10.1016/j.soncn.2019.08.006
18. Alice, M., Willrich, V.: Therapeutic monoclonal antibodies in the clinical laboratory. J. Appl. Lab. Medi. **2**(3) (2017)
19. Redig, A.J., McAllister, S.S.: :Breast cancer as a systemic disease: a view of metastasis. J. Intern. Med. **274**(2), 113–126 (2013)
20. Ref. Cynthia Lynch, MD, Medical Oncologist, Cancer Treatment Center of America, Phoenix, Arizona)

Chapter 6
Modeling Strategies to Win the War Against Breast Cancer

Abstract This is the central part of the monograph. Mathematically, we have tried to model the present treatment methodologies for various forms of breast cancer. Here we got excellent qualitative agreements with existing data. Data were analyzed and presented in tabular forms, in graphs, and through computer simulations.

6.1 Rationale

Our objective is now to study mathematical models for standard treatment for breast cancer. We will analyze the models computationally.

The body becomes a battlefield when cancer attacks. As we have seen before, cancer cells use every strategy that most vehement attackers apply in any battlefield to defeat the defenders. So the defenders must adopt more superb tactics to fight back and bring the body back to normalcy. This is indeed a game of strategy. Doctors and all medical professionals adopt very meticulous strategies every now and then and attempt to make the body fight back successfully. The first author made the first attempt to look into this phenomenon in [14]. Here we will study this in more detail.

The biggest challenge is: Deaths of tumors do not, by any means, guarantee the death of cancer. Cancer cells play a game of hide-and-seek with the immune system, manage to survive with less oxygen (hypoxic), run away from their sites during surgeries and radiation, even learn to survive injections of poison (chemotherapy) while metastasizing, and can reappear practically at any instant without notice. So we need to introduce Super Combat Drugs (SCD's) to fight back. And finally an artificial intelligence (AI) guided Skilled Killer Drug (SKD) to destroy all malignancies at the end of the book.

At the present time, surgery, radiation, and chemotherapies (cut, burn, and poison) are still the standardized techniques to treat most cancers including breast cancer. These protocols are the primary topics of our mathematical simulation now. We have included in our model a parameter *ISN* which stands for immune stimulation/suppression number with an attempt to mathematize the role of the state of mind of a patient during treatments.

© Springer Nature Singapore Pte Ltd. 2021
S. Dey and C. Dey, *Mathematical and Computational Studies on Progress, Prognosis, Prevention & Panacea of Breast Cancer*, Forum for Interdisciplinary Mathematics, https://doi.org/10.1007/978-981-16-6077-1_6

It has been observed that [1] when stress goes down, medications become more effective, and when stress goes up, they are less effective. Our fundamental assumption is that psychology definitely plays a significant role in the treatment of any ailment especially in breast cancer. "Mind–body symbiosis has its roots in antique Greece through the writings of Hippocrates and Galen who observed an increase in breast cancer incidence among the melancholic compared with the sanguine women. The connection between "psyche" and cancer remains mostly anecdotal till the end of nineteenth century, when Dr. Snow [13] reported in his paper that 156 out of 200 women with breast cancer had suffered a traumatic life event, usually the loss of a dear person" [1].

Sometimes by blood test, doctors can detect the presence of cancer cells in the body. According to the Web site [10], there are several markers, some specific proteins, which cancer cells release in the blood that doctors look for in blood tests. The presence of the proteins CA15.3, TRU-QUANT, CA27.29, and CA125 indicates the possibility of the presence of breast cancer. Also blood tests could show even the presence of cancer cells circulating in the blood.

We have discussed here several mathematical models dealing with a large number of patients in a theoretical setup with an attempt to simulate conditions of real patients. In fact, mathematicians could develop hundreds of such models. But one primary question is: Do these models or can these models simulate the real-world scenarios? In order to answer this question, computational results must be presented through some form of simulation and compared with the equivalent physiological environments observed by medical professionals investigating cancer. Having a great deal of discussions with a number of medical professionals about breast cancer patients, our models with varieties of inputs have been set up. Also, simulations of the microenvironments inside the body generated by cancer and simulations obtained by numerical solutions of the mathematical models have been checked.

6.2 The Burning Question on Validation of Computational Findings

A big challenge to mathematicians is as follows: The body has over 10^{14} (one hundred trillion) cells, we are limited to using a much smaller field in a computer model. How effective that could be? We should analyze it briefly.

In our mathematical models, we have considered, in general, a computational field consisting only $101 \times 101 \times 101 = 1,030,301$ points and solved ten time-dependent nonlinear reaction–diffusion equations in three dimensions. Obviously, in comparison with the number of cells in the body, this size is significantly small. So a question may be raised regarding how such small fragment of cells could describe the functions of the cancer cells and the dynamics of the immune response in the entire body which consists of 10^{14} number of cells. Mathematics has an answer.

In this regard, we would like to remind our readers that in order to check blood glucose, doctors certainly do not check the entire amount of blood in the body, they take a small sample. That means on average, or statistically, all blood cells are carrying the same amount of glucose.

6.2.1 Definition: Surjective, Injective, and Bijective Mappings

Let us consider an example: Let $A =$ The set of all real numbers. Let A be mapped onto B, where $B = \{1,2,3\}$. So, $B \subset A$. However the cardinal number of the set A is transfinite, the cardinal number of B is finite. Let us consider an example:

Ex. 1. Let $f(x) = 1, \forall x \in R^1$, (both x and $f(x)$ are real) where R^1 is the set of all real numbers, and let A be the set of all integers. Then, mathematically we may consider a mapping $f : A \rightarrow B$ (onto but not one to one). This is a surjective mapping, meaning an infinite number of elements are all projected into one element, which is 1.

Ex. 2. Let $f(x) = 2$ if $0 \leq x < 1$, and $f(x) = x$ if $x \geq 1$. Here for the entire domain of $0 \leq x < 1$, $f(x)$ is surjective.

But if we consider only, $f(x) = x$ for $x \geq 1$, it is bijective. Bijective mapping is one to one and onto.

So if $A =$ a set representing all the cells of a human body, and $B =$ all the elements of the grid points in the computational field, $n(A) > n(B)$, then we can think of a surjective mapping $f : A \rightarrow B$, such that B can have all (at least statistically) relevant properties of A. If our computational results match well with experimental data at least qualitatively, we will consider our assumption to be valid following the definition of a surjective mapping.

Injective mapping is simply one to one, and there is a point in the range of the dependent variable, which is not included in the mapping. In our field of computations, if certain parts of the body are not included, then we are dealing with injective mapping. This happens when cancer is in situ.

Let us understand these concepts with regard to the heading of this chapter. When we consider the field of computation representing the entire human body, then all cells are being projected into a small field of activities. This is surjective. However, inside the field of computation when we consider one cancer cell like $u(i, j, k)$ being attacked by one cell $v(i, j, k)$ of the defender at each time step t_n, this is one to one and onto (bijective) and when during the war a group of defenders are attacking an enemy soldier at a point (a lymph node in the field) while the others are fighting on a one-to-one basis that is injective mapping. So mathematically, the model has been set up properly.

6.2.2   Surjective Transformation in Set

Let us consider two sets A and B, and a mapping $F : A \to B$, where $\mu(B) < \mu(A)$ where $\mu(A)$ and $\mu(B)$ are the measures of A and B, respectively.

Here every element of A has an image in B. This is an *onto* mapping but not *one to one*. This is a surjective transformation in sets.

A simple example is as follows: A tiny drop of blood describes the glucose level of the entire amount of blood in the body. Most adults will have approximately 1.2–1.5 gallons of blood in the body.

Our finite difference model defined within a computational field will be a surjective transformation from the geometry of the human body to the computational cube.

6.2.3   Mathematical Models on Breast Cancer Treatments. The Attacker–Defender Models

Cells perform various kinds of biochemical tasks. One major task is releasing proteins. When leukocytes attack cancer cells, both release chemicals which are proteins to fight against each other. We are primarily interested in these specific functions of the cells.

In this work, all cells are considered as biochemical elements. So treatment of cancer represents a war between biochemicals. Our bodies are activated by trillions of biochemicals. But in the computational fields, we are considering only a small fraction of these biochemical activities to make some meaningful predictions about what is happening in the real world. In the biology of cancer cells, there are several war zones. These are the sites of the tumors and locations where cancer cells are present. These biological war zones should be mapped mathematically into the computational zones. This is a surjective mapping.

In Chap. 1, we discussed some of the general properties of the immune system. Here we plan to analyze them from a mathematical/computational standpoint. Treatments of cancer patients often require drugs to stimulate T-cells, B-cells, NK (natural killer cells), dendritic cells, and macrophages to get them all lined up and organized to fight against the invading antigens (cancer cells). Drugs are given (i) to inhibit the hormone (like estrogen) which helps the invaders, (ii) to minimize glucose intake by the cancer cells, and (iii) angiogenesis, which are some of the supply lines for food and oxygen to the enemies. Cancer cells often build these tiny blood vessels all around them so methodically, so that they can draw blood, but the immune soldiers could not enter into those tiny vessels. So these supply lines must be destroyed by medications.

The dosages depend upon the results of regular blood tests showing regressive and/or aggressive properties of the disease and most importantly the state of health of the patient. This is obviously an individualized treatment that changes as treatments proceed. Indeed, most cancer patients require individualized treatments which often vary from day to day. Every case is unique in general. So one standardized treatment cannot fit all. Statistical generalizations of treatments are not quite appropriate because the data are generally grossly heterogeneous. That is another big reason why we have considered here a large number of different inputs. We reiterate that our mathematical models have been set up based upon a broad class of individualized treatments. Each set of inputs is related to a live story of a patient.

Mathematical description of the entire course of treatment is presented as a direct war between cancer cells as aggressors and the immune system and medications as defenders.

The general model is

$$\partial q_i / \partial t = r_i q_1 - r_{i1} q_i + \sum_{j=1, j \neq i}^{N} r_{ij} q_i q_j + \kappa_i \nabla^2 q_i \tag{6.1}$$

$(i = 1, 2, \ldots, N, \text{ where } N \text{ is the total number of equations.})$

This is a set of nonlinear parabolic partial differential equation. The initial conditions are $q_i(t = 0) = Q(x, y, z)$ (will be assigned). Boundary conditions depend upon some particular treatment procedure. If boundary conditions are set as q_i (at all boundary points) $= 0$, for all $i = 1, 2, \ldots, N$ under certain treatment criterions, then at some point during computations all variables will totally diffuse (going to be null) as treatments proceed. Only under such conditions, this model can be called a set of nonlinear reaction–diffusion equations, because of the properties of parabolic equations. In most practical cases, this will not happen. Doctors keep on treating their cancer patients long after cancer is virtually cured. Because dead cancer cells are not dead forever. They can rise from their graves and become more forceful and resourceful. They can become hypoxic. So we have carefully chosen the boundary conditions such that we could see a more realistic scenario. That is, treatments should continue for sometimes even when cancer is virtually conquered. Also, from a very practical (unfortunate) experience, we know that cancer cells may take shelter in some tissues to avoid any detection by the defense. So we have extrapolated computed values from the computational fields at most points on the boundary, and at some points, we simply used an injective projection.

In the Eq. (6.1), q_1 is the antigen (malignant cell), $q_j's(j = 2, 3, \ldots N)$ are the defending agents, $r_i's$ are the rates of growths/supplies of $q_i's$, r_{i1} are the rates of natural decay of $q_i's$, and $r_{ij}'s$ are the rate constants in biochemical interactions simulating the fight (which means releases of interacting proteins). All $q_i's$ are treated like biochemicals. It should be noted that $r_{ij} > 0$ if q_j is a collaborator of q_i, $r_{ij} < 0$ if q_j is an adversary or antagonist of q_i, and $r_{ij} = 0$ if q_j does not have any effect on q_i. In an individual equation, some small changes were made sometimes according to the special properties of the biochemical that the individual equation represents. After studying many different treatments, these reaction coefficients have been estimated through statistical analysis.

It must be noted that $\forall i$, $q_i \geq 0$. As time goes, a necessary condition that q_1 will be damped out is

$$\sigma = \left(r_1 - r_{11} - \sum_{j=2}^{N} r_{1j} q_j \right) < 0 \tag{6.2}$$

If boundary conditions could be made, by applying some drugs, as $q_1 = 0$, then the condition (6.2) is both necessary and sufficient to damp out cancer from the body. From (6.1) and (6.2), it is evident that a treatment is mathematically strong, if

$$\left(r_1/r_j \right) < q_j/q_1 \text{ and } r_{1j}/r_{i1}) > (q_1/q_j) \quad \forall i, j \tag{6.2.1}$$

But depending upon the physical conditions of the body, physicians should determine whether they may be applicable to a patient.

For these biochemical/chemical interactions, rate constants are assumed to be constants in this chapter. Although there are drugs, known as monoclonal antibodies, which could be used to alter these rate constants. So in some of our models, they may be time dependent. Especially applying medications, r_1 is often being reduced as time progresses. If under certain chemical interactions,

$$|q_i(x, y, z, t_n) - q_i(x, y, z, t_{n+1})| < \varepsilon \quad \forall n > N \tag{6.2.1}$$

then the chemical q_i has reached a steady state. For such a condition, at the boundaries $q_i \neq 0$. (That chemical did not diffuse at the boundary).

This model (6.1) is similar to a set of nonlinear reaction–diffusion equations. It is a set of nonlinear reaction–dispersion equations, better it is an attacker–defender model or an activator–inhibitor model.

From the point of view of biophysics, what we see is as follows: At some point inside the body, cancer is growing and from outside some cancer cells which are flowing into that area. This picture is certainly valid, but this is not the complete story. Some cells also leave that infected area and move away.

6.2.4 A Measure of Success of Treatment

All medical professionals are rather eager to ascertain how they could measure the success of their methods of treatment. The value of σ given by the Eq. (6.2) has a partial answer. So σ is called the measure of success of the treatment. The condition $\sigma < 0$ is a necessary condition, so that treatment will lead to success. If on top of that a schedule of treatment is added such that $\kappa_i \to 0$, (no incoming cancer cells), $\sigma < 0$ is both necessary and sufficient condition, so that the treatment must succeed. We have already studied this in Chap. 4, we will look into more later.

6.2.5 Condition for Non-replicability of Malignant Cells

Another immediate observation is if a medication could be found such that it will make $r_1 \to 0$ at all points wherever q_1, if the antigen is present, then the foreign antigen (cancer) will start growing more and more slowly. And finally, it will not grow anymore. But that does not mean those which are already circulating in blood and in lymph will go away. They will keep on fighting till the therapy destroys them all. What $r_1 \to 0$ will guarantee that the necrotic malignant cells which do create new cancer cells will not do it anymore.

6.2.6 Conditions for the Rates of Growths of Chemicals/Biochemicals Fighting Cancer

Next we focus on the rates of growths/changes of all other biochemicals. It is noticeable that the first term of the right side is $r_i q_1$. That suggests that growths of all biochemicals fighting cancer depend on q_1, the number of cancer cells at a given point in the field which could be estimated using various medical tests. This principle is based upon the fundamental law in a war zone. The defenders must size their army according to the size of the army of the attackers. Sometimes, however if doctors want to apply a particular dosage of a particular medicine at each time interval, then the first term in each equation will be $r_i q_{i0} (i \neq 1)$ where q_{i0} is the preselected dosage of the drug to be used at each time step. In that case, the codes are altered accordingly which we will see later.

The set of Eq. (6.1) are nonlinear partial differential equations. Numerical solutions have been done here by approximating the differential equations which are defined in a continuous space by a set of nonlinear difference equations in a finite discrete space. We solve these difference equations in that discrete space. Two forms of errors are generated because of these approximations: (i) truncation errors and (ii) rounding off errors. So we must follow some specific algorithms for solutions such that these two errors remain bounded during computations. For more details, we request readers to look into some text book on numerical analysis [8].

6.2.7 Mathematical Model for Standard Breast Cancer Treatment

A large number of medications listed in [9] are at present being used to treat breast cancer. We have considered only five including local radiation at the tumor site. However, because this is a mathematical model, these medications may all be considered as different forms of chemos which may not be administered at the same time.

In all, we have used $N = 10$, the total number of equations. So our model consists of the following ten nonlinear reaction–diffusion equations.

6.2.8   Variables Used in the Equations

We have used ten variables for ten reaction–diffusion equations, which are as follows:

(ag) = the antigen to be destroyed which is a cancer cell. It is being inhibited by all the lymphocytes, radiation, and various chemos.

(ct) = the cytotoxic T-cells; the fighting lymphocytes. Its inhibitors are (ag) and (rd). Helpers are (vc), (dm), (st), and (ch).

(dm) = the dendritic and macrophages which inform and attract the T-cells and they inform B-cells to produce antibodies (immunoglobulins) to destroy the antigen, while macrophages serve as phagocytic killers and debris cleaners when the war is over. Its inhibitors are (ag) and (rd). Helpers are (vc) and (ch).

(b) = the B-cells another relentless antigen-specific fighter. Both T-cells and B-cells are lymphocytes. Inhibitors of B-cells are (ag) and (rd). Its helpers are (vc), (st), and (ch).

(nk) = the natural killer cells, another kind of lymphocytes that attack all tumors, activated in response to interferons or cytokines released by macrophages. They are a part of our innate defense system. Its inhibitors are (ag) and (rd). Its helpers are (vc), (st), and (ch).

(vc) = a drug that enhances the power of the defense. May be considered as the vitamin C given intravenously. In several clinical trials (2), it is found that the amount of (vc) in the blood is reduced as other biochemicals like (ct), (dm), (b), and (nk) start using it. However, (rd) reduces its effects, and (ch) has a negative interaction with it.

(ei) = estrogen inhibitor. It is negatively affected only by (ag), (rd), and (ch).

There are several types of hormone therapy for breast cancer. Most types of hormone therapy either lower estrogen levels or stop estrogen from acting on breast cancer cells.

(rd) = radiation therapy. Radiation mainly focuses toward the tumors and burns them. However, when radiation starts, many cancer cells leave their camps in groups and try to move somewhere else. The effects of radiation move inward (inward dispersion) to burn incoming cells at a much faster rate.

(st) = a drug that strengthens the immune system, especially the T-cells.

(ch) = a drug which inhibits angiogenesis thereby enhancing capabilities of the immune system to destroy the enemies and cutting off the supply lines of the enemies. The two medications (st) and (ch) improve low counts of lymphocytes. The term $\nabla^2 q_i$ is a measure of diffusion of q_i. As cancer cells grow, the body produces more defensive forces, in general, to fight back. In fact, this is how our defense mechanisms work. However, whereas the growth of cancer is unrestricted, the number of leukocytes (immune forces, our white blood cells) must be under control. It should not exceed about 11,000 per microliter of blood. So in the codes, some mathematical restrictions have been imposed on their growths. In some models, we have considered delivery of fixed predetermined amounts of drugs both intravenously and intratumorally. We should remember that in a one centimeter tumor, there are about 100 million cells. Also we should note that values of $r_{1j} > 0$ determine the strength of the defense and the medications that destroy the cancer cells. Computationally, we have noticed that it is not the number of the defensive forces, rather the amount of their strength and strategy that help them defeat cancer. The first terms of the right sides of the Eqs. (6.3) to (6.14) indicate that the values of the inhibitors are adjusted at each unit of time according to the states of aggressive nature of (ag). Doctors generally do that according to various test results. In these models, radiation is primarily focused on the tumor, except when we use the properties of nanoparticles carrying radiation to destroy all malignant cells all over the body. However, effects and remnants of radiation are also carried by blood all over the body.

6.2.9 The Equations Representing the Model

All cancer cells carry foreign antigens, which are protein molecules that cause our immune system to release antibodies, if and when they are recognized as foreign by the body's immune system.

1. **Antigen (Cancer)**

$$\partial(ag)/\partial t = \alpha_1(ag) + \kappa_1 \nabla^2(ag) \tag{6.3}$$

where

$$\alpha_1 = r_1 - r_{11} - r_{12}(ct) - r_{13}(b) - r_{14}(dm)$$
$$- r_{15}(nk) - r_{16}(vc) - r_{17}(ei) - r_{18}(rd) - r_{19}(ch) + r_{110}\theta(ag) \tag{6.4}$$

Here, (ag) = the foreign antigen (cancer) to be destroyed.

A notation: $\alpha_1(ag) = \alpha_1 \times (ag)$ etc. Not α_1 as a function of (ag). This notation is true for other terms also and has been maintained throughout this chapter.

Meaning of θ

$r_{12} = the\ rate\ at\ which(ag)\ is\ annihilated\ by\ each\ unit\ of(ct)$. This holds true for all other terms. The parameter θ is used to quantify aggressiveness of the aggressor. It is defined as follows:

$$\theta = (ag)/((ct) + (b) + (dm) + (nk)). \tag{6.5}$$

In any war, the aggressor attempts to have more strength of aggression to fight to overcome the enemies. By definition, θ is a scalar.

θ cannot be negative because all the variables defining θ are positive. However, for a favorable result we need

$$0 < \theta < 1 \tag{6.6}$$

Also $\theta(ag)$ is:

$$\theta \times ag \tag{6.7}$$

Larger value of θ means the more powerful ag will be and less strong will be the defense of the body.

2. Cytotoxic T-cells

The equations for the lymphocytes are as follows:

$$\partial(ct)/\partial t = r_2(ag) - \alpha_2(ct) + \kappa_2 \nabla^2(ct) \tag{6.8}$$

where

$$\alpha_2 = r_{21} + r_{22}(ag) - r_{23}(vc) + r_{24}(rd) - r_{25}(dm) - r_{26}(st) - r_{27}(ch)$$
$$-(ISN)(ag) \tag{6.9}$$

(ct) = cytotoxic T-cells, a group of antigen-specific ruthless lymphocytes. Their rate of growth is triggered by the rate of growth of (ag). The chemo (st) improves the performance of the immune system. It could be considered as a drug that inhibits angiogenesis. The chemo (ch) kills both cancer cells and helps lymphocytes. It may be considered as a glucose inhibitor. Glucose and fat supply cancer cells with more energy. Too much glucose in the blood causes angiogenesis in diabetic patients [16]. We have assumed that this is true in general. This scenario is represented by the terms $r_{26}(ct)(st)$ and $r_{27}(ct)(ch)$. We have assumed that chemically these two drugs interact with each other negatively. While fighting with cancer, it has been assumed that (ct) will lose some of its army due to the fight and also due to radiation. The terms $r_{22}(ag)(ct)$ and $r_{24}(ct)(rd)$ represent this concept.

Dendritic cells and macrophages help cytotoxic T-cells to find and destroy (ag). So they are helpers. That is why the term $r_{25}(ct)(dm)$ is positive. Macrophages are also phagotizers. They phagocytize debris of the body. They have the ability to locate where the debris is. They also engulf bacteria, viruses, parasites, and cancer cells. However, unfortunately, cancer cells can trick both dendritic cells and macrophages and make them build new blood vessels, so that they can get fresh supplies of nutrients from blood and grow and move toward other sites of the body. In general, it has been assumed that radiation represented by (rd) often damages the performance of these cells. This concept is included in the Eq. (6.10) which gives the dynamics of both dendritic cells and macrophages jointly in the term (dm).

3. **Dendritic Cells and Macrophages**

$$\partial(dm)/\partial t = r_3(ag) - \alpha_3(dm) + \kappa_3\nabla^2(dm) \tag{6.10}$$

where

$$\alpha_3 = r_{31} + r_{32}(ag) - r_{33}(vc) + r_{34}(rd) - r_{35}(ch) - (ISN)\,(ag) \tag{6.11}$$

(dm) = dendritic cells. (vc) represents vitamin C often given to patients intravenously to increase the strength of the immune system. However, it could be any drug that strengthens the immune system.

It has been assumed that while (ch) improves the performance of (dm), (st) has no effect on it.

4. **The B-cells**

The dynamics of the B-cells, another powerful antigen-specific lymphocytes, which secrete immunoglobulins (antibodies) are represented by the Eq. (6.12).

$$\partial(b)/\partial t = r_4(ag) - \alpha_4(b) + \kappa_3\nabla^2(b) \tag{6.12}$$

where

$$\alpha_4 = r_{41} + r_{42}(ag) - r_{43}(vc) + r_{44}(rd) - r_{45}(dm) - r_{46}(st)$$
$$- r_{47}(ch) - (ISN)\,(ag) \tag{6.13}$$

While (vc), (dm), (st), and (ch) help B-cells, (rd) affects it negatively.

5. **The Natural Killer Cells**

The next equation represents the dynamics of the natural killer cells (nk).

$$\partial(nk)/\partial t = r_5(ag) - \alpha_5(nk) + \kappa_5\nabla^2(nk) \tag{6.13}$$

where

$$\alpha_5 = r_{51} + r_{52}(ag) - r_{53}(vc) + r_{54}(rd) - r_{55}(dm) - r_{56}(st) - r_{57}(ch)$$
$$-(ISN)(ag) \tag{6.14}$$

It is noticeable that we have separate equations for the lymphocytes fighting cancer cells. The primary reason for that is cancer cells may confuse some of the defenders but not all at the same time. For instance, if T-cells are tricked by cancer, that trick may not be applicable to NK-cells or B-cells.

The parameters r_2, r_3, r_4, r_5 are truly the rate at which chemotaxis is taking place.

6. Estrogen Inhibitors

If (ei) = estrogen inhibitors, the equation for that is

$$\partial(ei)/\partial t = r_6(ag) - \alpha_6(ei) + \kappa_6 \nabla^2(ei) \tag{6.15}$$

$$\text{where } \alpha_6 = r_{61} + r_{62}(ag) + r_{63}(rd) + r_{64}(ch) \tag{6.16}$$

Both radiation (rd) and the drug (ch) affect (ei) negatively.

For triple-negative breast cancer patients, (ei) is not needed. So it could be considered as a chemo for them. In general, two out of three breast cancers are hormone receptor positive. Their cells have receptors (protein molecules) for the hormones estrogen (ER-positive cancers) and/or progesterone (PR-positive cancers) which help cancer cells grow and spread. It is negatively affected only by (ag), (rd), and (ch) due to drug interactions.

7. Vitamin C, an Immunostimulator

The next is the equation for (vc):

$$\partial(vc)/\partial t = r_7(ag) - \alpha_7(vc) + \kappa_7 \nabla^2(vc) \tag{6.17}$$

$$\alpha_7 = r_{71} + r_{72}(ag) + r_{73}(ct) + r_{74}(dm) + r_{75}(b) + r_{76}(nk) + r_{77}(ch) + r_{78}(rd) \tag{6.18}$$

(vc) is vitamin C. Several clinical trials [2, 3], found:" ...vitamin C breaks down easily, generating hydrogen peroxide, a so-called reactive oxygen species that can damage tissue and DNA. Normal cells remove the damaging hydrogen peroxide. The study shows that tumor cells are much less capable of removing the damaging hydrogen peroxide than normal cells. Vitamin C also increases collagen production and enhances the immune system activity. It has been demonstrated that high dose intravenous vitamin C has given many cancer patients the opportunity to improve their quality of life. This explains how the very, very high levels of vitamin C used in our clinical trials do not affect normal tissue, but can be damaging to tumor tissue." says Prof. Buettner, a professor of radiation oncology and a member of the Holden Comprehensive Cancer Center at The University of Iowa. Also vitamin C strengthens the immune system.

Normally, cancer cells have an increased requirement for glucose, and therefore, there is an increase in glucose transporters in cancer cell membranes. This action enhances and favors the entrance of vitamin C into the cancer cell and facilitates the action of this ascorbate acid as a selective, non-toxic chemotherapeutic agent that slows tumor growth [3]. So, we have included in our model (vc) as an inhibitor of cancer. The amount of (vc) in the blood is reduced as other biochemicals like (ct), (dm), (b), and (nk) start using it. However, (rd) reduces its effects, and (ch) has a negative interaction with it. In case vitamin C is not used, it could be considered as a drug that enhances strength of the defense.

8. **Radiation**

The next equation is for (rd), the radiation therapy

$$\partial(rd)/\partial t = r_8(ag) - \alpha_8(rd) + \kappa_8 \nabla^2(rd) \tag{6.19}$$

where

$$\alpha_8 = r_{81} + r_{82}(ag) + r_{83}((ct) + (dm) + (b) + (nk)) \\ + r_{84}(ei) + r_{85}(vc) + r_{86}(st) + r_{87}(ch) \tag{6.20}$$

In general, it damages both normal and cancer cells. We have assumed that it interacts negatively with all the chemos. In case it represents nanoparticles carrying radioactive drugs to deliver at the cancer sites, reaction coefficients should be adjusted. Radiation mainly focuses toward the tumors and burns them. However, when radiation starts, many cancer cells leave their camps in groups and try to move somewhere else. Mentioning the side effects of radiation, in [6] it is stated: "It is well known that tumor cells migrate from the primary lesion to distant sites to form metastases and that these lesions limit patient outcome in a majority of cases." Radiation also damages the immune cells. This radiotherapy is external beam radiation. The beam is highly focused targeting the cancerous area for two to three minutes. Internal radiation or *partial breast radiation* is also sometimes done during a treatment. Oncologists or surgeons insert a radioactive liquid using needles, wires, or a catheter in order to target the area where the cancer originally began to grow and tissues closest to the tumor site, to eliminate malignant cells. We have assumed that the rate of dispersion of radiation is higher than that of other biochemicals.

9. **An Immunostimulant**

Next is the equation for an immunostimulant (st). It is a chemo negatively affected by (rd) and (ch).

$$\partial(st)/\partial t = r_9(ag) - \alpha_9(st) + \kappa_9 \nabla^2(st) \tag{6.21}$$

where

$$\alpha_9 = r_{91} + r_{92}((ct) + (b) + (nk)) + r_{93}(rd) + r_{94}(ch) \tag{6.22}$$

As (st) gets absorbed by (ct), (b), and (nk), it loses its potency. (dm) is not being affected (an assumption). That is represented by the second term of (6.22).

10. **Angiogenesis Inhibitor**

Finally, we consider another drug (ch) which is an angiogenesis inhibitor. It is an anticancer chemo. It helps defense forces, and while doing so, we have assumed that it loses some of its own potency.

$$\partial(ch)/\partial t = r_{10}(ag) - \alpha_{101}(ch) + \kappa_{10}\nabla^2(ch) \qquad (6.23)$$

where

$$\alpha_{101} = r_{101} + r_{102}(ag) + r_{103}((ct) + (dm) + (b) + (nk))$$
$$+ r_{104}(rd) + r_{105}(ei) + r_{106}(vc) + r_{107}(st) \qquad (6.24)$$

6.2.10   Measure of Aggression of Cancer

An aggressor always attempts to overpower the defenders by increasing its power of aggression. θ, defined in (6.5), represents an attempt to quantify that tendency. Because cancer cells always display a strong tendency to overrule the immune system. So it has been assumed in this model that they are trying to gain the upper hand in the fight very aggressively by increasing their number which must exceed the number of soldiers of the defense. This is a standard strategy of the aggressors in a war. θ is called the measure of aggression.

If $\theta > 1$, cancer is winning over the lymphocytes.
If $0 < \theta < 1$, cancer is getting defeated by the lymphocytes.
If $\theta = 0$, cancer is cured. But this is hypothetical.

It must be noted that θ is always positive, because (ag), (ct), (dm), (b), and (nk) are all positive. If θ becomes negative during computation, the model should be corrected. These conditions demonstrate the overall strength of cancer cells as demonstrated by them during the fight.

6.2.11   Immunostimulation/Immunosuppression Parameter ISN

The parameter (ISN) is the immunostimulation/suppression parameter. If it is positive, it is a stimulant, and if it is negative, it is a suppressant. We have assumed that $-1 \leq ISN \leq 1$. If $ISN = 1$, it means the patient is 100% stress free, and if $ISN = -1$, it means that the patient has no feelings of happiness, excessively stressed. Stress weakens the body's immune system and by the same token strengthens the power of attack of the attackers. So, ISN is the rate of increase or decrease of the strength of per unit of (ag) according to the state of mind of the patient. So its dimension is the same as r_1 which is $1/T$, and the dimension of the term $ISN(ag)$ is M/T. If the patient feels optimistic about the treatment, it is positive. Immune system gets stronger. That is very helpful to her. That makes treatment more effective. For exactly the opposite scenario, $ISN < 0$. A chronic state of depression is a cause for breast cancer because it weakens the immune system. So, if the state of depression could be reversed, that should act like a form of immunotherapy. Interestingly, if and when $(ag) \to 0$, immunostimulation is turned off. Similarly, if and when (ag) gets stronger and (ct) etc. get weaker, a negative (ISN) makes the immune system much more weak.

6.2.12 Dimensional Analysis

If T, M, and L represent time, mass, and length, respectively, for dimensional analysis, then the dimensions of r_1, r_{11}, and r_{110} are $1/T$. However, the dimensions of r_{ij} are $1/(T \times M)$, $(for \ i = 1, 2, \ldots, 10 \ and \ j = 2, 3, \ldots, 10 \ for \ each \ i)$. For instance, $r_{12} = rate \ of \ destruction \ of \ (ag) \ per \ unit \ of \ (ct)$. The dimension of $\kappa_i = L^2/T$ and that of $\frac{\partial^2 (ag)}{\partial x^2}$ is (M/L^2). So, the dimension of $\kappa_1 \nabla^2 (ag)$ is M/T. θ is a scalar. This dimensional analysis perfectly balances dimensions of both sides of the Eq. (6.3).

We should also note that the larger the values of $r_{1j} (j = 2, 3, \ldots, 9)$, the higher the strengths of the corresponding defenders. Pharmaceutical companies who manufacture drugs, doctors, and biochemists determine what strengths of medications should be appropriate for a particular patient.

Another point to notice is that, if and when all the inhibitors reach a steady state denoted by (CT), (DM), (B), (NK), (EI), (VC), (RD), (ST), and (CH), respectively, then if,

$$\sigma = r_1 - r_{11} - r_{12}(CT) - r_{13}(B) - r_{14}(DM) - r_{15}(NK)$$
$$- r_{16}(VC) - r_{17}(EI) - r_{18}(RD) - r_{19}(CH) + r_{110}\theta$$

its dimension is 1/T.

6.2.13 Defeat of Cancer Cells

As discussed before, a necessary condition that (ag) will be defeated/destroyed in the war in due course of time by the treatment is

$$\sigma < 0$$

If applying medication, the values $(ag)_{ijk}^{n} \to 0$ at all points on the boundary (6.7) will be both necessary and sufficient that the patient will be cancer free. σ is called the parameter of success or a measure of success of the treatment.

6.2.14 The Law of Physiology on the Replication of the Defenders Used in the Model

We know from physiology that an increase of lymphocytes in the blood is caused by viral/bacterial infections. Accordingly, we have assumed that lymphocytes replicate themselves in accordance with the severity of the attack of the aggressor (ag) and will aggressively move where they need to fight against the invading pathogens. This is called the chemotaxis. This is an integral aspect of our physiological structure. So in any model for treatment, this part must be done and that is present in our reaction terms on the right side of the Eq. (6.3).

In our previous discussions, we have noted that if u is the pathogen, the first term on the right side represents the rate of growth of u, may be given by $a_1 u$. This is shown by the term $r_1(ag)$. This reveals the aggressiveness of the cancer. Possibly, pathologists find r_1 using biopsy of a tumor. Oncologists can possibly figure it out too while conducting regular blood tests, almost every day, to check the state of their treatment, analyze the results, and accordingly, figure out how treatment must proceed. Their strategies often change every now and then. So, all medications were adjusted and applied accordingly. We thus conclude that the rate of growth of the aggressor triggers the rates of growth of the lymphocytes and other medicines. This rate often changes during the progress of cancer and its treatment. However, we have considered it to be fixed in this chapter.

6.2.15 A Note on the Rate Constants r_{ij} and the Coefficients of Dispersions κ_i

These rate constants should be known prior to application of any mathematical model related to chemical interactions. Our research studies on radiolysis of water [15] at UC, Berkeley and at NASA Ames [18] helped us in this respect very much. Also we have observed a large number of treatments, consulted with the doctors/surgeons/nurses about the dosages of medications and applied statistical analysis to estimate these values. Accordingly, we have set up these values. One may consider them as our best educated guess. These should be made available by the pharmaceutical companies supplying the drugs. κ_i 's are the coefficients of dispersion. In this model, we have assumed that all dispersions are carried out by blood, so to simulate this situation we have considered $\kappa_i = 10^{-7}$, the kinematic coefficient close to the value of the kinematic coefficient of viscosity. The exception is for radiation (rd). Because it disperses much faster. However, when cancer cells move through tissues, they can disperse much faster. So, values of κ_i 's could be much higher that will be considered later.

6.2.16   The CDey-Simpson Method for Numerical Solution

Numerical solution of these ten nonlinear reaction–dispersion equations of chemical kinetics has been done by the CDey-Simpson forward difference method. We already have discussed this numerical method in detail in Chap. 4. Here the application of this technique has been done in a little different way. First we have approximated all the equations by the forward Euler method. It is well known that this method has a very poor stability property. So, it must be corrected at each time step. First correction was done by Charlie's method, and the second correction was done by Simpson's rule. An analysis of stability has been done in the Appendix A.

There are ten nonlinear partial differential equations. When we approximate the space derivatives by finite differences on a computational field $101 \times 101 \times 101$, we generate 10,303,010 numbers of time-dependent ordinary differential equations. If we solve each of them numerically for 500 time steps, it means in total we are solving 5,151,505,000 (fifty-one hundred fifty-one million, five hundred five thousand) nonlinear algebraic equations in total. Furthermore, these equations are stiff. In a gaming Dell Laptop with an i5 processor, numerical solution by Charlie-Simpson took about 10 min and ten seconds for each time step computation. This numerical algorithm is fully vectorized, so it could be parallelized as done in [18]. This was discussed for massively large-scale computations in acoustic parabolic models [20].

The algorithmic procedure is as follows: At each nth time step denoted by t_n, we do the following: (1) First take a half-time step move using Euler forward prediction. (2) Correct this scheme with Charlie's corrector. (3) From there, take a half-time step forward with Euler. (4) Correct it again with Charlie. (5) Now we have values of all the biochemicals at t_n, $t_{n+1/2}$, and t_{n+1}. All these values have been corrected by Simpson's rule twice to make a move from t_n to t_{n+1}. Computations move forward explicitly as time goes. There is no inner computational loop as done for implicit finite difference methods when we have used perturbed functional iterations (PFI) in the past. One more point is, just the way PFI has been parallelized [20], CDey-Simpson being a vectorized scheme, it may also be parallelized exactly in the same way. So in a high-speed supercomputer, results could be found in a few seconds.

There is no package for this algorithm, so we wrote our own codes. The graphs indicate that the equations are extremely stiff, especially when we compare them with our previous works on chemical kinetics [16].

6.2.17 A Special Note on Graphing

In all our graphs, the largest values of all the parameters (biochemicals) over the entire computational field have been plotted. For instance, if at $t = 0$, $(ct) = 20$ at the tumor site, and it is 200 outside the tumor, then the max value of (ct) at $t = 0$ over the entire field is 200. All values have been plotted on a $log_{10}(10 \times ag)$ or $1 + log_{10}(ag)$, etc., base scale. We felt that since these are no experimental data, no medical professional will use the actual values. They may be used as educational guesses or as some guidelines and suggestions regarding treatments. So, on the vertical axis, printing values of the biochemicals are sometimes omitted intentionally. We have printed the initial values of the parameters and their final values on the figures just to indicate what the model reveals as time changes. On the horizontal axis, representing time, no point has been allocated to show the value of t. This is done intentionally to refrain readers from thinking exactly at what particular time what they will expect to see. These are NOT the exact values in real-time treatments. However, in the future, these preliminary models may lead to developing models for real-time simulation. Notably, when the values of all the inhibitors go into steady states or start decreasing, meaning their slopes are zero or negative, it will indicate that the performance of the model is correct. Our primary objective is to show that if the slope of a graph of a particular chemical is as follows: (i) positive, when its value is increasing, (ii) negative, when its value is decreasing, and (iii) zero, *as* $t \to \infty$ when it has gone to a steady state.

On the vertical axis, the color of a chemical matches with the color of the graph it represents.

A Note: The graphs must be read with much care. Sometimes we had to reorganize the logarithmic scale to get better looking graphs. Please note that graphs reveal the pattern of treatment only. Sometimes we have intentionally omitted the descriptions of the inputs. Readers could see the inputs from the graphs and analyze from their values all the ramifications associated with them from prior discussions about them.

6.2.18 Computational Studies of the Model

In the first computer run, the number of grid points $= 101 \times 101 \times 101$. $\Delta x = \Delta y = \Delta z = 0.02$, $\Delta t = 0.015$. There is just one tumor extending from $(0,0,0)$ to $(5,5,5)$. There are three lymph nodes at $(8,8,8)$, $(15,15,15)$, and $(30,30,30)$ which are affected by cancer. At the tumor, $(ag) = 15000$ at each grid point. Outside this tumor and at the three lymph nodes, we have considered various values of (ag) for various patients. Also at the tumor site, the number of T-cells, dendritic cells, B-cells, and NK-cells has different values for different patients. In the field of computation, their upper limits were set at 3000 at each grid point. However, the number of cancer cells is unrestricted because their growth knows no constraint. We have assumed the existence of three affected lymph nodes. They are defined as having a very small number of lymphocytes. Although, at these locations cancer cells may not be large in number, they successfully block the activities of lymphocytes. "Cancer cells can travel through the bloodstream to reach distant organs. If they travel through the lymph system, the cancer cells may end up in lymph nodes. Either way, most of the escaped cancer cells die or are killed before they can start growing somewhere else. But one or two might settle in a new area, begin to grow, and form new tumors. This spread of cancer to a new part of the body is called metastasis….When cancer has spread to lymph nodes, there's a higher risk that the cancer might come back after surgery. This information helps the doctor decide whether more treatment, like chemo or radiation, might be needed after surgery." [Ref. American Cancer Society]. There could be a small number of cancer cells in a lymph node. "…if there are only a small number of cancer cells in the lymph nodes, they may feel normal. It's only possible to tell whether a cancer is present by removing part or all of the lymph node and examining the cells in a laboratory." [The Center for Cancer and Blood Disorder, Fortworth, Texas].

This scenario represents a typical patient, who has discovered a lump in the breast, often feels very tired, gets headaches practically for no apparent reason, and feels weak at the same time. The first author saw several patients like this. Their ages are between 40 and 60. In all computer runs in this Chapter, the values of $r_i, i = 2, 3, \ldots, 10$ and $r_{i1}, i = 1, 2, 3, \ldots 10$ and $\kappa_j, j = 1, 2 \ldots 10$ remain unchanged as follows: $r_{11} = 10^{-7}$ and $r_{21} = r_{31} = r_{41} = r_{51} = 10^{-6}$ and $r_{61} = r_{71} = r_{81} = r_{91} = r_{101} = 10^{-5}$ and $\kappa_1 = 10^{-7}$, $\kappa_2 = \kappa_3 = \kappa_4 = \kappa_5 = 10^{-7}$, $\kappa_6 = \kappa_7 = 10^{-7}$, $\kappa_8 = 10^{-3}$ (assuming that radiation has a faster dispersion), $\kappa_9 = \kappa_{10} = 10^{-7}$. In general, the temperatures of the body and blood change. At about ninety degree fahrenheit, diffusion coefficient of blood is about 1.62×10^{-7}. In all of our codes, we have used 10^{-7}. In the code, MAXAGO represents the largest value of (ag) in the field outside the primary site and outside the affected lymph nodes.

We have considered: $r_1 = 2.5$

That means the rate of growth of cancer is 150%. Most breast cancers about 70% to 80% start in the milk ducts and slowly become invasive. These are invasive ductal carcinoma. 150% is a relatively slow growth.

The values of r_{ij} 's show how strong or weak the antagonists are. The strengths of these values mean how strongly they are fighting.

Inputs for Fig. 6.1.

```
** INPUTS FOR THE FIG.6.1      ISN= .001   DX= .02  DY= .02   DZ= .02   DT= .015   FIELD 101x101x101 # OF TIME STEPS=   100  **
R1= 2.5 R11= .0000001  R12= .00025  R13= .00025  R14= .00005  R15= .00025  R16= .00001  R17= .0001
R18= .015 R19= .0001  R110= .00001  KAPPA1= .00000001
R2=  1 R21= .000001  R22= .00001  R23= .0005  R24= .0001  R25= .00025  R26= .0005  R27= .0001  KAPPA2= .0000001
R3=  1 R31= .000001  R32= .00002  R33= .00005  R34= .0001  R35= .0001  KAPPA3= .0000001
R4=  1 R41= .000001  R42= .00002  R43= .00005  R44= .0001  R45= .00025  R46= .0005  R47= .0001  KAPPA4= .0000001
R5=  1 R51= .000001  R52= .00002  R53= .00005  R54= .0001  R55= .00025  R56= .0001  R57= .0001  KAPPA5= .0000001
R6=  1 R61= .00001  R62= .00005  R63= .0005  R64= .0001  KAPPA6= .000001
R7=  1 R71= .00001  R72= .00006  R73= .00005  R74= .00005  R75= .00005  R76= .00005
R77= .0001  R78= .005  KAPPA7= .000001
R8=  1 R81= .00001  R82= .0005  R83= .0001  R84= .0001  R85= .0001  R86= .0001  R87= .0001  KAPPA8= .001
R9=  1 R91= .00001  R92= .000025  R93= .0001  R94= .0001  KAPPA9= .00001
R10= 1 R101= .00001  R102= .00005  R103= .00001  R104= .00005  R105= .00001  R106= .00001
R107= .00001  KAPPA10= .00001

CONDITIONS AT THE PRIMARY SITE AT T=0
AG0= 15000   CT0= 20  DM0= 50  B0= 20
NK0=  20   EI0= 100  VC0= 100  RD0= 250  ST0= 250  CH0= 150
AT THE THREE AFFECTED LYMPH NODES AG0,CT0,DM0,B0,NK0=   10  20  500  20  20
CONDITIONS OUTSIDE THE SITES, AT T=0
AG0=  10  CT0= 200  DM0= 500  BM0= 200  NK0= 200
EI0= 100  VC0= 100  RD0= 0  ST0= 250  CH0= 150
```

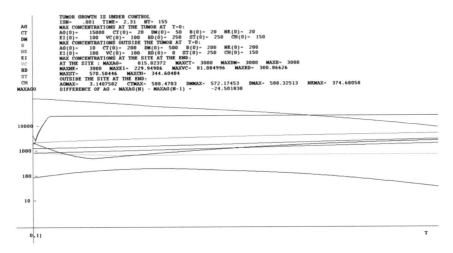

Fig. 6.1 Radiation, chemos, immunotherapy, & inhibition of estrogen and glucose. Treatment worked. *A **Note**:* There is a mistake in this graph. The lower blue line represents the values of MAXAGO as time changes. However, on the vertical axis, MAXAGO is written in black. It should be blue. We apologize to the readers

r_{ij} 's are the rate constants, and AG0,CT0,DM0,B00,NK0,EI0,VC0,RD0,ST0, and CH0 are the initial values of (ag), (ct), (dm), (b), (nk), (ei), (vc), (rd), (st), and (ch), respectively. MAXAGO (this is the letter O, not the numeral 0) is the largest value of (ag) in the field outside the tumor. It is very important to observe the values of MAXAGO. If this increases, it indicates that outside the primary tumor, somewhere in the field cancer is growing, giving a strong indication of metastasization. In all the figures, the max values of (ag), (ct), (dm), (b), (nk), (ei), (vc), (rd), (st), and (ch) in the field of computations are plotted against time. In some cases, max value of (ag) at the tumor site is denoted by MAXAGP. All drugs, except radiation (rd), have been administered intravenously (injected in the blood), and radiation is focused at the tumor (outside the tumor site $rd(t = 0) = 0$).

In Fig. 6.1, we have also assumed that the patient is in a moderately good spirit and $(ISN) = 0.001$. That also has a positive impact on the results showing that not only the tumor is shrinking, the number of cancer cells elsewhere is also diminishing after they attempt to increase (the bottom blue line in Fig. 6.1). The treatment is a success. The medications are working, and the patient is moving toward a cure. It is clear from the model that the entire defense system adjusts itself at each time step as the antigen (ag) changes. That is why the coefficients of $r_j (j = 2, 3, \ldots 10)$ are always (ag). Most oncologists do it routinely by conducting regular blood tests. This could also become a part of immunotherapy. Because in the laboratory, biochemists could simulate a system like this.

One other important point is that wherever and whenever the antigen (ag) appears and/or increases, the defensive forces, (ct), (b), (dm), $and (nk)$ increase and attack (ag). This is the dynamics of chemotaxis.

From Fig. 6.1, it is clear that initially, at the tumor site $(ag) = 15{,}000$ at each point of the tumor which extended from $(1,1,1)$ to $(5,5,5)$ and $(ct) = 20$, $(dm) = 50$, $(b) = 20$ and $(nk) = 20$ at each point of the tumor. So the defense is very poor. Outside the tumor, defense is not that strong either. $(ct) = 200$, $(dm) = 500$, $(b) = (nk) = 200$. The drugs $(ei) = 100$, $(vc) = 100$, $(st) = 250$, and $(ch) = 150$ at each point injected intravenously. It is assumed that a drug injected intravenously goes into the bloodstream and moves all over the body. However, radiation is given only at the site (intratumoral). $(rd) = 250$ at each point of the tumor. So $(rd) = 0$ elsewhere. Once radiation is given at the tumor, its effects do not stay bounded at the sight. It affects the entire body, often causing neutropenia (low white blood cell count), anemia (low red blood cell count), thrombocytopenia (low platelet count), and many more. The reason is it affects the entire body. So, even when it is directed toward the tumor, it goes all over. All treatments, regardless how good they work at the outset, must be examined most carefully analyzing what the long-term effects could be. So later in the conclusion, we have tried to look into this.

Our next computational study is with $r_1 = 4.5$. That means aggressiveness of cancer is significantly increased. The rate of growth is 350%. This type of cancer is an invasive ductal carcinoma (IDC).

With the same rate constants, and exactly the same initial conditions, the treatment seemed to have failed. The tumor is getting bigger and spreading as shown in Fig. 6.2. As we have mentioned earlier, for each patient, treatment must be individualized. Here tumors are growing and metastasizing, apparent from the blue line representing MAXAGO (growth of cancer cells outside the tumor). It has been assumed that at the tumor site, defense is very poor, maybe because the foreign antigens already changed some normal cells into cancer cells. So the number of fighting (ct), (dm), (b), and (nk) is very small in that scenario, whereas the number

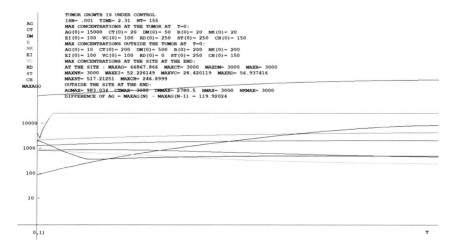

Fig. 6.2 Radiation, chemos, immunotherapy, & inhibition of estrogen and glucose. Treatment failed

of cancer cells is very large. The number of lymphocytes outside the tumor is relatively small too. That has increased θ, the measure of aggressiveness. We know that many cancer patients develop low white blood counts (WBC). Often chemotherapy also lowers the number of leucocytes. Sometimes, computational results show that initially the tumor is shrinking but later it starts increasing. So the treatment ultimately fails. It may be noted that radiation $(rd) = 0$ outside the tumor at t $= 0$. But because of faster diffusion, as time goes, its effects move all over the body. Treatments for such cases are discussed later with more examples. The graph shows that the tumor growth is not under control. Also notably, the increasing values of MAXAGO the largest value of (ag) outside the primary site recorded by the lower blue line show that cancer is metastasizing outside, and the values of all the defenders have reached steady states. So no more medications should be added. However, one must keep in mind that this is a mathematical model and may not show exactly what should be done by the medical practitioners.

The graph shows that as cancer cells are fast replicating aggressively by a factor of $r_1 = 4.5$, (350%), the treatment which worked nicely for $r_1 = 2.5$ totally fails (with the same inputs for antagonists). However, there are a few points to note: (1) The tumor grew from 15,000 at each point to 66,868 (at the site means at the site of the tumor), whereas the fighting leukocytes have attained their maximum value. (2) The value of cancer cells outside the tumor was only 10 at the beginning and that rose sharply to 983. (3) However, the fighting leukocytes came to a steady state, and medications decreased.

So each treatment should be tailored for each individual patient. In conclusion, a long-term study of this method of treatment has been looked into.

Now we will observe the same (IDC) case where the defense has been strengthened. Here the medicines have played a major role in saving the life of the patient. This is shown in Fig. 6.3 The inputs are ISN $= -0.01$.

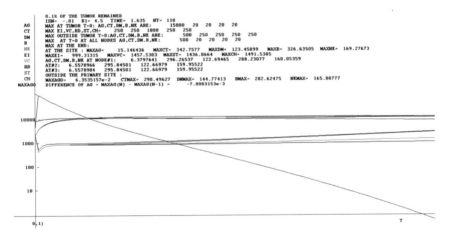

Fig. 6.3 A powerful treatment

Inputs for Fig. 6.3.

```
•••••••••••••••••••••••••••••••••••••••••••••••••••••••••••••••••••••••••••••••
  •• INPUTS FOR THE FIG.6.2      ISN=-.01   DX=  .02   DY=  .02   DZ=  .02   DT=  .015   FIELD 101x101x101 # OF TIME STEPS-    151  ••
  R1=   4.5  R11=  .0000001   R12=  .0065  R13=  .0065  R14=  .0001  R15=  .0065  R16=  .0001  R17=  .0005
  R18=   .015  R19=  .0001   R110=  .000001   KAPPA1=  .0000001
  R2=   1  R21=  .000001   R22=  .00001   R23=  .00005  R24=  .0001  R25=  .00025  R26=  .0005   R27=  .0001   KAPPA2=  .0000001
  R3=   1  R31=  .000001   R32=  .00002   R33=  .00005  R34=  .0001  R35=  .0001  KAPPA3=  .0000001
  R4=   1  R41=  .000001   R42=  .00005   R43=  .00005  R44=  .0001  R45=  .00025  R46=  .0005   R47=  .0001   KAPPA4=  .0000001
  R5=   1  R51=  .000001   R52=  .00001   R53=  .0001  R54=  .0001  R55=  .00025  R56=  .0001   R57=  .0001   KAPPA5=  .0000001
  R6=   1  R61=  .000001   R62=  .00005   R63=  .0005  R64=  .00001  KAPPA6=  .0000001
  R7=   1  R71=  .000001   R72=  .00005   R73=  .000005  R74=  .000009  R75=  .000005  R76=  .000005
  R77=  .000001   R78=  .0001   KAPPA7=  .0000001
  R8=   1  R81=  .000001   R82=  .0005   R83=  .000001  R84=  .000001  R85=  .000001  R86=  .000001  R87=  .000001  KAPPA8=  .001
  R9=   1  R91=  .00001   R92=  .000025   R93=  .0001  R4=  .000001  KAPPA9=  .0000001
  R10=  1  R101=  .00001   R102=  .00001   R103=  .00001  R104=  .0001  R105=  .00001  R106=  .00001
  R107=  .000001   KAPPA10=  .0000001
  CONDITIONS AT THE PRIMARY SITE AT T=0:
  AG0=  15000   CT0=  20   DM0=  20   B0=  20  NK0=  20   EI0=  250   VC0=  250   RD0=  1000   ST0=  250  CH0=  250
  AT THE THREE AFFECTED LYMPH NODES AG0,CT0,DM0,B0,NK0=       5000   20    20    20    20
  CONDITIONS OUTSIDE THE SITES, AT T=0:
  AG0=  500   CT0=  250   DM0=  250   BM0=  250  NK0=  250  EI0=  250   VC0=  250   RD0=  0   ST0=  250  CH0=  250
  IMMUNE SYSTEM IS VERY WEAK. ONLY RD WAS GIVEN INTRATUMORAL,ALL OTHERS WERE CONDUCTED INTRAVENOUSLY.
•••••••••••••••••••••••••••••••••••••••••••••••••••••••••••••••••••••••••••••••
```

Note: When we state that radiation (rd) is affected by the chemo (ch), we mean that observations regarding the "late effects" have revealed the information from past similar treatments. The same is true when we have stated that (ei) is not affected by (st). We have assumed that these late effects really started when the treatment was done and were noticed by the doctors while observing the after effects of the treatment during some course of time. The code converged at the time NT = 110, long before NT = 151, because the tumor and the value of (ag) all over the field became very small.

• Here we notice that the strengths of (vc) and (ei) are significantly increased which are represented by the values of r_{16} and r_{17} in comparison with their previous values and the value of θ, measuring the strength of aggressiveness of cancer is decreasing. Also here (ei) being less affected by (ch) ($r_{64} = 0.00001$ against previously $r_{64} = 0.0001$) which strengthens lymphocytes, is much stronger than what it was previously. More importantly, here, values of r_{12}, r_{13}, and r_{15} are each 0.00065, considerably higher than their previous values. This means the cytotoxic T-cells, the B-cells and the NK-cells are fighting hard. However, $r_{14} = 0.0001$ implies that the dendritic cells and macrophages are not that active. The underlying assumption is that once tumors are formed, the malignant cells try to make friends with dendritic cells and macrophages to help them grow more blood vessels for fresh supplies of nutrients. So overall, the defense is more aggressive, and medications are more powerful than the previous case. Even though the patient is depressed (ISN = −0.01), treatment is successful.

6.2.19   Radiation is Administered Intravenously Using Nanoparticles as Vectors

In all the above cases whereas (ei). (vc), (st), and (ch) were all administered intravenously, only (rd) which may be considered as radiation treatment, was administered intratumoral. Reports from MD Anderson Cancer Center reveal "Enhancement of radioresponse with nanoparticles is attributed to increased drug delivery, synergist

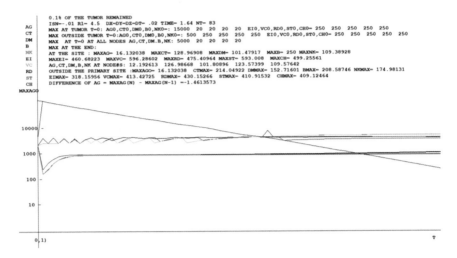

Fig. 6.4 All medications including RD have been administered intravenously

interaction with different modes of cell killing, and/or increased cellular exposure to ionizing radiation beams" [BMC, Part of Springer Nature, May 2, 2018]. Recent studies by the National Cancer Institute [13] on the applications of nanotechnology for cancer treatment revealed the success of using nanoparticles for delivery of drugs. "Cancer therapies are currently limited to surgery, radiation, and chemotherapy. All three methods risk damage to normal tissues or incomplete eradication of the cancer. Nanotechnology offers the means to target chemotherapies directly and selectively to cancerous cells and neoplasms, guide in surgical resection of tumors, and enhance the therapeutic efficacy of radiation-based and other current treatment modalities. All of this can add up to a decreased risk to the patient and an increased probability of survival." On the basis of our studies on the process of metastasization and effectiveness on intravenous delivery of medications, we have decided to look into the computational results when (rd) is administered intravenously. The results are amazingly interesting. In the previous case, we have used $(rd0)$ = the initial intratumoral dose of (rd) = 1000 at each point. Now we used $(rd0)$ = 250 at each point in the entire field. The inputs are as follows.

Inputs for Fig. 6.4.

```
••••••••••••••••••••••••••••••••••••••••••••••••••••••••••••••••••••••••••••••••••••••••••
** INPUTS FOR THE FIG.6.4 :ISN=-.01  DX= .02  DY= .02  DZ= .02  DT= .02  FIELD 101x101x101 # OF TIME STEPS= 151 **
R1= 4.5  R11= .0000001  R12= .0065  R13= .0065  R14= .0001  R15= .0065  R16= .00005  R17= .0005
R18= .015  R19= .0001  R110= .000001  KAPPA1= .0000001
R2= 1   R21= .000001  R22= .00001  R23= .00005  R24= .0001  R25= .00025  R26= .0005  R27= .0001  KAPPA2= .0000001
R3= 1   R31= .000001  R32= .00002  R33= .00005  R34= .0001  R35= .0001   KAPPA3= .0000001
R4= 1   R41= .000001  R42= .00001  R43= .00005  R44= .0001  R45= .00025  R46= .0005  R47= .0001  KAPPA4= .0000001
R5= 1   R51= .000001  R52= .00001  R53= .0001   R54= .0001  R55= .00025  R56= .0001  R57= .0001  KAPPA5= .0000001
R6= 1   R61= .000001  R62= .00005  R63= .0005   R64= .00001 KAPPA6= .0000001
R7= 1   R71= .000001  R72= .000005 R73= .000005 R74= .000005 R75= .000005 R76= .000005
R77= .000001  R78= .0001  KAPPA7= .0000001
R8= 1   R81= .000001  R82= .0005  R83= .000001  R84= .000001  R85= .000001  R86= .000001  R87= .000001  KAPPA8= .001
R9= 1   R91= .00001  R92= .000025 R93= .0001   R4= .000001  KAPPA9= .0000001
R10= 1  R101= .00001  R102= .00001  R103= .00001  R104= .0001  R105= .00001  R106= .00001
R107= .000001  KAPPA10= .0000001


CONDITIONS AT THE PRIMARY SITE AT T=0:
AG0= 15000  CT0= 20  DM0= 20  B0= 20  NK0= 20  EI0= 250  VC0= 250  RD0= 250  ST0= 250  CH0= 250
NK0= 20  EI0= 250  VC0= 250  RD0= 250  ST0= 250  CH0= 250
AT THE THREE AFFECTED LYMPH NODES AG0,CT0,DM0,B0,NK0= 5000   20  20  20  20


CONDITIONS OUTSIDE THE SITES, AT T=0:
AG0= 500  CT0= 250  DM0= 250  BM0= 250  NK0= 250  EI0= 250  VC0= 250  RD0= 250  ST0= 250  CH0= 250


ALL MEDICATIONS,INCLUDING RD ARE ADMINISTERED INTRAVENOUSLY.
CT,DM,B,NK ALL CANNOT EXCEED 3000 AND ALL OTHER MEDICATIONS CANNOT FALL BELOW 200 AND EXCEED 500.
IF THEY DO THEY ARE SET EQUAL TO 300.
••••••••••••••••••••••••••••••••••••••••••••••••••••••••••••••••••••••••••••••••••••••••••
```

Looking at the values of $r_{12}, r_{13}, r_{15} = 0.0065$, it is evident that the immune system is giving a relatively strong fight.

The graph of (rd) reveals some ups and downs. That means every time (rd) rises over its limits, it is brought down by the drug monitoring device. The restrictions were not applied at the boundaries where cancer cells can hide. In the real world, these are the locations in the body where cancer cells may hide, like adipose tissues, under the skin, etc.

6.2.20   The Confused T-Cells and An Inactive Chemo

There are some cases where cancer cells could totally confuse the T-cells by releasing certain cytokines which make immune cells confused. That indicates that although T-cells are in circulation, they do not attack the foreign antigens. We have also assumed that the drug (vc) (which may not be the vitamin C) is also in circulation and is inactive. So, $r_{12} = r_{16} = 0$. Under such a condition what could be done to save a patient? We will discuss such a case now. Here we have considered $r_{13} = r_{15} = 0.0065$, which implies that the B-cells and NK-cells are fighting very very hard. Results are given in Fig. 6.5. The inputs are as follows: ISN $= -0.01$ (a sign of depression), $r_1 = 4.5$ (cancer replicating very fast). Here we are solving the entire set of equation as before assuming that the T-cells are fighting and the chemo (vc) is doing its job. But actually both have no effect on (ag).

Here $rd(t = 0) = 0$, outside the tumor meaning that radiation is directed only toward the tumor.

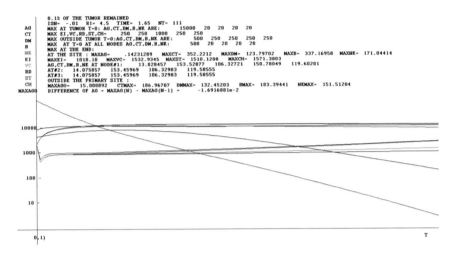

Fig. 6.5 CT AND VC are doing no harm to cancer cells. Cancer could still be controlled, if other drugs are more effective

Inputs for Fig. 6.5.

```
** INPUTS FOR THE FIG.6.5 :    ISN=-.01  DX= .02  DY= .02  DZ= .02  DT= .015  FIELD 101x101x101 # OF TIME STEPS=   153 **
R1=  4.5  R11=.0000001  R12=  0  R13= .0065  R14= .0001  R15= .0065  R16=  0  R17= .0005
R10= .015  R19= .0001  R110= .00001  KAPPA1= .0000001
R2=  1  R21= .000001  R22= .000001  R23= .0005  R24= .0001  R25= .00025  R26= .0005  R27= .0001  KAPPA2= .0000001
R3=  1  R31= .000001  R32= .00002  R33= .00005  R34= .0001  R35= .0001  KAPPA3= .0000001
R4=  1  R41= .000001  R42= .000001  R43= .00005  R44= .0001  R45= .00025  R46= .0005  R47= .0001  KAPPA4= .0000001
R5=  1  R51= .000001  R52= .000001  R53= .0001  R54= .0001  R55= .00025  R56= .0001  R57= .0001  KAPPA5= .0000001
R6=  1  R61= .000001  R62= .00005  R63= .0005  R64= .0001  KAPPA6= .0000001
R7=  1  R71= .000001  R72= .000005  R73= .000005  R74= .000005  R75= .000005  R76= .000005
R77= .000001  R78= .0001  KAPPA7= .0000001
R8=  1  R81= .000001  R82= .0005  R83= .000001  R84= .000001  R85= .000001  R86= .000001  R87= .000001  KAPPA8= .001
R9=  1  R91= .00001  R92= .000025  R93= .0001  R4= .000001  KAPPA9= .0000001
R10=  1  R101= .00001  R102= .00001  R103= .00001  R104= .0001  R105= .00001  R106= .00001
R107= .000001  KAPPA10= .0000001

CONDITIONS AT THE PRIMARY SITE AT T=0:
AG0=  15000  CT0= 20  DM0= 20  B0= 20  NK0= 20  EI0= 250  VC0= 250  RD0= 1000  ST0= 250  CH0= 250
NK0=  20  EI0= 250  VC0= 250  RD0= 1000  ST0= 250  CH0= 250
AT THE THREE AFFECTED LYMPH NODES AG0,CT0,DM0,B0,NK0=      500  20  20  20  20

CONDITIONS OUTSIDE THE SITES, AT T=0:
AG0= 500  CT0= 250  DM0= 250  BM0= 250  NK0= 250  EI0= 250  VC0= 250  RD0= 0  ST0= 250  CH0= 250

ALL MEDICATIONS ARE ADMINISTERED INTRAVENOUSLY. RD IS MONITORED INTRATUMORAL
CT,DM,B,NK ALL CANNOT EXCEED 3000.
```

6.2.21   Use of Fixed Dosages of Medications

Our next topic is the standardized treatment with fixed amounts of chemos and radiation. In most treatments, doctors use fixed dosages of medication after some fixed intervals of time. But once these medications move in the bloodstream, they follow the dynamics of the reaction–dispersion equations. The model may now be expressed as follows:

$$\partial(ei)/\partial t = r_6(ei0) - r_{61}(ei) - r_{62}(ag)(ei) - r_{63}(ei)(rd)$$
$$- r_{64}(ei)(ch) + \kappa_6 \nabla^2(ei) \tag{6.15}$$

$$\partial(vc)/\partial t = r_7(vc0) - r_{71}(vc) - r_{72}(ag)(vc) - r_{73}(ct)(vc)$$
$$- r_{74}(dm)(vc) - r_{75}(b)(vc) - r_{76}(nk)(vc)$$
$$- r_{77}(ch)(vc) - r_{78}(rd)(vc) + \kappa_7 \nabla^2(vc) \tag{6.16}$$

$$\partial(rd)/\partial t = r_8(rd0) - r_{81}(rd) - r_{82}(ag)(rd)$$
$$- r_{83}((ct) + (dm) + (b) + (nk))(rd) - r_{84}(ei)(rd)$$
$$- r_{85}(vc)(rd) - r_{86}(st)(rd) - r_{87}(ch)(rd) + \kappa_8 \nabla^2(rd) \tag{6.17}$$

$$\partial(st)/\partial t = r_9(st0) - r_{91}(st) - r_{92}((ct) + (b) + (nk))(st)$$
$$- r_{93}(rd)(st) - r_{94}(ch)(st) + \kappa_9 \nabla^2(st) \tag{6.18}$$

$$\partial(ch)/\partial t = r_{10}(ch0) - r_{101}(ch) - r_{102}(ag)(ch) - r_{103}((ct) + (dm)$$
$$+ (b) + (nk)))(ch) - r_{104}(rd)(ch) - r_{105}(ei)(ch)$$
$$- r_{106}(vc)(ch) - r_{107}(st)(ch) + \kappa_{10} \nabla^2(ch) \tag{6.19}$$

where $(ei0)$, $(vc0)$, $(rd0)$, $(st0)$, and $(ch0)$ are the fixed dosages of medications given to the patient at each time interval. The first five equations for (ag), (ct), (dm), (b), and (nk) all remain unchanged.

It has to be noticed that although a fixed amount of medications is infused in the blood, when they get mixed up, then they start changing with time due to chemical interactions.

Inputs for Fig. 6.6.

```
•• INPUTS FOR THE FIG.6.6   :    ISN- -.005  DX-  .02  DY-  .02  DZ-  .02  DT-  .02  FIELD 101x101x101 # OF TIME STEPS-    2 ••
R1-  4.5  R11- .0000001  R12-  .0025  R13-  .0025  R14-  .0025  R15-  .0025  R16-  .00025  R17-  .0001
R18-  .015  R19-  .0065  R110-  .00001  KAPPA1-  .0000001
R2-   1  R21-  .000001  R22-  .00001  R23-  .00005  R24-  .0001  R25-  .00025  R26-  .0005  R27-  .0001  KAPPA2-  .0000001
R3-   1  R31-  .000001  R32-  .00002  R33-  .00005  R34-  .0001  R35-  .0001  KAPPA3-  .0000001
R4-   1  R41-  .000001  R42-  .00002  R43-  .00005  R44-  .0001  R45-  .00025  R46-  .0005  R47-  .0001  KAPPA4-  .0000001
R5-   1  R51-  .000001  R52-  .00002  R53-  .00005  R54-  .0001  R55-  .00025  R56-  .0001  R57-  .0001  KAPPA5-  .0000001
R6-   1  R61-  .00001  R62-  .00005  R63-  .0005  R64-  .0001  KAPPA6-  .0000001
R7-   1  R71-  .00001  R72-  .00006  R73-  .00005  R74-  .00005  R75-  .00005  R76-  .00005
R77-  .0001  R78-  .005  KAPPA7-  .0000001
R8-   1  R81-  .00001  R82-  .0005  R83-  .0001  R84-  .0001  R85-  .0001  R86-  .0001  R87-  .0001  KAPPA8-  .001
R9-   1  R91-  .00001  R92-  .000025  R93-  .0001  R4-  .0001  KAPPA9-  .0000001
R10-  1  R101- .00001  R102-  .00005  R103-  .00001  R104-  .00005  R105-  .00001  R106-  .00005
R107-  .00001  KAPPA10-  .0000001

CONDITIONS AT THE PRIMARY SITE AT I=0
AG0-  15000  CT0-  20  DM0-  20  B0-  20  NK0-  20
EI0-  200  VC0-  200  RD0-  400  ST0-  200  CH0-  200
AT THE THREE AFFECTED LYMPH NODES AG0,CT0,DM0,B0,NK0-    2000   20   20   20   20
CONDITIONS OUTSIDE THE SITE:
AG0-  500  CT0-  100  DM0-  100  B0-  100  NK0-  100
EI0-  200  VC0-  200  RD0-  0  ST0-  200  CH0-  200
```

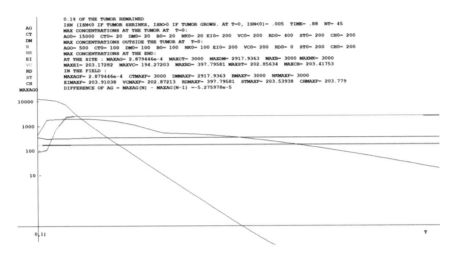

Fig. 6.6 With fixed amount of EI, VC, RD, ST & CH

With these inputs, we get the following results from the Fig. 6.6:

1. The primary tumor is shrinking fast. However, the number of cancer cells outside the tumor first started increasing fast and then slowly decreased.
2. Also the defense went into a steady state. This happens in general when a treatment is successful. So our results are very realistic.

6.3 Immunotherapy for Breast Cancer Treatment

In this section, we have considered a stronger immune system. No other drug has been used. BBc.com, January 20, 2020 reported that there is a new discovery by "The Cardiff University team" which mentions: "A newly-discovered part of our immune system could be harnessed to treat all cancers…" [7]. They discovered a T-cell in the blood which can scan the entire body to "assess whether there is a threat that needs to be eliminated." The researchers believe that this could attack a wide range of cancers including breast cancer while leaving all normal cells untouched. However, we have one big concern. Unless this finding is tested in vivo, in cancer patients, the results shall remain uncertain. Because the cancer cells may release some new proteins to overcome these new obstacles and confuse the T-cells. So we considered immunotherapy which will strengthen all the lymphocytes which launch antigen-specific attacks in the entire body. We hope that cancer cells may not be able to confuse all of them.

Our principle of immunotherapy is based upon the fact that there is a strong rapport between all the cells of our immune system. They together work to recognize proteins released by the malignant cells and fight accordingly. This model consists of just five nonlinear reaction–diffusion equations.

$$\partial(ag)/\partial t = r_1(ag) - r_{11}(ag) - r_{12}(ag)(ct) - r_{13}(ag)(b)$$
$$- r_{14}(ag)(dm) - r_{15}(ag)(nk) + r_{110}(ag)\theta + \kappa_1\nabla^2(ag). \quad (6.20)$$

$$\partial(ct)/\partial t = r_2(ag) - r_{21}(ct) - r_{22}(ag)(ct)$$
$$+ r_{25}(ct)(dm) + (ISN)(ct)(ag) + \kappa_2\nabla^2(ct). \quad (6.21)$$

$$\partial(dm)/\partial t = r_3(ag) - r_{31}(dm) - r_{32}(ag)(dm)$$
$$+ (ISN)(dm)(ag) + \kappa_3\nabla^2(ct). \quad (6.22)$$

$$\partial(b)/\partial t = r_4(ag) - r_{41}(b) - r_{42}(ag)(b)$$
$$+ r_{45}(b)(dm) + (ISN)(b)(ag) + \kappa_4\nabla^2(b). \quad (6.23)$$

$$\partial(nk)/\partial t = r_5(ag) - r_{51}(nk) - r_{52}(ag)(nk)$$
$$+ r_{55}(nk)(dm) + (ISN) + \kappa_5\nabla^2(nk). \quad (6.24)$$

At the very beginning, the stress level of a patient is generally high. So we have assumed ISN $= -0.01$. During the treatment, if the blood test reveals better results, ISN is positive. So the value of ISN is negative if the patient is depressed, and it is positive if prognosis is favorable.

We have considered three cases, namely Figs. 6.7, 6.8, and 6.9. For all these cases, the initial values are $(ct0) = 1200$, $(dm0) = 700$, $(b0) = 900$, and $(nk0) = 1000$ at each point outside the tumor. At the cancer site where the tumor is, as well as at each three affected lymph nodes, the value of each of them is 20 at each point. The initial

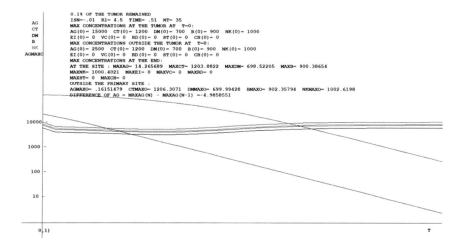

Fig. 6.7 Immunotherapy for CT, DM, B & NK, they change AS AG changes

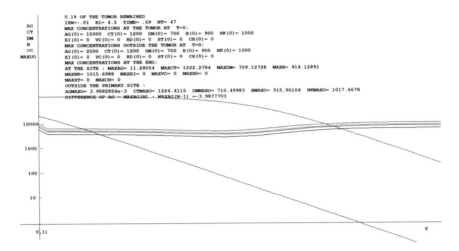

Fig. 6.8 Immunotherapy for CT, DM, B & NK, fixed amounts used at each time step

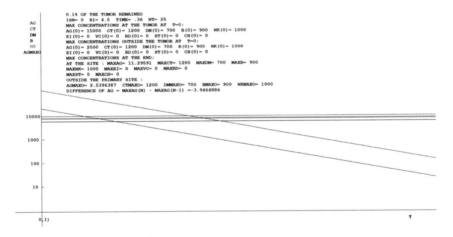

Fig. 6.9 Validation of the resuts from Figs. 6.7 and 6.8

value of $(ag0) = 15000$ at each point at the cancer site and 2500 at the affected lymph nodes. Cancer is aggressive $r_1 = 4.5$, a very high speed of replication. As before, $\kappa_1 = \kappa_2 = \kappa_3 = \kappa_4 = \kappa_5 = 10^{-7}$. Here the reaction coefficients are as follows.

$r_{12} = 6.5E - 03, r_{13} = r_{14} = r_{15} = r_{16} = r_{12}, r_{17} = r_{18} = r_{19} = 0, r_{110} = 1.0E\text{-}06$. All other reaction constants are mentioned in the inputs for Fig. 6.7.

No radiation and no other medications have been applied.

Inputs for Fig. 6.7.

```
•• INPUTS FOR THE FIG.6.7      ISN=-.01  DX=  .02  DY=  .02  DZ=  .02  DT=  .015  FIELD 101x101x101 # OF TIME STEPS=    300  ••
R1=   4.5  R11=  .0000001  R12=  .0065  R13=  .0065  R14=  .0065  R15=  .0065  R16=  0  R17=  0
R18=   0  R19=  0  R110=  .000001   KAPPA1=  .0000001
R2=   1  R21=  .000001  R22=  .000001  R23=  0  R24=  0  R25=  .0005  R26=  0  R27=  0  KAPPA2=  .0000001
R3=   1  R31=  .000001  R32=  .000001  R33=  0  R34=  0  R35=  0  KAPPA3=  .0000001
R4=   1  R41=  .000001  R42=  .000001  R43=  0  R44=  0  R45=  .00025  R46=  0  R47=  0  KAPPA4=  .0000001
R5=   1  R51=  .000001  R52=  .000001  R53=  0  R54=  0  R55=  .00025  R56=  0  R57=  0  KAPPA5=  .0000001
R6=   0  R61=  0  R62=  0  R63=  0  R64=  0  KAPPA6=  0
R7=   0  R71=  0  R72=  0  R73=  0  R74=  0  R75=  0  R76=  0
R77=   0  R78=  0  KAPPA7=  0
R8=   0  R81=  0  R82=  0  R83=  0  R84=  0  R85=  0  R86=  0  R87=  0  KAPPA8=  0
R9=   0  R91=  0  R92=  0  R93=  0  R4=  0  KAPPA9=  0
R10=   0  R101=  0  R102=  0  R103=  0  R104=  0  R105=  0  R106=  0
R107=   0  KAPPA10=  0

CONDITIONS AT THE PRIMARY SITE AT T-0
AG0=  15000  CT0=  1200  DM0=  700  B0=  900  NK0=  1000
EI0=   0  VC0=  0  RD0=  0  ST0=  0  CH0=  0
AT THE THREE AFFECTED LYMPH NODES AG0,CT0,DM0,B0,NK0=          2500   20   20   20   20
CONDITIONS OUTSIDE THE SITES, AT T-0
AG0=  2500  CT0=  1200  DM0=  700  BM0=  900  NK0=  1000
EI0=   0  VC0=  0  RD0=  0  ST0=  0  CH0=  0
```

The initial dosages of (ct), (dm), (b), and (nk) given as immunotherapy intravenously are, respectively, 1200, 700, 900, and 1000 at each point in the entire field. The dosages of the immunotherapy change in the blood with time. The results are given in Fig. 6.7.

It is evident from the figure that the therapy first faced a huge blow that downgraded its effectiveness and later sped up and defeated the enemy.

Our next effort was to see that instead of adjusting the values of this therapeutic procedure in accordance with the values of (ag) if we maintain the same dosages of (ct), (dm), (b), and (nk) at each time step. However, we considered that when these biochemicals enter into the blood system intravenously, they change following the Eqs. (6.14), (6.15), (6.16), and (6.17) as the dynamics of the war change. All other inputs are kept the same.

The new model is

$$\partial(ag)/\partial t = r_1(ag) - r_{11}(ag) - r_{12}(ag)(ct) - r_{13}(ag)(b)$$
$$- r_{14}(ag)(dm) - r_{15}(ag)(nk) + r_{110}(ag)\theta + \kappa_1\nabla^2(ag). \quad (6.25)$$

$$\partial(ct)/\partial t = r_2(ct0) - r_{21}(ct) - r_{22}(ag)(ct)$$
$$+ r_{25}(ct)(dm) + (ISN)(ct)(ag) + \kappa_2\nabla^2(ct). \quad (6.26)$$

$$\partial(dm)/\partial t = r_3(dm0) - r_{31}(dm) - r_{32}(ag)(dm)$$
$$+ (ISN)(dm)(ag) + \kappa_3\nabla^2(ct). \quad (6.27)$$

$$\partial(b)/\partial t = r_4(b0) - r_{41}(b) - r_{42}(ag)(b)$$
$$+ r_{45}(b)(dm) + (ISN)(b)(ag) + \kappa_4\nabla^2(b) \quad (6.28)$$

$$\partial(nk)/\partial t = r_5(nk0) - r_{51}(nk) - r_{52}(ag)(nk)$$
$$+ r_{55}(nk)(dm) + (ISN) + \kappa_5\nabla^2(nk). \quad (6.29)$$

We again need to understand that once these drugs enter into the bloodstream, they behave like any other variable biochemicals, or in other words, they do not maintain exactly their same strengths as they initially had. Their chemical strengths change. The results are given in Fig. 6.8. The input data are exactly the same as in Fig. 6.7.

The inputs for Fig. 6.8 are the same as before, the only difference is that a constant amount of medications was infused in the blood in this case. Those dosages are the same as the initial dosages in the case shown in Fig. 6.7.

Interestingly, it may be observed as follows: (The Results).

In Fig. 6.7, at the steady state, $(ct) = 1204$, $(dm) = 699$, $(b) = 900$, and $(nk) = 1000$.

In Fig. 6.8, at the steady state, $(ct) = 1222$, $(dm) = 709$, $(b) = 914$, and $(nk) = 1015$.

These two results are very similar, which was expected. The initial values of the parameters are $(ct) = 1200$, $(dm) = 700$, $(b) = 900$, and $(nk) = 1000$. This reveals that the oncologist, at the outset of the treatment made an excellent choice of treatment. Let us briefly discuss these outcomes.

Mathematically, these results validate that our numerical method (Charlie-Simpson) is truly an excellent stiff system solver because looking at the graphs we notice that the nonlinear system is stiff.

So, we considered a next computer experiment where we have traced the values of (ag) keeping the values of $(ct) = 1200$, $(dm) = 700$, $(b) = 900$, and $(nk) = 1000$ fixed all through. This is just an assumption.

The model is

$$\partial(ag)/\partial t = r_1(ag) - r_{11}(ag) - r_{12}(ag)(ct0) - r_{13}(ag)(b0)$$
$$- r_{14}(ag)(dm0) - r_{15}(ag)(nk0) + r_{110}(ag)\theta + \kappa_1 \nabla^2(ag).$$

where $\theta = (ag)/(ct0 + b0 + dm0 + nk0)$.

So, we may reorganize this by substituting the value of θ equation as follows:

$$\partial(ag)/\partial t = (r_1 - r_{11} - r_{12}(ct0) - r_{13}(b0) - r_{14}(dm0) - r_{15}(nk0)$$
$$+ r_{110}(ag)/(ct0 + b0 + dm0 + nk0))(ag) + \kappa_1 \nabla^2(ag).$$

Then

$$\sigma = (r_1 - r_{11} - r_{12}(ct0) - r_{13}(b0) - r_{14}(dm0)$$
$$- r_{15}(nk0) + r_{110}(ag0)/(ct0 + b0 + dm0 + nk0))$$

Using the values of the reaction coefficients and the initial values $(ag0) = 15000$, $(ct0) = 1200$, $(b0) = 900$, $(dm0) = 700$, $(nk0) = 100$, we get $\sigma = -20.20 < 0$. So, the necessary condition that (ag) should decrease is satisfied.

This is what we observe in Fig. 6.9. Here the reaction coefficients are as follows:

$$r_{12} = 6.5E - 03, r_{13} = r_{14} = r_{15} = r_{16} = r_{12}, r_{17} = r_{18} = r_{19} = 0, r_{110} = 1.0E - 06.$$

This is another way to conduct immunotherapy mathematically and validate its outcome.

Only medical professionals could interpret these results better. But mathematically, the results appear to be very consistent as shown before.

6.4 Modeling a Stark Tragedy in Cancer Treatment

Sometimes false hopes overwhelm both doctors and patients. It could appear from blood tests, MRIs, ultrasounds, etc., that the primary tumor is shrinking very well. All tests reveal very clearly that the patient is progressing toward a cure. However, secretly cancer cells could slowly grow and metastasize and pose a threat to the life of the patient. We have considered a case ISN $= -0.001$. The rate contents for the immune system are relatively small. $r_{12} = r_{13} = r_{14} = r_{15} = 0.00005$. These show at what rates leukocytes are destroying cancer cells. In comparison with these $r_{22} = r_{42} = r_{52} = 0.00001$ and $r_{32} = 0.00002$, the rate constants are giving the rates at which cancer cells are destroying them. So element wise, the two sets $(r_{12}, r_{13}, r_{14}, r_{15})$ and $(r_{22}, r_{32}, r_{42}, r_{52})$ do not differ very much. That obviously indicates that the immune system is not very aggressive. At first, MAXAG (in Fig. 6.10) was getting smaller (tumor was shrinking) while MAXAGO was rising. Later both started getting larger and larger. This seems to be a disaster. First there was a ray of hope, and later it was just all darkness!

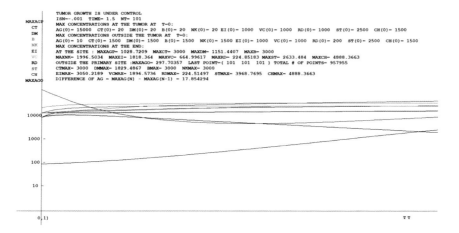

Fig. 6.10 A tragedy: tumor metastasizing

Inputs for Fig. 6.10.

```
*** INITIAL CONDITIONS FOR MODEL ON MATHEMATICIZATION OF A TALE OF TRAGEDY.FIG.6.10 ***
ISN- -.001   DX- .02   DY- .02   DZ- .02   DT- .015   FIELD SIZE- 101x101x101   # OF TIME STEPS-  100
R1-  2.5   R11- .0000001   R12- .00005   R13- .00005   R14- .00005   R15- .00005   R16- .00001   R17- .0001
R18- .015   R19- .0001   R110- .000001   KAPPA1- .0000001
R2-   1   R21- .000001   R22- .00001   R23- .00005   R24- .0001   R25- .00025   R26- .0005   R27- .0001   KAPPA2- .0000001
R3-   1   R31- .000001   R32- .00002   R33- .00005   R34- .0001   R35- .0001   KAPPA3- .0000001
R4-   1   R41- .000001   R42- .00001   R43- .00005   R44- .0001   R45- .00025   R46- .0005   R47- .0001   KAPPA4- .0000001
R5-   1   R51- .000001   R52- .00001   R53- .00005   R54- .0001   R55- .00025   R56- .0001   R57- .0001   KAPPA5- .0000001
R6-   1   R61- .000001   R62- .00005   R63- .0005   R64- .0001   KAPPA6- .0000001
R7-   1   R71- .00001   R72- .00006   R73- .00005   R74- .00005   R75- .00005   R76- .00005
R77- .0001   R78- .005   KAPPA7- .0000001
R8-   1   R81- .00001   R82- .0005   R83- .0001   R84- .0001   R85- .0001   R86- .0001   R87- .0001   KAPPA8- .001
R9-   1   R91- .00001   R92- .000025   R93- .0001   R94- .0001   KAPPA9- .0000001
R10-  1   R101- .00001   R102- .00001   R103- .00001   R104- .0005   R105- .00001   R106- .00001
R107- .00001   KAPPA10- .0000001

CONDITIONS AT THE PRIMARY SITE AT T=0:
AGMAX0-   15000   CTMAX0-   20   DMMAX0-   20   BMAX0-   20   NKMAX0-   20
EIMAX0-   1000   VCMAX0-   1000   RDMAX0-   1000   STMAX0-   2500   CHMAX0-   1500
THE THREE LYMPH NODES AGO, CT0,DM0,B0,NK0-       10   20   20   20   20
CONDITIONS OUTSIDE THE SITE:
AGMAX0-   10   CTMAX0-   1500   DMMAX0-   1500   BMAX0-   1500   NKMAX0-   1500
EIMAX0-   1000   VCMAX0-   1000   RDMAX0-   200   STMAX0-   2500   CHMAX0-   1500
```

Here $r_1 = 2.5$. Cancer is not growing very fast at all. Outside the tumor site, in comparison with other models, the immune system is relatively strong, yet the rate constants reveal that they are not fighting very strongly.

The results, shown in Fig. 6.10, show a rather grim situation. Although the tumor is shrinking, defense seems to be working well, and silently cancer is spreading all over the body.

We took a shorter run to check what was going on, while the tumor was shrinking and cancer was growing outside. This could be because more cells were leaving the tumor site through the flow of blood and lymph. We have used the same inputs as before in Fig. 6.11.

The first author saw such scenarios while observing cancer patients in hospitals for almost two years. When the immune system was made four times stronger and the patient was able to become less stressful having ISN $= 0.01$, the situation did not improve as may be seen in Fig. 6.12.

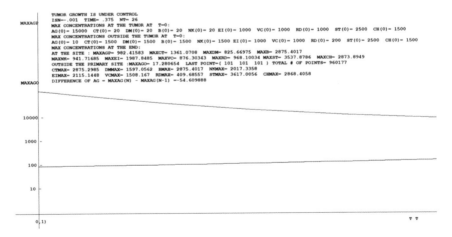

Fig. 6.11 A tragedy: primary tumor shrinking but cancer is metastasizing

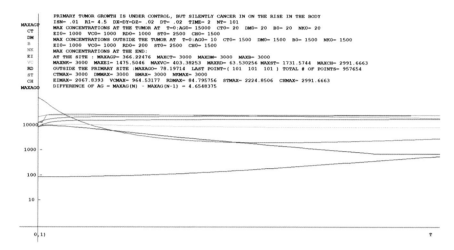

Fig. 6.12 Primary tumor shrinking but cancer cells increasing very slowly in the field

Inputs for Fig. 6.12.

```
•••••••••••••••••••••••••••••••••••••••••••••••••••••••••••••••••••••••••••••••••••••
 ••• INITIAL CONDITIONS FOR MODEL ON MATHEMATICIZATION OF A TALE OF TRAGEDY.FIG.6.12  •••
 ISN-    .01   DX-  .02   DY-  .02  DZ-  .02   DT-  .02   FIELD SIZE- 101x101x101   # OF TIME STEPS-    100
 R1-   4.5   R11-  .0000001   R12-  .0002  R13-  .0002   R14-  .0002   R15-  .0002   R16-  .00001   R17-  .0001
 R18-   .015   R19-  .0001   R110-  .000001   KAPPA1-  .0000001
 R2-   1   R21-  .000001   R22-  .00001   R23-  .00005   R24-  .0001   R25-  .00025   R26-  .0005   R27-  .0001   KAPPA2-  .0000001
 R3-   1   R31-  .000001   R32-  .00002   R33-  .0005   R34-  .0001   R35-  .0001   KAPPA3-  .0000001
 R4-   1   R41-  .000001   R42-  .00001   R43-  .00005   R44-  .0001   R45-  .00025   R46-  .0005   R47-  .0001   KAPPA4-  .0000001
 R5-   1   R51-  .000001   R52-  .00001   R53-  .0005   R54-  .0001   R55-  .00025   R56-  .0001   R57-  .0001   NU5-  .0000001
 R6-   1   R61-  .00001   R62-  .00005   R63-  .0005   R64-  .0001   KAPPA6-  .0000001
 R7-   1   R71-  .00001   R72-  .00006   R73-  .00005   R74-  .00005   R75-  .00005   R76-  .00005
 R77-   .0001   R78-  .005   KAPPA7-  .0000001
 R8-   1   R81-  .00001   R82-  .0005   R83-  .0001   R84-  .0001   R85-  .0001   R86-  .0001   R87-  .0001   KAPPA8-  .001
 R9-   1   R91-  .00001   R92-  .000025   R93-  .0001   R94-  .0001   KAPPA9-  .0000001
 R10-   1   R101-  .00001   R102-  .00001   R103-  .00001   R104-  .00005   R105-  .00001   R106-  .00001
 R107-   .00001   KAPPA10-  .0000001
 CONDITIONS AT THE PRIMARY SITE AT T-0:
 AGMAXO-    15000   CTMAXO-   20   DMMAXO-   20   BMAXO-   20   NKMAXO-   20
 EIMAXO-    1000   VCMAXO-   1000   RDMAXO-  1000   STMAXO-  2500   CHMAXO-  1500
 CT,DM,B,NK,AG AT EACH AFFECTED NODE -       20   20   20   20   5000
 CONDITIONS OUTSIDE THE SITE:
 AGMAXO-    10   CTMAXO-   1500   DMMAXO-  1500   BMAXO-  1500   NKMAXO-  1500
 EIMAXO-    1000   VCMAXO-   1000   RDMAXO-  200   STMAXO-  2500   CHMAXO-  1500
•••••••••••••••••••••••••••••••••••••••••••••••••••••••••••••••••••••••••••••••••••••
```

Here (rd) was injected (using radiation-carrying drugs) both at the tumor site and intravenously. To look for a better solution, we have increased the strengths of drugs especially enhancing the immune system. At the same time, radiation has been concentrated only at the tumor site. The results are given by Fig. 6.13. Rate constants were increased to destroy cancer as may be noticed in the inputs for Fig. 6.13.

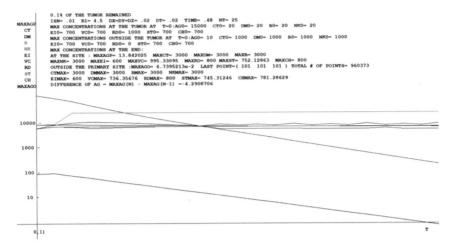

Fig. 6.13 Medications are always adjusted for a successful treatment

Inputs for Fig. 6.13.

```
•••••••••••••••••••••••••••••••••••••••••••••••••••••••••••••••••••••••••••••••••••••••••••••

*** INITIAL CONDITIONS FOR MODEL ON MATHEMATICIZATION OF A TALE OF TRAGEDY CORRECTED.FIG.6.13  ***

ISN- .01  DX- .02  DY- .02  DZ- .02  DT- .02  FIELD SIZE- 101x101x101 # OF TIME STEPS- 100
R1- 4.5  R11- .0000001 R12- .00025 R13- .00025 R14- .00025 R15- .00025 R16- .00025 R17- .00025
R18- .0185  R19- .00025  R110- .000001  KAPPA1- .0000001
R2- 1  R21- .000001 R22- .00001 R23- .00005 R24- .0001 R25- .00025 R26- .0005 R27- .0001 KAPPA2- .0000001
R3- 1  R31- .000001 R32- .00002 R33- .00005 R34- .0001 R35- .0001 KAPPA3- .0000001
R4- 1  R41- .000001 R42- .00001 R43- .00005 R44- .0001 R45- .00025 R46- .0005 R47- .0001 KAPPA4- .0000001
R5- 1  R51- .000001 R52- .00001 R53- .00005 R54- .0001 R55- .00025 R56- .0001 R57- .0001 NU5- .0000001
R6- 1  R61- .00001 R62- .00005 R63- .0005 R64- .0001 KAPPA6- .0000001
R7- 1  R71- .00001 R72- .00006 R73- .00005 R74- .00005 R75- .00005 R76- .00005
R77- .0001  R78- .005  KAPPA7- .0000001
R8- 1  R81- .00001 R82- .0005 R83- .0001 R84- .0001 R85- .0001 R86- .0001 R87- .0001 KAPPA8- .001
R9- 1  R91- .00001 R92- .000025 R93- .0001 R94- .0001 KAPPA9- .0000001
R10- 1  R101- .00001 R102- .00001 R103- .00001 R104- .00005 R105- .00005 R106- .00001
R107- .00001  KAPPA10- .0000001
CONDITIONS AT THE PRIMARY SITE AT T-0: AGMAX0- 15000  CTMAX0- 20  DMMAX0- 20  BMAX0- 20  NKMAX0- 20
EIMAX0- 700  VCMAX0- 700  RDMAX0- 1000  STMAX0- 700  CHMAX0- 700
THE THREE LYMPH NODES AG0, CT0,DM0,B0,NK0- 5000   20   20   20   20
CONDITIONS OUTSIDE THE SITE:AGMAX0- 10  CTMAX0- 1000  DMMAX0- 1000  BMAX0- 1000  NKMAX0- 1000
EIMAX0- 700  VCMAX0- 700  RDMAX0- 0  STMAX0- 700  CHMAX0- 700
DURING COMPUTATIONS THE VALUES OF ALL PARAMETERS ARE ADJUSTED IF THEY FALL OR GO UP AT ANY TIME.

•••••••••••••••••••••••••••••••••••••••••••••••••••••••••••••••••••••••••••••••••••••••••••••
```

In Fig. 6.13, we see some welcoming results. Cancer is not spreading anymore. The lower blue line, representing the spread of cancer cells outside the tumor site, has a strong negative slope. *The graphs must be read with much care. Sometimes, we had to reorganize the logarithmic scale to get graphs which display results easy for the readers to comprehend the conditions of patients, who are truly virtual.*

This is what happens if the immune system is strengthened.

6.4.1   The Necessity for Adjuvant Therapy

After the regular treatment of a cancer patient is over with the removal of tumors and reduction of cancer cells in blood, she is not let free. Oncologists start adjuvant therapy, therapy after surgeries, and examine her often, generally, every three to six months, using MRI, mammograms, etc. Tumors may not be there but remnants of cancer cells, at the site of the tumor and elsewhere, could be present completely undetected circulating in the blood. Later, after months or even years, they could form tumors and be a threat to life.

We will now attempt to study this phenomenon mathematically. We have the following inputs:

Inputs for Fig. 6.14.

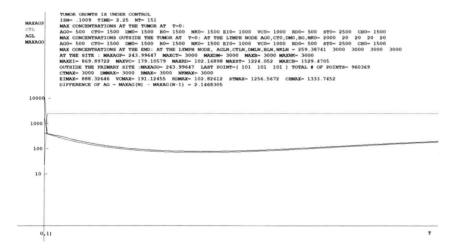

Fig. 6.14   Affected remote lymph node is not removed

ISN $= 0.1009$, meaning that the patient is psychologically in good spirit, having less stress. Tumor is gone. However, cancer cells were not totally eliminated at three points. These could be three lymph nodes. At these three points, $ag = 2000$, $(ct), (dm), (b), and (nk)$ are each 20. Also some cancer cells are still circulating in blood. Tumor being gone, here strengths of medications have been diminished. So we have considered: $r_{12} = r_{13} = r_{14} = r_{15} = 0.00015$ as recorded in the inputs for Fig. 6.14. Here MAXAGP = the largest value of cancer cells at the primary site of the tumor, (ct), (cytotoxic T-cells) at the site where initially it was 20 (the same is true for $(dm), (b)$, and (nk) cells). Now they are strongly present there. Each is 1500. At t $= 0$, the amounts of drugs used are as follows: $(ei0) = (vc0) = 1000$; $(rd0) = 500$; $(st0) = 2500$ and $(ch0) = 1500$. They change as (ag) changes. As (ag) started decreasing, they started decreasing, and similarly they started increasing as (ag) started increasing. AGL = the value of (ag) at the site where cancer cells were hiding, not detected earlier. At these points, $(ag0) = 2000$. These points could be some undetected lymph nodes. There are three such points at three different points in the computational field. MAXAGO = the largest number of (ag) outside the primary site. The treatment continues though. Values of the drugs were adjusted as the number of (ag) changes as done before. The graph shows that at first MAXAGP, AGL, and MAXAGO all started getting smaller. AGL went down from 2000 to 259.39 at the end. But prior to that, it was much less. This is true for MAXAGP and MAXAGO. At some point before the $NT = 150$, they all started increasing. This is a disaster for the treatment procedure. What this reveals is that in the case of cancer treatment, all good news should be taken with very much care and caution. Or in other words, both patients and the teams of medical professionals should remain cautiously optimistic. (Note: AGL = the largest value of (ag)).

Now we consider a case where by surgery or chemo all cancerous locations were removed. The following inputs for Fig. 6.15 have been used in the code.

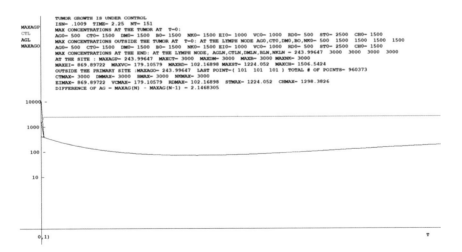

Fig. 6.15 Tumor and all affected lymph nodes removed

Inputs for Fig. 6.15.

```
•••••••••••••••••••••••••••••••••••••••••••••••••••••••••••••••••••••••••••••••••••••••••••
*** INITIAL CONDITIONS FOR THE FIG.6.15    . ALL AFFECTED LYMPH NODES WERE REMOVED ***
ISN- .1009  DX- .02  DY- .02  DZ- .02  DT- .015  FIELD SIZE- 101x101x101 # OF TIME STEPS- 150
R1- 4.5  R11- .0000001  R12- .00015  R13- .00015  R14- .00015  R15- .00015  R16- .00001  R17- .0001
R18- .015  R19- .0001  R110- .000001  KAPPA1- .0000001
R2- 1  R21- .000001  R22- .00001  R23- .00005  R24- .0001  R25- .00025  R26- .0005  R27- .0001  KAPPA2- .0000001
R3- 1  R31- .000001  R32- .00002  R33- .00005  R34- .0001  R35- .0001  KAPPA3- .0000001
R4- 1  R41- .000001  R42- .00001  R43- .00005  R44- .0001  R45- .00025  R46- .0005  R47- .0001  KAPPA4- .0000001
R5- 1  R51- .000001  R52- .0001  R53- .00005  R54- .0001  R55- .00025  R56- .0001  R57- .0001  KAPPA5- .0000001
R6- 1  R61- .00001  R62- .00005  R63- .0005  R64- .0001  KAPPA6- .0000001
R7- 1  R71- .00001  R72- .00006  R73- .00005  R74- .00005  R75- .00005  R76- .00005
R77- .0001  R78- .005  KAPPA7- .0000001
R8- 1  R81- .00001  R82- .0005  R83- .0001  R84- .0001  R85- .0001  R86- .0001  R87- .0001  KAPPA8- .001
R9- 1  R91- .00001  R92- .000025  R93- .0001  R94- .0001  KAPPA9- .0000001
R10- 1  R101- .00001  R102- .00001  R103- .00001  R104- .00005  R105- .00001  R106- .00001
R107- .00001  KAPPA10- .0000001
CONDITIONS AT THE PRIMARY SITE AT T-0:
AGMAX0- 500  CTMAX0- 1500  DMMAX0- 1500  BMAX0- 1500  NKMAX0- 1500
EIMAX0- 1000  VCMAX0- 1000  RDMAX0- 500  STMAX0- 2500  CHMAX0- 1500
THE THREE LYMPH NODES AG0, CT0,DM0,B0,NK0- 500   1500   1500   1500   1500
CONDITIONS OUTSIDE THE SITE:
AGMAX0- 500  CTMAX0- 1500  DMMAX0- 1500  BMAX0- 1500  NKMAX0- 1500
EIMAX0- 1000  VCMAX0- 1000  RDMAX0- 500  STMAX0- 2500  CHMAX0- 1500
•••••••••••••••••••••••••••••••••••••••••••••••••••••••••••••••••••••••••••••••••••••
```

The results did not change very much when the very same treatments were used for the case when all affected lymph nodes were removed and cancer cells were removed from all their hiding points. Only, they are still in the blood flow and in lymph. Noting that $r_1 = 4.5$ is a fast growing cancer. Cancer is replicating at the rate of 350%. The results are shown in Fig. 6.15. It could be noticed that in the entire field, lymphocytes came back with full force as computations began. Even then, as treatment began, first cancer cells swimming in groups in the streams of blood and lymph started decreasing and the amount of drugs for the therapy were decreasing in keeping with the decrease of cancer cells. However, later cancer cells meticulously took advantage of that situation and started increasing again. The lesson is as follows: Protocols for the treatments must be vigilantly observed at every instant. Amount of drugs may be reduced as cancer cells keep on decreasing; however, the moment they take advantage of this scenario and show a tendency to increase, and the method of delivery of drugs must be amended.

There are several procedures to correct such a sad situation. First thought should be application of more drugs if it is possible. That has been done for Fig. 6.16. Inputs for (ei), (vc), (rd), (st), and (ch) were increased. Results got better, although a major question is: Can the patient tolerate this treatment? In fact this question arises for all cancer patients.

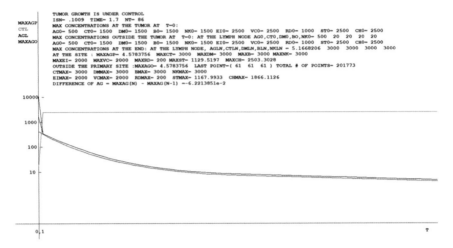

Fig. 6.16 Therapy made stronger

Inputs for Fig. 6.16.

```
*** INITIAL CONDITIONS FOR THE FIG.6.16    . THERAPY MADE STRONGER ***
ISN- .1009  DX- .02  DY- .02  DZ- .02  DT- .02  FIELD SIZE- 101x101x101 # OF TIME STEPS- 150
R1- 4.5  R11- .0000001  R12- .00015  R13- .00015  R14- .00015  R15- .00015  R16- .00001  R17- .0001
R18- .015  R19- .0001  R110- .000001  KAPPA1- .0000001
R2- 1  R21- .000001  R22- .00001  R23- .00005  R24- .0001  R25- .00025  R26- .0005  R27- .0001  KAPPA2- .0000001
R3- 1  R31- .000001  R32- .00002  R33- .00005  R34- .0001  R35- .0001  KAPPA3- .0000001
R4- 1  R41- .000001  R42- .00001  R43- .00005  R44- .0001  R45- .00025  R46- .0005  R47- .0001  KAPPA4- .0000001
R5- 1  R51- .000001  R52- .00001  R53- .00005  R54- .0001  R55- .00025  R56- .0001  R57- .0001  KAPPA5- .0000001
R6- 1  R61- .00001  R62- .00005  R63- .0005  R64- .0001  KAPPA6- .0000001
R7- 1  R71- .00001  R72- .00006  R73- .00005  R74- .00005  R75- .00005  R76- .00005
R77- .0001  R78- .005  KAPPA7- .0000001
R8- 1  R81- .00001  R82- .0005  R83- .0001  R84- .0001  R85- .0001  R86- .0001  R87- .0001  KAPPA8- .001
R9- 1  R91- .00001  R92- .000025  R93- .0001  R94- .0001  KAPPA9- .0000001
R10- 1  R101- .00001  R102- .00001  R103- .00001  R104- .00005  R105- .00001  R106- .00001
R107- .00001  KAPPA10- .0000001
CONDITIONS AT THE PRIMARY SITE AT T=0:
AGMAX0- 500  CTMAX0- 1500  DMMAX0- 1500  BMAX0- 1500  NKMAX0- 1500
EIMAX0- 2500  VCMAX0- 2500  RDMAX0- 1000  STMAX0- 2500  CHMAX0- 2500
THE THREE LYMPH NODES AG0, CT0,DM0,B0,NK0- 500  20  20  20  20
CONDITIONS OUTSIDE THE SITE:
AGMAX0- 500  CTMAX0- 1500  DMMAX0- 1500  BMAX0- 1500  NKMAX0- 1500
EIMAX0- 2500  VCMAX0- 2500  RDMAX0- 1000  STMAX0- 2500  CHMAX0- 2500
```

6.5 More on Adjuvant Therapy

We will now discuss some more details on the adjuvant therapy assuming that a surgical procedure has been completed and the patient is getting more therapy to eliminate all remnants of cancer cells still in circulation in the blood and lymph. This is the standard adjuvant therapy. Some of the medications generally used are as follows: (1) For estrogen positive breast cancer Tamoxifen is used preventing estrogen helping cancer cells to replicate; (2) aromatase is an enzyme released by the adrenal glands located on top of our kidneys that convert androgen into estrogen. So aromatase inhibitors (AIs) are used during adjuvant therapy, and (3) HER2 is an epidermal growth factor which is a protein. A gene mutation causes too much of these proteins that help growth of cancer cells in almost 20% of breast cancer patients. During adjuvant therapy, doctors use some of these and more as needed. These treatments are also sometimes given like neoadjuvant therapy before the primary treatment.

In the first model in this section, we have considered a form of immunotherapy together with a fixed amount of $(ei0)$ and $(vc0)$. We must note that during surgery, some cancer cells will run away from the site and remain in the circulation. Later they could form tumors again. This is also true during radiation treatment [11, 12]. In our models, we have considered all these possibilities. A few more points to remember are when a drug is given intravenously or by mouth and it goes into the bloodstream, then as it interacts with other biochemicals it changes. For instance, the first term like $r_6(ei0)$ denotes the rate of change of $(ei0)$ which is a constant. So, mathematically this term should be zero. But in reality, that is not correct. Because the moment it gets mixed with blood, it becomes a variable and it changes. At this point, we have assumed that radiation therapy has been stopped. Only some form of chemo continues.

6.5.1 The Mathematical Model for Adjuvant Therapy

$$\partial(ag)/\partial t = r_1(ag) - r_{11}(ag) - r_{12}(ag)(ct)$$
$$- r_{13}(ag)(b) - r_{14}(ag)(dm) - r_{15}(ag)(nk)$$
$$- r_{16}(ag)(vc) - r_{17}(ag)(ei) + r_{110}(ag)\theta + \kappa_1 \nabla^2(ag) \qquad (6.25)$$

$$\partial(ct)/\partial t = r_2(ag) - r_{21}(ct) - r_{22}(ag)(ct) + r_{23}(ct)(vc)$$
$$+ r_{25}(ct)(dm) + (ISN)(ct)(ag) + \kappa_2 \nabla^2(ct) \qquad (6.26)$$

$$\partial(dm)/\partial t = r_3(ag) - r_{31}(dm) - r_{32}(ag)(dm)$$
$$+ r_{33}(dm)(vc) + (ISN)(dm)(ag) + \kappa_3 \nabla^2(dm) \qquad (6.27)$$

$$\partial(b)/\partial t = r_4(ag) - r_{41}(b) - r_{42}(ag)(b) + r_{43}(b)(vc)$$
$$+ r_{45}(b)(dm) + (ISN)(b)(ag) + \kappa_3 \nabla^2(b) \qquad (6.28)$$

$$\partial(nk)/\partial t = r_5(ag) - r_{51}(nk) - r_{52}(ag)(nk) + r_{53}(nk)(vc)$$
$$- r_{54}(nk)(rd) + r_{55}(nk)(dm) + r_{56}(nk)(st)$$
$$+ r_{57}(nk)(ch) + (ISN)(nk)(ag) + \kappa_5 \nabla^2(nk) \qquad (6.29)$$

$$\partial(ei)/\partial t = r_6(ei0) - r_{61}(ei) - r_{62}(ag)(ei)$$

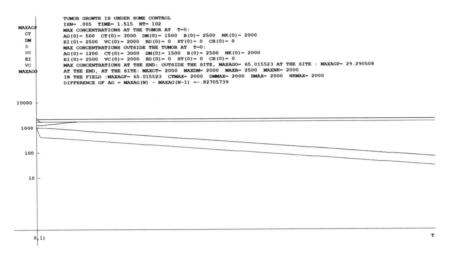

Fig. 6.17 Adjuvant therapy. No tumor. Fixed EI, VC only

$$- r_{63}(ei)(rd) - r_{64}(ei)(ch) + \kappa_6 \nabla^2(ei)$$
$$\partial(vc)/\partial t = r_7(vc0) - r_{71}(vc) - r_{72}(ag)(vc)$$
$$- r_{73}(ct)(vc) - r_{74}(dm)(vc) - r_{75}(b)(vc) + \kappa_7 \nabla^2(vc) \qquad (6.30)$$

Inputs for Fig. 6.17.

```
*** INITIAL CONDITIONS FOR ADJUVANT THERAPY.TUMOR IS REMOVED AND AFFECTED LYMPH NODES WERE REMOVED .FIG.6.17 ***
ISN= -.005   DX=  .02  DY=  .02  DZ=  .02  DT=  .015  FIELD SIZE= 101x101x101   # OF TIME STEPS=   10
R1=  4.5  R11= .0000001  R12= .00065  R13= .00065  R14= .00065  R15= .00065  R16= .00025  R17= .00025
R18=  0  R19= 0  R110= .00005  KAPPA1= .0000001
R2=  1  R21= .000001  R22= .000001  R23= .00005  R24= 0  R25= .00025  R26= 0  R27= 0  KAPPA2= .0000001
R3=  1  R31= .000001  R32= .00002  R33= .00005  R34= 0  R35= 0  KAPPA3= .0000001
R4=  1  R41= .000001  R42= .00002  R43= .00005  R44= 0  R45= .00025  R46= 0  R47= 0  KAPPA4= .0000001
R5=  1  R51= .000001  R52= .00002  R53= .00005  R54= 0  R55= .00025  R56= 0  R57= 0  KAPPA5= .0000001
R6=  1  R61= .00001  R62= .00005  R63= 0  R64= 0  KAPPA6= .0000001
R7=  1  R71= .00001  R72= .00006  R73= .00005  R74= .00005  R75= .00005  R76= .00005
R77=  0  R78= 0  KAPPA7= .0000001
R8=  0  R81= 0  R82= 0  R83= 0  R84= 0  R85= 0  R86= 0  R87= 0  KAPPA8= 0
R9=  0  R91= 0  R92= 0  R93= 0  R94= 0  KAPPA9= 0
R10= 0  R101= 0  R102= 0  R103= 0  R104= 0  R105= 0  R106= 0
R107= 0  KAPPA10= 0

CONDITIONS AT THE PRIMARY SITE ( WHERE TUMOR WAS FOUND AND REMOVED ) AT T=0:
AGMAX0=  500  CTMAX0=  3000  DMMAX0=  1500  BMAX0=  2500  NKMAX0=  2000
EIMAX0=  2400  VCMAX0=  2000  RDMAX0=  0  STMAX0=  0  CHMAX0=  0
CONDITIONS OUTSIDE THE SITE:
AGMAX0=  1200  CTMAX0=  3000  DMMAX0=  1500  BMAX0=  2500  NKMAX0=  2000
EIMAX0=  2400  VCMAX0=  2000  RDMAX0=  0  STMAX0=  0  CHMAX0=  0
```

Together with $(ei0)$ and $(vc0)$, we have considered a strengthened immune system with $r_{12} = r_{13} = r_{14} = r_{15} = 0.00065$. Initially, at the site $(ag0) = 500$, and elsewhere in the field $(ag0) = 1200$. These patients could do much better, even if they are a bit depressed. This is our first example. We also considered that as the patient feels better and better, immunosuppression is changed into immunostimulation and vice versa. So ISN changes the sign accordingly. This means if prognosis gets better, it is positive, else it is negative. The results given in Fig. 6.17 show steady decay of cancer cells in the entire computational field.

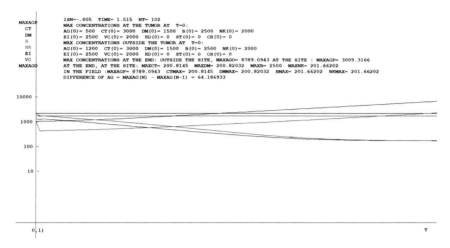

Fig. 6.18 Adjuvant therapy. No tumor. Fixed EI, VC with more strength

If the immune system is not kept strong for such cases, adjuvant therapy will not work. At first, they may look nice but as time goes on they may fail. In Fig. 6.18, that is what is revealed. Here we have started with the same initial inputs, except $r_{12} = r_{13} = r_{14} = r_{15} = 0.00015$.

Inputs for Fig. 6.18.

```
••• INITIAL CONDITIONS FOR ADJUVANT THERAPY.TUMOR IS REMOVED AND AFFECTED LYMPH NODES WERE REMOVED .FIG.6.18 •••
ISN- -.005   DX- .02  DY- .02   DZ- .02   DT- .015   FIELD SIZE- 101x101x101   # OF TIME STEPS-   5
R1-   4.5   R11- .0000001   R12- .00015   R13- .00015   R14- .00015   R15- .00015   R16- .00065   R17- .00065
R18-   0   R19-  0   R110- .00005   KAPPA1- .0000001
R2-   1   R21- .000001   R22- .000001   R23- .00005   R24-  0   R25- .00025   R26-  0   R27-  0   KAPPA2- .0000001
R3-   1   R31- .000001   R32- .000001   R33- .00005   R34-  0   R35-  0   KAPPA3- .0000001
R4-   1   R41- .000001   R42- .00002   R43- .00005   R44-  0   R45- .00025   R46-  0   R47-  0   KAPPA4- .0000001
R5-   1   R51- .000001   R52- .00002   R53- .00005   R54-  0   R55- .00025   R56-  0   R57-  0   KAPPA5- .0000001
R6-   1   R61- .00001   R62- .0005   R63-  0   R64-  0   KAPPA6- .0000001
R7-   1   R71- .00001   R72- .00006   R73-  0   R74- .00005   R75- .00005   R76- .00005
R77-   0   R78-  0   KAPPA7- .0000001
R8-   0   R81-  0   R82-  0   R83-  0   R84-  0   R85-  0   R86-  0   R87-  0   KAPPA8-  0
R9-   0   R91-  0   R92-  0   R93-  0   R94-  0   KAPPA9-  0
R10-   0   R101-  0   R102-  0   R103-  0   R104-  0   R105-  0   R106-  0
R107-   0   KAPPA10-  0

CONDITIONS AT THE PRIMARY SITE ( WHERE TUMOR WAS FOUND AND REMOVED ) AT T=0:
AGMAX0-    500   CTMAX0-   3000   DMMAX0-   1500   BMAX0-   2500   NKMAX0-   2000
EIMAX0-   2500   VCMAX0-   2000   RDMAX0-   0   STMAX0-   0   CHMAX0-   0
CONDITIONS OUTSIDE THE SITE:
AGMAX0-   1200   CTMAX0-   3000   DMMAX0-   1500   BMAX0-   2500   NKMAX0-   2000
EIMAX0-   2500   VCMAX0-   2000   RDMAX0-   0   STMAX0-   0   CHMAX0-   0
```

After surgery when the tumor is totally removed, still there could be cancer cells around that area and some remain in the blood. Total annihilation of cancer cells is practically impossible. Furthermore after surgery, patients become very weak in general and their immune systems become weak too. So cancer cells remaining in the blood circulation often take full advantage of this and thrive.

We notice by comparing the values of the lymphocytes at $t = 0$ and at the end that they were diminished alarmingly in the field. Now the drugs for (ei) and (vc) have been made much stronger with larger rate constants $r_{16} = r_{17} = 0.0015$. The results are given in Fig. 6.19.

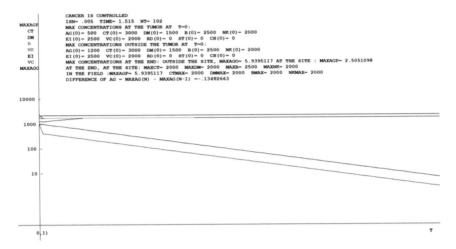

Fig. 6.19 Adjuvant therapy. Fixed EI, VC have been made much stronger

Inputs for Fig. 6.19.

```
••••••••••••••••••••••••••••••••••••••••••••••••••••••••••••••••••••••••••••••••••••••••
  ••• INITIAL CONDITIONS FOR ADJUVANT THERAPY.TUMOR IS REMOVED AND AFFECTED LYMPH NODES WERE REMOVED .FIG.6.19  •••
   ISN= -.005   DX- .02  DY- .02   DZ- .02   DT- .015   FIELD SIZE- 101x101x101    # OF TIME STEPS-   101
   R1-   4.5   R11-  .0000001   R12-  .00015   R13-  .00015   R14-  .00015   R15-  .00015   R16-  .0015   R17-  .0015
   R18-   0   R19-   0   R110-  .00005   KAPPA1-  .0000001
   R2-   1   R21-  .000001   R22-  .000001   R23-  .00005   R24-   0   R25-  .00025   R26-   0   R27-   0   KAPPA2-  .0000001
   R3-   1   R31-  .000001   R32-  .00002   R33-  .00002   R34-   0   R35-   0   KAPPA3-  .0000001
   R4-   1   R41-  .000001   R42-  .00002   R43-  .00005   R44-   0   R45-  .00025   R46-   0   R47-   0   KAPPA4-  .0000001
   R5-   1   R51-  .000001   R52-  .00002   R53-  .00005   R54-   0   R55-  .00025   R56-   0   R57-   0   KAPPA5-  .0000001
   R6-   1   R61-  .00001   R62-  .00005   R63-   0   R64-   0   KAPPA6-  .0000001
   R7-   1   R71-  .00001   R72-  .00006   R73-  .00005   R74-  .00005   R75-  .00005   R76-  .00005
   R77-   0   R78-   0   KAPPA7-  .0000001
   R8-   0   R81-   0   R82-   0   R83-   0   R84-   0   R85-   0   R86-   0   R87-   0   KAPPA8-   0
   R9-   0   R91-   0   R92-   0   R93-   0   R94-   0   KAPPA9-   0
   R10-   0   R101-   0   R102-   0   R103-   0   R104-   0   R105-   0   R106-   0
   R107-   0   KAPPA10-   0

   CONDITIONS AT THE PRIMARY SITE ( WHERE TUMOR WAS FOUND AND REMOVED ) AT T-0:
   AGMAX0-   500   CTMAX0-  3000   DMMAX0-  1500   BMAX0-  2500   NKMAX0-  2000
   EIMAX0-   2500   VCMAX0-  2000   RDMAX0-   0   STMAX0-   0   CHMAX0-   0
   CONDITIONS OUTSIDE THE SITE:
   AGMAX0-   1200   CTMAX0-  3000   DMMAX0-  1500   BMAX0-  2500   NKMAX0-  2000
   EIMAX0-   2500   VCMAX0-  2000   RDMAX0-   0   STMAX0-   0   CHMAX0-   0
••••••••••••••••••••••••••••••••••••••••••••••••••••••••••••••••••••••••••••••••••••••••••
```

Here we will see simulations of our computational results in graphics. These simulations have been done by the second author exclusively.

6.6 The Computer Visualization of Numerical Solutions

All mathematical modelings should be validated by experimental data. In this case, these data should be collected from the hospital records, which by law cannot be obtained. So we have to search for data available on the Internet. However, the first author has got some first-hand knowledge on this being with many teams of oncologists at various hospitals in the midwest of the USA for almost two years. Many oncologists, internationally, checked our computer outcomes and made very favorable comments. Some of the graphics below were some slides presented at NASA conferences at NASA Ames, Moffett Field, California and NASA Johnson Space Center, Houston, Texas. We can state for sure that our results have been validated qualitatively through computer graphics. Only a few have been given here. These all are graphic outputs of numerical solutions of our mathematical models.

We have noted before that the computational field is a cube. This cube is subdivided into many smaller cubes of the same size. That helps us see what is happening at different parts of the body as time goes on and treatment continues. Medical professionals can understand these figures relatively easily and they did. They saw these results matching quite well with what they saw in the real world. All these results are numerical solutions scaled and simulated in computer graphics. Figure 6.20 shows the development of cancer at the beginning. Tumor is still in situ. In the computational field, this is at the corner stretching from $(1,1,1)$ to $(5,5,5)$. As cancer started metastasizing and treatments began, the defenders started attacking. We notice that they have launched their attack primarily at those points where there are more cancer cells. For better visualization, we decided to show the fights, generated by numerical solutions in separate figures, showing the engagements between cancer cells (red dots) and different defenders.

In Fig. 6.21, cytotoxic T-cells (blue dots) are on the attack wherever cancer cells (red dots) are moving all over the field. They are surrounding the cancer cells (red dots). Likewise, other fighters are also assembling all around. For simplification, those are not shown in the same figure.

In Fig. 6.22, cancer cells (red dots) are not shown. Instead we plotted the deep pink dots representing estrogen inhibitors massively assembling all around cancer cells flagging the assemblage of T-cells (blue dots). At the center of this figure there are lots of blue dots, the tumor is surrounded and massively attacked by T-cells. The big cube representing the computational field in Fig. 6.20 is rotated in such a way that the corner where the tumor is present is now placed at the point little left from the center where many blue dots are packed together. Estrogen inhibitors (pink dots) have surrounded the cancer cells so that they cannot metastasize. They have formed shields around cancer cells (Fig. 6.22).

In the Fig. 6.23 fighting is seen to be very intense. Cancer cells have been completely overpowered and overrun by the defensive forces described in the model.

The tumor (placed at the lower left corner) is literally engulfed by T-cells (the blue dots). The grayish blue dots represent the presence of drugs enhancing the immune system. They have also surrounded the cancer cells. T-cells (blue dots) are significantly strengthened. All have surrounded the red dots in the entire field which are not much visible. The war is at its climax. Sadly, even under these conditions, cancer sometimes wins, and the reason is the patient gets weaker and more stressed. So she fails to tolerate the side effects of those strong medications. The solution is to find less strong inputs which could still work for the patient. We will look into that in Chap. 8.

In Fig. 6.24, we kept only the red dots representing cancer cells. The box presenting the computational field has been rotated in such a way that at the corner where the

Fig. 6.20 Growth of cancer shown in the lower left side box (Tumor in situ)

tumor appeared first is now at a point identified by a black circle. We notice that cancer is metastasizing at six other points (Fig. 6.24).

The figure represents the vanishing of the primary tumor from the primary site and reappearing at a different location. In fact, it can appear at several locations in the body. There are many cases like this that the first author witnessed and heard from friends and oncologists. In one case, cancer from lungs was all gone, and it appeared later in the bone. In another case, cancer from the liver vanished after a lengthy treatment, and later it appeared in the bone.

Fig. 6.21 Cytotoxic
T-cells (blue dots)
overpowering cancer (red) as
they try to metastasize

Fig. 6.22 Cancer at the
center. Overpowered by
T-cells. Estrogen (darkish
pink dots) moved all around
the breast helping T-cells
(blue dots)

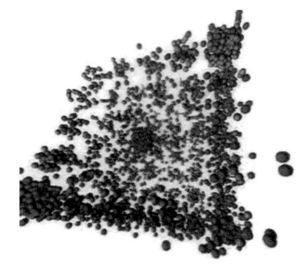

Figure 6.26 displays another success of the mathematical model. Here the semi-darkish blue dots represent radiation. From the lower box (the primary site of the tumor), where radiation was first given, cancer disappeared totally and started appearing at other locations. Interestingly, computational results showed that the effects of radiation are still present at the primary site. And the injected radiation is attacking the metastatic sites. The other defensive forces are also present, but these are not shown in the Fig. 6.26.

Here we notice as the numerical solution shows no cancer at the primary site, estrogen (pink) dots vanish from the site too. However, it is evident from the figure that they are present where cancer cells (red dots) are trying to metastasize.

Enemy cannot be underestimated.

Figure 6.28 shows that at the cancer site (upper left corner), although cancer totally vanished, (as seen in Fig. 6.27), the site is still covered by cytotoxic T-cells. They are

Fig. 6.23 Very intense fight
between cancer (red dots)
and T-cells and
immunotherapy drugs.
Defeat of cancer totally
overpowered by the defense

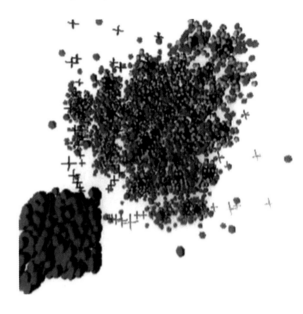

Fig. 6.24 Metastasization.
Tumors are being formed at
several locations

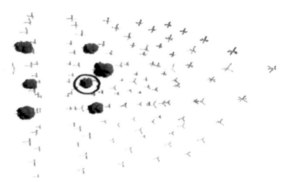

also fighting wherever cancer cells are present in the field. Estrogen inhibitors (pink
dots) are also present where cancer cells (red dots) are.

Here cytotoxic T-cells (deep blue dots) are fighting cancer cells (red dots), and
dendritic cells and macrophages (green dots)are present to help T-cells. All of them
are massively present around the red dots. This is a scenario that all oncologists want
to see in actual treatments.

Here the fighters are the T-cells (deep blue dots) and radiation (lightly sky-blue
dots). We may compare the Figures 6.28, 6.29, and 6.30. However, estrogen is
strongly present everywhere in the field, dendritic cells, macrophages, and radiation, especially the cytotoxic T-cells are mostly present where cancer cells (red dots)
are. Fighters have surrounded the enemies. This is precisely what we expected to
see. Successful chemotaxis.

Fig. 6.25 Cancer vanished from the primary site (bottom first box). But reappearing at another location

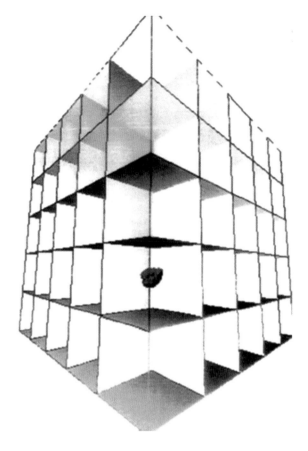

Fig. 6.26 While attacking the metastatic states, radiation is still present at the primary site (lower front box)

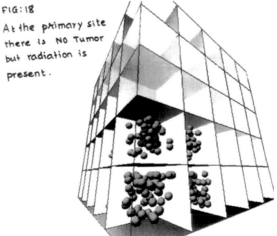

Fig. 6.27 Once tumor
vanished at the primary site
(lowest front box), estrogen
is no longer there

Fig. 6.28 Cytotoxic T-cells
are still massively present at
the primary site where the
tumor vanished (as noticed
in the computer output)

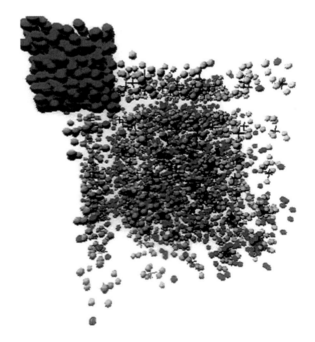

6.6.1   *Graphical Results on Long Term Effects of Cancer Treatment*

In graphs, we do not directly see these simulations. However, if we analyze the graphs, then visions of such simulations and more could be very clear. We need to remember that these works are not really just theoretical studies. Following our past observations and discussions with many international experts, we have considered a vast number of

Fig. 6.29 As T-cells launch fight, B-cells, dendritic cells and macrophages (green dots) assist them at every location. Chemotaxis vividly presented by numerical outputs

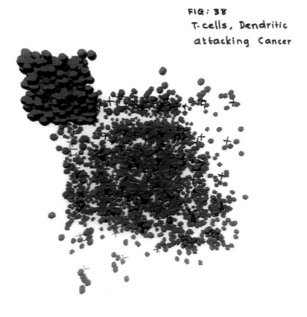

Fig. 6.30 As cancer (red dots) spreads, at every location they face massive attacks by the defense (Here only T-cells and radiation have been shown)

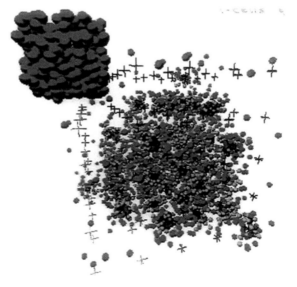

realistic conditions and selected a few in this monograph. Our mathematical findings matched quite well with them. For example, in one case, cancer cells were fast replicating and very aggressive. Some doctors proposed whether a particular method of adjuvant therapy could work. We have simulated these conditions mathematically by the inputs of Fig. 6.31. Certainly, we do not claim to know all the details of this treatment. Our simulation shows that the doctors were possibly right.

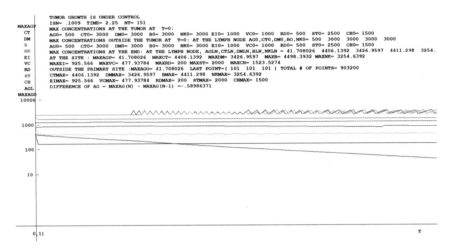

Fig. 6.31 Drugs adjusted as needed to keep cancer under control

Here, the patient is in relatively good spirits. ISN = 0.1009. Immune system is not very aggressive because $r_{12} = r_{13} = r_{14} = r_{15} = 0.00015$. (ct), (dm), (b), and (nk) are restricted. They cannot be less than 2500 and cannot exceed 3500. On the average, they should be about 3000. Also the amount of drugs are all restricted not to fall below certain limits as given in the INPUT for Fig. 6.31.

Inputs for Fig. 6.31.

```
...........................................................................................
*** INITIAL CONDITIONS FOR THE CODE 6.5A.DRUGS WERE ADJUSTED AS NEEDED TO KEEP CANCER UNDER CONTROL ***
ISN- .1009  DX- .02  DY- .02  DZ- .02  DT- .015  FIELD SIZE- 101x101x101 # OF TIME STEPS- 150
R1- 4.5  R11- .0000001  R12- .00015  R13- .00015  R14- .00015  R15- .00015  R16- .00001  R17- .0001
R18- .015  R19- .0001  R110- .000001  KAPPA1- .0000001
R2- 1  R21- .000001  R22- .00001  R23- .00005  R24- .0001  R25- .00025  R26- .0005  R27- .0001  KAPPA2- .0000001
R3- 1  R31- .000001  R32- .00002  R33- .00005  R34- .0001  R35- .0001  KAPPA3- .0000001
R4- 1  R41- .000001  R42- .00001  R43- .00005  R44- .0001  R45- .00025  R46- .0005  R47- .0001  KAPPA4- .0000001
R5- 1  R51- .000001  R52- .00001  R53- .00005  R54- .0001  R55- .00025  R56- .0001  R57- .0001  KAPPA5- .0000001
R6- 1  R61- .00001  R62- .00005  R63- .0005  R64- .0001  KAPPA6- .0000001
R7- 1  R71- .00001  R72- .00005  R73- .00005  R74- .00005  R75- .00005  R76- .00005
R77- .0001  R78- .005  KAPPA7- .0000001
R8- 1  R81- .00001  R82- .0005  R83- .0001  R84- .0001  R85- .0001  R86- .0001  R87- .0001  KAPPA8- .001
R9- 1  R91- .00001  R92- .000025  R93- .0001  R94- .0001  KAPPA9- .0000001
R10- 1  R101- .00001  R102- .00001  R103- .00001  R104- .00005  R105- .00001  R106- .00001
R107- .00001  KAPPA10- .0000001
CONDITIONS AT THE PRIMARY SITE AT T-0:
AGMAX0- 500  CTMAX0- 3000  DMMAX0- 3000  BMAX0- 3000  NKMAX0- 3000
EIMAX0- 1000  VCMAX0- 1000  RDMAX0- 500  STMAX0- 2500  CHMAX0- 1500
THE THREE LYMPH NODES AG0, CT0,DM0,B0,NK0- 500  3000  3000  3000  3000
CONDITIONS OUTSIDE THE SITE:
AGMAX0- 500  CTMAX0- 3000  DMMAX0- 3000  BMAX0- 3000  NKMAX0- 3000
EIMAX0- 1000  VCMAX0- 1000  RDMAX0- 500  STMAX0- 2500  CHMAX0- 1500

RESTRICTIONS ARE :
IF CT(I,J,K), OR DM(I,J,K) OR B(I,J,K) OR NK(I,J,K) < 2500 OR > 3500 THEN THEY ARE SET EQUAL TO 3000
IF CH(I,J,K) < 1500 THEN CH(I,J,K) - 1500, IF EI(I,J,K) < 885 THEN EI(I,J,K) - 1000.
IF VC(I,J,K) < 500 THEN VC(I,J,K) - 500,  IF RD(I,J,K) < 200 THEN RD(I,J,K) - 200.
IF ST(I,J,K) < 2000 THEN ST(I,J,K) - 2000.
...........................................................................................
```

The results are given in Fig. 6.31.

If we look into the values of MAXAGO, at the beginning, MAXAGO(0) = 500 and at the end, it is about 42. Numerically, it is a big jump. But on the $1 + log_{10}(500) = 3.7$ and $1 + log_{10}(42) = 2.6$. The same is valid for the values of (ag) at the undetected

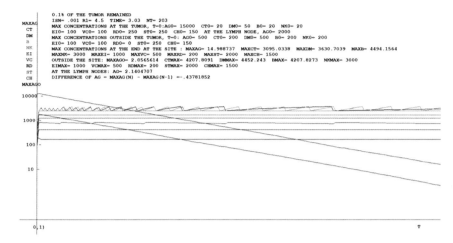

Fig. 6.32 This is a modified Fig. 6.31. Fixed dosages of drug have been used at each time step

lymph node. So those three lines merged with one another and that is the redline with a negative slope. Since no line has a positive slope, it signifies the success of the treatment. The values of (b) and (nk) increased and decreased in a wave, and their amplitudes are slowly dying moving toward a steady state.

Our next attempt is to check on a long-term study of this treatment. Here at each time step, fixed amount of drugs were given intravenously. As they go into the bloodstream and mix with other elements, they change.

Inputs Fig. 6.32.

```
••••••••••••••••••••••••••••••••••••••••••••••••••••••••••••••••••••••••••••••••••••••••••••••
*** INITIAL CONDITIONS FOR THE CODE 6.5B.FIXED VALUES OF DRUGS WERE USED AT EACH TIME STEP. ***
ISN= .001  DX= .02  DY= .02  DZ= .02  DT= .015  FIELD SIZE= 101x101x101 # OF TIME STEPS= 400
R1= 4.5  R11= .0000001  R12= .00025  R13= .00025  R14= .00025  R15= .00025  R16= .00001  R17= .0001
R18= .015  R19= .0001  R110= .000001  KAPPA1= .0000001
R2= 1  R21= .000001  R22= .00001  R23= .00005  R24= .0001  R25= .00025  R26= .0005  R27= .0001  KAPPA2= .0000001
R3= 1  R31= .000001  R32= .00002  R33= .00005  R34= .0001  R35= .0001  KAPPA3= .0000001
R4= 1  R41= .000001  R42= .00001  R43= .00005  R44= .0001  R45= .00025  R46= .0005  R47= .0001  KAPPA4= .0000001
R5= 1  R51= .000001  R52= .00001  R53= .00005  R54= .0001  R55= .00025  R56= .0001  R57= .0001  KAPPA5= .0000001
R6= 1  R61= .00001  R62= .00005  R63= .0005  R64= .0001  KAPPA6= .0000001
R7= 1  R71= .00001  R72= .00006  R73= .00005  R74= .00005  R75= .00005  R76= .00005
R77= .0001  R78= .005  KAPPA7= .0000001
R8= 1  R81= .00001  R82= .0005  R83= .0001  R84= .0001  R85= .0001  R86= .0001  R87= .0001  KAPPA8= .001
R9= 1  R91= .00001  R92= .000025  R93= .0001  R94= .0001  KAPPA9= .0000001
R10= 1  R101= .00001  R102= .00001  R103= .00001  R104= .00005  R105= .00001  R106= .00001
R107= .00001  KAPPA10= .0000001
CONDITIONS AT THE PRIMARY SITE AT T=0:
AGMAX0= 15000  CTMAX0= 20  DMMAX0= 50  BMAX0= 20  NKMAX0= 20
EIMAX0= 100  VCMAX0= 100  RDMAX0= 100  STMAX0= 250  CHMAX0= 150
THE THREE LYMPH NODES AG0, CT0,DM0,B0,NK0= 2000  20  20  20  20
CONDITIONS OUTSIDE THE SITE:
AGMAX0= 500  CTMAX0= 200  DMMAX0= 500  BMAX0= 200  NKMAX0= 200
EIMAX0= 100  VCMAX0= 100  RDMAX0= 0  STMAX0= 250  CHMAX0= 150

RESTRICTIONS ARE :
IF CT(I,J,K), OR DM(I,J,K) OR B(I,J,K) OR NK(I,J,K) < 2500 OR > 3500 THEN THEY ARE SET EQUAL TO 3000
IF CH(I,J,K) < 1500 THEN CH(I,J,K) = 1500,  IF EI(I,J,K) < 885 THEN EI(I,J,K) = 1000.
IF VC(I,J,K) < 500 THEN VC(I,J,K) = 500,  IF RD(I,J,K) < 200 THEN RD(I,J,K) = 200.
IF ST(I,J,K) < 2000 THEN ST(I,J,K) = 2000.
••••••••••••••••••••••••••••••••••••••••••••••••••••••••••••••••••••••••••••••••••••••••••••••
```

Here the rate constants are identically the same, but this is a regular treatment of a breast cancer patient. Here at $t = 0$, at the primary site, $(ag(t = 0)) = 15000$, at the lymph nodes, it is 2000, and outside the site it is 500 spreading all over

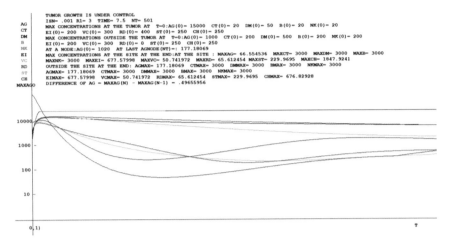

Fig. 6.33 Drugs are self adjusted as ag changes (In the code a different figure # has been used)

the computational field. Here ISN = 0.001. As time progresses, whereas all the medications along with leukocytes are going toward the steady state, (ag) at the site and (ag) outside the site both have sharp negative slopes, they are decreasing. That marks the success of the treatment.

Short-term progress of treatments sometimes does not match with long-term results. We will consider two such cases here. First we will extend our findings after NT=155 from the Fig. 6.1 making very small changes in the inputs.

For most computer runs, we have considered NT = total number of time steps = 150 on the average. In some cases, we have found results earlier as the numerical solutions converged. However, we need to study some cases more deeply looking into some long-term solution of the model. In this regard, we have selected two cases. The first one is shown in Fig. 6.33, where we saw cancer cells start decreasing both at the tumor site and outside. We decided to make some small changes to match some of our observations.

Inputs for Fig. 6.33 (Fig. 6.1 is extended).

```
** INPUTS FOR THE FIG.6.1A:ISN- .001  DX- .02  DY- .02  DZ- .02  DT- .015  FIELD 101x101x101 # OF TIME STEPS- 500  **
R1- 3  R11- .0000001  R12- .00015  R13- .00015  R14- .00015  R15- .00015  R16- .0001  R17- .0001
R18- .015  R19- .0001  R110- .00001  KAPPA1- .00000001
R2- 1  R21- .000001  R22- .00001  R23- .00001  R24- .0001  R25- .00025  R26- .0005  R27- .0001  KAPPA2- .0000001
R3- 1  R31- .000001  R32- .00002  R33- .00005  R34- .0001  R35- .0001  KAPPA3- .0000001
R4- 1  R41- .000001  R42- .00002  R43- .00005  R44- .0001  R45- .00025  R46- .0005  R47- .0001  KAPPA4- .0000001
R5- 1  R51- .000001  R52- .00002  R53- .00005  R54- .0001  R55- .00025  R56- .0001  R57- .0001  KAPPA5- .0000001
R6- 1  R61- .00001  R62- .00005  R63- .0005  R64- .0001  KAPPA6- .000001
R7- 1  R71- .00001  R72- .00006  R73- .00005  R74- .00005  R75- .00005  R76- .00005
R77- .0001  R78- .005  KAPPA7- .000001
R8- 1  R81- .00001  R82- .0005  R83- .0001  R84- .0001  R85- .0001  R86- .0001  R87- .0001  KAPPA8- .001
R9- 1  R91- .00001  R92- .000025  R93- .0001  R94- .0001  KAPPA9- .00001
R10- 1  R101- .00001  R102- .00005  R103- .00001  R104- .00005  R105- .00001  R106- .00001
R107- .00001  KAPPA10- .00001

CONDITIONS AT THE PRIMARY SITE AT T-0
AG0- 15000  CT0- 20  DM0- 50  B0- 20
NK0- 20  EI0- 200  VC0- 300  RD0- 400  ST0- 250  CHO- 250
AT THE THREE AFFECTED LYMPH NODES AG0,CT0,DM0,B0,NK0- 1020  20  30  20  20
CONDITIONS OUTSIDE THE SITES, AT T-0
AG0- 1000  CT0- 200  DM0- 200  BM0- 200  NK0- 200
EI0- 200  VC0- 300  RD0- 0  ST0- 250  CHO- 250
```

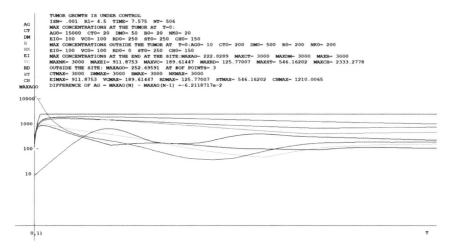

Fig. 6.34   No modification of conditions for Fig. 6.2

Here cancer is a bit more aggressive than in the Fig. 6.1. Previously, r_1 was 2.5; here, $r_1 = 3$. Interesting is at $NT = 501$, outside the site $(ag) = 177.18$, whereas in Fig. 6.1 at $NT = 155$, it was only 3.14. That indicates that after a short term treatment good prognosis may not be much meaningful. Cancer practically vanished in Fig. 6.1 and now reappearing in due time.

During the first part of the figure, both max value of (ag), shown by the upper most redline, and the value of MAXAGO, the black line, have negative slopes. But as time moves, both started increasing and then decreasing and increasing again, whereas other medications are moving into steady states or getting smaller. So the treatment which was earlier very promising is finally a total failure.

We have decided to look into another case. Regarding this treatment, neuro oncologists predicted that after surgery is completed, a small pump could be attached to the chest wall of the patient, and she could pump the medication herself (Ref. www.mskcc.org) as needed. Another oncologist predicted that for this particular patient, who has already gone through two surgeries, this process of chemo infusion will not be ultimately fruitful. Because by the time it will start working, that could be a long time for her to survive. The patient finally did not survive. All hopes ended in smoke. That grim picture is mathematically painted in Fig. 6.34 with the inputs given in the inputs for Fig. 6.34.

Inputs for Fig. 6.34. (These are exactly the same inputs used in the Fig. 6.2 which are same as in Fig. 6.1).

```
••••••••••••••••••••••••••••••••••••••••••••••••••••••••••••••••••••••••••••••••••••••••••••••
*** INITIAL CONDITIONS FOR THE CODE 6.5C.NO DRUG ADJUSTMENT THE SAME AS IN THE FIG.6.1B ***
ISN- .001  DX- .02  DY- .02  DZ- .02  DT- .015  FIELD SIZE- 101x101x101 # OF TIME STEPS- 505
R1- 4.5  R11- .0000001  R12- .00025  R13- .00025  R14- .00005  R15- .00025  R16- .00001  R17- .0001
R18- .015  R19- .0001  R110- .000001  KAPPA1- .0000001
R2- 1  R21- .000001  R22- .00001  R23- .00005  R24- .0001  R25- .00025  R26- .0005  R27- .0001  KAPPA2- .0000001
R3- 1  R31- .000001  R32- .00002  R33- .00005  R34- .0001  R35- .0001  KAPPA3- .0000001
R4- 1  R41- .000001  R42- .00001  R43- .00005  R44- .0001  R45- .00025  R46- .0005  R47- .0001  KAPPA4- .0000001
R5- 1  R51- .000001  R52- .00001  R53- .00005  R54- .0001  R55- .00025  R56- .0001  R57- .0001  KAPPA5- .0000001
R6- 1  R61- .00001  R62- .00005  R63- .0005  R64- .0001  KAPPA6- .0000001
R7- 1  R71- .00001  R72- .00006  R73- .00005  R74- .00005  R75- .00005  R76- .00005
R77- .0001  R78- .005  KAPPA7- .0000001
R8- 1  R81- .00001  R82- .0005  R83- .0001  R84- .0001  R85- .0001  R86- .0001  R87- .0001  KAPPA8- .001
R9- 1  R91- .00001  R92- .000025  R93- .0001  R94- .0001  KAPPA9- .0000001
R10- 1  R101- .00001  R102- .00001  R103- .00001  R104- .00005  R105- .00001  R106- .00001
R107- .00001  KAPPA10- .0000001
CONDITIONS AT THE PRIMARY SITE AT T-0:
AGMAX0- 15000  CTMAX0- 20  DMMAX0- 50  BMAX0- 20  NKMAX0- 20
EIMAX0- 100  VCMAX0- 100  RDMAX0- 250  STMAX0- 250  CHMAX0- 150
THE THREE LYMPH NODES AGL0, CTL0,DML0,BL0,NKL0- 10   20   20   20   20
CONDITIONS OUTSIDE THE SITE:
AGMAX0- 10  CTMAX0- 200  DMMAX0- 500  BMAX0- 200  NKMAX0- 200
EIMAX0- 100  VCMAX0- 100  RDMAX0- 0  STMAX0- 250  CHMAX0- 150

RESTRICTIONS ARE :
IF CT(I,J,K), OR DM(I,J,K) OR B(I,J,K) OR NK(I,J,K) < 3000  THEN THEY ARE SET EQUAL TO 3000
••••••••••••••••••••••••••••••••••••••••••••••••••••••••••••••••••••••••••••••••••••••••••••••
```

There are very small changes in the inputs taken from Fig. 6.2. For instance, earlier, (dm) at t = 0 was 500 at an affected lymph node, and here it is 20.

What is obvious from the graphs is tumor was first shrinking (the upper redline); however, silently the number of cancer cells was increasing elsewhere (the blue line below). After some time, both lines started having negative slopes. Both doctors and patients will surely like this. Please check. A truly great news of apparent success. But soon after that they both started rising (positive slopes) requiring more drugs (self-adjusted in the code) for treatment, so that both can come down. Theoretically, this is quite nice but not in reality. More drugs mean more poison in the body which an exhausted body may not tolerate. Although all other biochemicals seem to have reached steady states and theoretically patients should be able to live with cancer, sadly that did not happen to the patient whom I knew. So treatment worked but the patient expired. This implies that mathematical modeling should be analyzed with care and caution. Yet since in most cases, these models have predicted some real-world scenarios, medical students studying bioengineering dealing with malignancy should look into these models and work for the betterment.

6.6.2 Application of Skilled Killer Drug (SKD)

The Skilled Killer Drug (SKD) works inside the body against all cancer cells wherever they may be like a targeted guided missile. This is achieved by using nanoparticles carrying the medications only to the cancer cells. Scientists are finding that increasingly the cancer cells are becoming drug resistant. Even radiation is not working against them. We have mentioned here a review article [24]: "This review highlights design considerations to develop nanoparticle-based approaches for overcoming physiological hurdles in cancer treatment, as well as emerging research in engineering advanced delivery systems for the treatment of primary, metastatic, and multidrug resistant cancers. A growing understanding of cancer biology will continue to foster development of intelligent nanoparticle-based therapeutics that take into account diverse physiological contexts of changing disease states to improve treatment outcomes." We will discuss more on this later. Here we will end our chapter with two examples. We have considered primarily the inputs from Fig. 6.34. In fact, we have considered a much worse scenario. Here $r_1 = 5.5$, meaning that the rate of growth of cancer is much bigger than before where $r_1 = 4.5$. The strengths of the immune system are less, meaning that the immune system is much weaker than before. Because in Fig. 6.34, $r_{12} = r_{13} = r_{15} = 0.00025$, whereas, here they are $r_{12} = r_{13} = r_{15} = 0.00015$. More importantly, (es), (vc), (rd), and (st) are not used at all. The drug SKD is carried by nanoparticle whose rate of growth is $r_{10} = 1$, and it is killing cancer cells at the rate $r_{102} = 0.00001$ which is not too powerful. But it is performing two jobs: (i) It is reducing the rate of cancer growth. The formula for that is.

$$dr_1/dt = -\lambda r_1 \times max(SKD(t)). \quad \text{where} \quad max(SKD(n)) = max_{ijk} SKD(i, j, k) \text{ at } t_n.$$

(more details are in Chap. 8).

Using $\lambda = 0.01$ and integrating from t_{n-1} to t_n, we get

$$r_1(n) = r_1(n - 1)exp(-0.01(DT/2)(max(SKD(n)) + max(SKD(n - 1))))$$

So at each time step, the rate of growth of cancer is being diminished.

It does one more job. At each time and at each location (i, j, k) in the field, it is getting attached to each cancer cell and shrinking it, meaning it is trying to destroy that cancer cell. The formula that we developed is $ag(i, j, k) = ag(i, j, k) - a_5 \times (ag(i, j, k) \times SKD(i, j, k))$. At each time step, we took $a_5 = 0.00075$.

Inputs for Fig. 6.35 The rate of growth of cancer, $r_1(n)$ is decreasing.

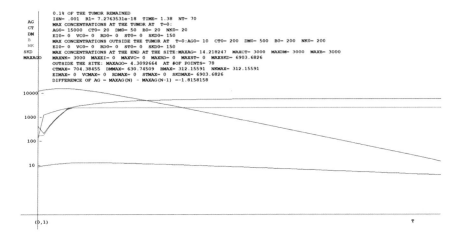

Fig. 6.35 Application of SKD on breast cancer model with no radiation or chemo

```
••••••••••••••••••••••••••••••••••••••••••••••••••••••••••••••••••••••••••••••••••••••••••••••••••••••••
*** INITIAL CONDITIONS FOR THE CODE 6.5D.APPLICATION OF INTRAVENOUS SKD ***
ISN= .001  DX= .02  DY= .02  DZ= .02  DT= .02  FIELD SIZE= 101x101x101 # OF TIME STEPS= 505
R1= 5.5  R11= .0000001  R12= .00015  R13= .00015  R14= .00005  R15= .00015  R16= 0  R17= 0
R18= 0  R19= .00065  R110= .000001  KAPPA1= .0000001  A5(REDUCTION OF AG AT EACH POINT IN THE FIELD) = .00075
R2= .5  R21= .000001  R22= .00001  R23= 0  R24= 0  R25= .00025  R26= 0  R27= .0005  KAPPA2= .0000001
R3= .5  R31= .000001  R32= .00002  R33= 0  R34= 0  R35= .0001  KAPPA3= .0000001
R4= .5  R41= .000001  R42= .00001  R43= 0  R44= 0  R45= .00025  R46= 0  R47= .0001  KAPPA4= .0000001
R5= .5  R51= .000001  R52= .00001  R53= 0  R54= 0  R55= .00025  R56= 0  R57= .0001  KAPPA5= .0000001
R6= 0  R61= 0  R62= 0  R63= 0  R64= 0  KAPPA6= 0
R7= 0  R71= 0  R72= 0  R73= 0  R74= 0  R75= 0  R76= 0
R77= 0  R78= 0  KAPPA7= 0
R8= 0  R81= 0  R82= 0  R83= 0  R84= 0  R85= 0  R86= 0  R87= 0  KAPPA8= 0
R9= 0  R91= 0  R92= 0  R93= 0  R94= 0  KAPPA9= 0
R10= 1  R101= .00001  R102= .00001  R103= 0  R104= 0  R105= 0  R106= 0
R107= 0  KAPPA10= .0000001
CONDITIONS AT THE PRIMARY SITE AT T=0:
AGMAX0= 15000  CTMAX0= 20  DMMAX0= 50  BMAX0= 20  NKMAX0= 20
EIMAX0= 0  VCMAX0= 0  RDMAX0= 0  SKDMAX0= 0
THE THREE LYMPH NODES AGL0, CTL0,DML0,BL0,NKL0= 10  20  20  20  20
CONDITIONS OUTSIDE THE SITE:
AGMAX0= 10  CTMAX0= 200  DMMAX0= 500  BMAX0= 200  NKMAX0= 200
EIMAX0= 0  VCMAX0= 0  RDMAX0= 0  CHMAX0= 150

RESTRICTIONS ARE : SKD(I,J,K) <= 7500 &
IF CT(I,J,K), OR DM(I,J,K) OR B(I,J,K) OR NK(I,J,K) < 3000  THEN THEY ARE SET EQUAL TO 3000
R1(N) = R1(N-1)* EXP(-0.01*(DT/2)*(MAXSKD(N) + MAXSKD(N-1)))
AG(I,J,K)= AG(I,J,K) - A5*AG(I,J,K)*SKD(I,J,K),REDUCTION OF AG AT EACH FIELD POINT
••••••••••••••••••••••••••••••••••••••••••••••••••••••••••••••••••••••••••••••••••••••••••••••••••••••••
```

This method worked well as could be seen in Fig. 6.35. We have allowed computations for 505 time steps as could be seen in the inputs for Fig. 6.35, but in 70 time steps, all values of $ag(i, j, k)$ throughout the computational field become less than $0.001 \times maxag(i, j, k)$, where $maxag(i, j, k) = 15000$. So the method worked well. Looking at the actual numerical computations given in Table 6.1, we notice that at the time step $t(n) = 58, r_1 = 2.87 \times 10^{-58}$. So the rate of growth is practically zero, yet the fight is going on which means those which are in circulation are fighting on. The lymphocytes $(ct), (dm), (b), and (nk)$ are strongly assembled at the primary site location, and (SKD) is also present with full force. Outside the primary site and at the undetected lymph nodes, affected by cancer, (SKD) is vigilantly fighting on. Initially, at $t = 0, (SKD) = 150$ as injected intravenously in the body.

Table 6.1 shows that medications are working quite well. Tumor is shrinking quite fast. The Skilled Killer Drug (SKD) has gone to a steady state by 57 time steps. r_1, the

Table 6.1 For Fig. 6.35. Massive fight is on with decreasing $r_1(n)$, the rate of growth of cancer.

```
SYSTEM TIME=19:14:40 TIME= 1.14  TIME STEP= 57  ISN= .001        NT= 505  R1(N)= 2.8722288e-14

AT THE SITE MAXAG= 60.132567  MAXCT= 3000  MAXDM= 3000  MAXB= 3000  MAXNK= 3000
MAXEI= 0  MAXVC= 0  MAXRD= 0  MAXSKD= 6896.6005  MAXST= 0
  AT THE NODE1 AG,CT,DM,B,NK,SKD : 5.3804822  57.543735  31.00804  30.636625  30.636625           1510.5659
  AT THE NODE2 AG,CT,DM,B,NK,SKD : 5.3804822  57.543735  31.00804  30.636625  30.636625  1510.5659
  AT THE NODE1 AG,CT,DM,B,NK,SKD : 5.3804822  57.543735  31.00804  30.636625  30.636625  1510.5659
MAX CONCENTRATIONS OUTSIDE THE PRIMARY SITE: MAXAGO= 5.7435934  AT #OF POINTS= 28
MAXCTO= 564.63457  MAXDM= 606.79629  MAXB= 288.96733  MAXNK= 288.96733
  MAXEI= 0  MAXVC= 0  MAXRD= 0  MAXST= 0  MAXSKD= 1513.0327
MAXAGO= 5.7435934  MAXCTO= 564.63457  MAXDMO= 606.79629  MAXBO= 288.96733  MAXSKDO= 1513.0327
DIFFERENCE IN AGMAX = MAXAG(N) - MAXAG(N-1)=-7.6751342


SYSTEM TIME=19:17:33 TIME= 1.16  TIME STEP= 58  ISN= .001        NT= 505  R1(N)= 1.4411311e-14

AT THE SITE MAXAG= 53.32561  MAXCT= 3000  MAXDM= 3000  MAXB= 3000  MAXNK= 3000
MAXEI= 0  MAXVC= 0  MAXRD= 0  MAXSKD= 6897.652  MAXST= 0
  AT THE NODE1 AG,CT,DM,B,NK,SKD : 5.2737775  58.503787  31.175695  30.798403  30.798403           1510.6705
  AT THE NODE2 AG,CT,DM,B,NK,SKD : 5.2737775  58.503787  31.175695  30.798403  30.798403  1510.6705
  AT THE NODE1 AG,CT,DM,B,NK,SKD : 5.2737775  58.503787  31.175695  30.798403  30.798403  1510.6705
MAX CONCENTRATIONS OUTSIDE THE PRIMARY SITE: MAXAGO= 5.6094286  AT #OF POINTS= 28
MAXCTO= 575.12164  MAXDM= 608.77199  MAXB= 290.82868  MAXNK= 290.82868
  MAXEI= 0  MAXVC= 0  MAXRD= 0  MAXST= 0  MAXSKD= 1513.1711
MAXAGO= 5.6094286  MAXCTO= 575.12164  MAXDMO= 608.77199  MAXBO= 290.82868  MAXSKDO= 1513.1711
DIFFERENCE IN AGMAX = MAXAG(N) - MAXAG(N-1)=-6.8069564
```

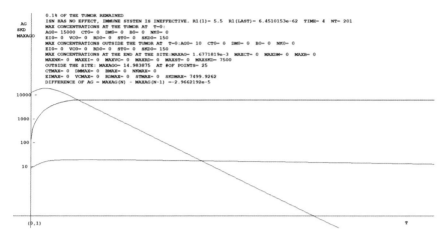

Fig. 6.36 AG versus SKD. SKD fighting alone

rate of growth of cancer cells, has been reduced from 5.5 to practically zero. Cancer cells outside the tumor site have first gone up, then started getting smaller. Although the rate of growth is zero, yet the presence of cancer cells is quite prominent. At the tumor site $MAXAG = 60.132567$. (Presence of this malignant mass is unacceptable).

Our final attempt was to use the Skilled Killer Drug (SKD) alone and turn off all other fighters, like the lymphocytes and all radiation and chemos. The inputs are the same as before, only all drugs are off, except (SKD). The results given by the graphs in Fig. 6.36 reveal some differences as compared with Fig. 6.35. To see the results in more details, we have the following.

Inputs for Fig. 6.36.

```
••••••••••••••••••••••••••••••••••••••••••••••••••••••••••••••••••••••••••••••••••••••••••••••
    *** INITIAL CONDITIONS FOR THE CODE 6.5E.APPLICATION OF INTRAVENOUS SKD. SKD FIGHTING ALONE. ***
    ISN= 0   DX= .02  DY= .02  DZ= .02  DT= .02  FIELD SIZE= 101x101x101 # OF TIME STEPS= 505
    R1= 5.5  R11= .0000001  R12= 0  R13= 0  R14= 0  R15= 0  R16= 0  R17= 0
    R18= 0   R19= .00065  R110= 0  KAPPA1= .0000001  A5(REDUCTION OF AG AT EACH POINT IN THE FIELD) = .00075
    R2= 0   R21= 0  R22= 0  R23= 0  R24= 0  R25= 0  R26= 0  R27= 0  KAPPA2= 0
    R3= 0   R31= 0  R32= 0  R33= 0  R34= 0  R35= 0  KAPPA3= 0
    R4= 0   R41= 0  R42= 0  R43= 0  R44= 0  R45= 0  R46= 0  R47= 0  KAPPA4= 0
    R5= 0   R51= 0  R52= 0  R53= 0  R54= 0  R55= 0  R56= 0  R57= 0  KAPPA5= 0
    R6= 0   R61= 0  R62= 0  R63= 0  R64= 0  KAPPA6= 0
    R7= 0   R71= 0  R72= 0  R73= 0  R74= 0  R75= 0  R76= 0
    R77= 0  R78= 0  KAPPA7= 0
    R8= 0   R81= 0  R82= 0  R83= 0  R84= 0  R85= 0  R86= 0  R87= 0  KAPPA8= 0
    R9= 0   R91= 0  R92= 0  R93= 0  R94= 0  KAPPA9= 0
    R10= 1  R101= .00001  R102= .00001  R103= 0  R104= 0  R105= 0  R106= 0
    R107= 0  KAPPA10= .0000001
    CONDITIONS AT THE PRIMARY SITE AT T=0:
    AGMAX0= 15000  CTMAX0= 0  DMMAX0= 0  BMAX0= 0  NKMAX0= 0
    EIMAX0= 0  VCMAX0= 0  RDMAX0= 0  STMAX0= 0  SKDMAX0= 150
    THE THREE LYMPH NODES AGL0, CTL0,DML0,BL0,NKL0= 10  0  0  0  0
    CONDITIONS OUTSIDE THE SITE:
    AGMAX0= 10  CTMAX0= 0  DMMAX0= 0  BMAX0= 0  NKMAX0= 0
    EIMAX0= 0  VCMAX0= 0  RDMAX0= 0  STMAX0= 0  SKDMAX0= 150

    RESTRICTIONS ARE : SKD(I,J,K) <= 7500 &
    R1(N) = R1(N-1)* EXP(-0.01*(DT/2)*(MAXSKD(N) + MAXSKD(N-1)))
    AG(I,J,K)= AG(I,J,K) - A5*AG(I,J,K)*SKD(I,J,K),REDUCTION OF AG AT EACH FIELD POINT
••••••••••••••••••••••••••••••••••••••••••••••••••••••••••••••••••••••••••••••••••••••••••••••
```

It is clear from the inputs that (SKD) alone is fighting against cancer. It is intravenously injected as before. At $t = 0$, $SKD(i, j, k) = 150$ at all points (i, j, k) in the entire computational field.

From Table 6.2, at $t_n = 85$, $r_1(n) = 1.8512 \times 10^{-24}$. So there is no growth of cancer. Therefore on the cancer, cells in circulation are fighting against (SKD). At the primary tumor site $max(ag(i, j, k)) = 18.22$, at an undetected lymph node, $ag(i, j, k) = 16.1715$ and out in the field $max(ag(i, j, k)) = 20.6141$. But when (SKD) was assisted by the lymphocytes, by $t_n = 70$, all over the field $max(ag(i, j, k)) < 15$. From both Tables 6.1 and 6.2, it is noticeable that the max concentrations of the defenders are at the primary site where the primary tumor was located. Even though the tumor is dying, the vigilance at this location remains at its highest level. In the codes, absolutely no special condition has been introduced which may have caused this phenomenon. The equations that we have solved for Fig. 6.36 are as follows:

$$\partial(ag)/\partial t = r_1(ag) - r_{11}(ag) - r_{19}(ag)(skd) + \kappa_1 \nabla^2(ag)$$

$$\partial(skd)/\partial t = r_{10}(ag) - r_{101}(skd) - r_{102}(ag)(skd) + \kappa_{10} \nabla^2(skd)$$

where all the values of the rate constants are given in the input for Fig. 6.36. Researchers in [28] have observed that "The conditions leading to cell necrosis, upon nutrient and oxygen deprivation, were recapitulated in-vitro and were used to generate samples for computational proteomic analysis. Under these conditions, we identified clusters of enriched pathways that may be involved in tumor resistance, leading to cancer recurrence. We show that the content of necrotic cells enhances angiogenesis and proliferation of endothelial cells, induces vasculature, as well as increases migration, invasion, and cell–cell interactions." So the presence of more defense near the necrotic tumors may be a good aspect of this model. It should be noticed that there are no roles for θ, the measure of aggressiveness and (ISN) because the immune system is turned off.

Table 6.2 for Fig. 6.36 shows the power of the drug SKD. It alone has brought down the growth of cancer. It may sound too good to be true, but we strongly believe this drug is forthcoming and shall make the future of cancer treatment much brighter.

We notice that the value of MAXAGO, which in this case is the largest value of (ag), is getting smaller very very slowly. In the logarithmic scale, this change is so small that the graph for MAXAGO shows hardly any change in Fig. 6.36.

In Chap. 8, we will discuss more on the Skilled Killer Drug (SKD).

Table 6.2 For Fig. 6.36

```
SYSTEM TIME=12:02:55 TIME= 1.7  TIME STEP= 85  ISN= 0              NT= 505  R1(N)= 1.8511715e-24

AT THE SITE MAXAG= 18.217224  MAXCT= 0  MAXDM= 0  MAXB= 0  MAXNK= 0
MAXEI= 0  MAXVC= 0  MAXRD= 0  MAXSKD= 7500  MAXST= 0
 AT THE NODE1 AG,CT,DM,B,NK,SKD : 16.175186  0  0  0  0              179.35145
 AT THE NODE2 AG,CT,DM,B,NK,SKD : 16.175186  0  0  0  0  179.35145
 AT THE NODE1 AG,CT,DM,B,NK,SKD : 16.175186  0  0  0  0  179.35145
MAX CONCENTRATIONS OUTSIDE THE PRIMARY SITE: MAXAGO= 20.614105  AT #OF POINTS= 25
MAXCTO= 0  MAXDM= 0  MAXB= 0  MAXNK= 0
 MAXEI= 0  MAXVC= 0  MAXRD= 0  MAXST= 0  MAXSKD= 189.13074
DIFFERENCE IN AGMAX = MAXAG(N) - MAXAG(N-1)=-1.8637175
```

```
SYSTEM TIME=12:05:38 TIME= 1.72  TIME STEP= 86  ISN= 0             NT= 505  R1(N)= 8.744315e-25

AT THE SITE MAXAG= 16.526482  MAXCT= 0  MAXDM= 0  MAXB= 0  MAXNK= 0
MAXEI= 0  MAXVC= 0  MAXRD= 0  MAXSKD= 7500  MAXST= 0
 AT THE NODE1 AG,CT,DM,B,NK,SKD : 16.137482  0  0  0  0              179.67396
 AT THE NODE2 AG,CT,DM,B,NK,SKD : 16.137482  0  0  0  0  179.67396
 AT THE NODE1 AG,CT,DM,B,NK,SKD : 16.137482  0  0  0  0  179.67396
MAX CONCENTRATIONS OUTSIDE THE PRIMARY SITE: MAXAGO= 20.563391  AT #OF POINTS= 25
MAXCTO= 0  MAXDM= 0  MAXB= 0  MAXNK= 0
 MAXEI= 0  MAXVC= 0  MAXRD= 0  MAXST= 0  MAXSKD= 189.5782
DIFFERENCE IN AGMAX = MAXAG(N) - MAXAG(N-1)=-1.6907417
```

```
SYSTEM TIME=12:08:20 TIME= 1.74  TIME STEP= 87  ISN= 0             NT= 505  R1(N)= 4.1305219e-25

AT THE SITE MAXAG= 14.992661  MAXCT= 0  MAXDM= 0  MAXB= 0  MAXNK= 0
MAXEI= 0  MAXVC= 0  MAXRD= 0  MAXSKD= 7500  MAXST= 0
 AT THE NODE1 AG,CT,DM,B,NK,SKD : 16.099799  0  0  0  0              179.99572
 AT THE NODE2 AG,CT,DM,B,NK,SKD : 16.099799  0  0  0  0  179.99572
 AT THE NODE1 AG,CT,DM,B,NK,SKD : 16.099799  0  0  0  0  179.99572
MAX CONCENTRATIONS OUTSIDE THE PRIMARY SITE: MAXAGO= 20.512667  AT #OF POINTS= 25
MAXCTO= 0  MAXDM= 0  MAXB= 0  MAXNK= 0
 MAXEI= 0  MAXVC= 0  MAXRD= 0  MAXST= 0  MAXSKD= 190.02465
DIFFERENCE IN AGMAX = MAXAG(N) - MAXAG(N-1)=-1.5338208
```

6.7 The Conclusion

The primary objective of these graphics is to show that our numerical results are properly simulating what is happening in the real time. However, these are not real-time computations. In the future, real-time reactive parallel computations should be conducted by researchers working on biocomputations. Then doctors will be able to visualize what is happening inside a patient's body as treatments continue. This was first suggested by Dr. John Ziebarth, former Chief of NAS division of NASA Ames who first sponsored this project. The first author did accomplish some mathematical works in this field, at Mississippi State University, Indian Statistical Institute, Calcutta, S. N. Bose Center for Basic Sciences, Calcutta, CNR, Naples, Italy, and Harvey Mudd College, Los Angeles, and received many good comments from Dr. Joe Thompson, from Mississippi State University, Prof. Ambarish Ghosh from ISI, from Amato Umberto from CNR, Italy. Dr. Chanchal Majumdar from the S. N. Bose Center gave me many valuable suggestions in this regard. The second author worked extremely hard on this part of the project. Oncologists and several scientists around the world liked our computed results because they were very meaningful to them.

References

1. Chiriac, V.-F., Baban, A., Dumitrascu, D.l.: Psychological Stress and Breast Cancer Incidence: a Systematic Review, Romania, Iuliu Hatieganu University of Medicine and Pharmacy Cluj-Napoca, Romania 2 Department of Psychology, Babes Bolyai University Cluj-Napoca (2017)
2. University of Iowa Health Care: Why high-dose vitamin C kills cancer cells (2017)
3. Gonzalez, M.J., Miranda-Massari, J.R., Mora, E.M., Guzmán, A., Riordan, N.H., Riordan, H.D., Casciari, J.J., Jackson, J.A., Román-Franco, A.: Orthomolecular oncology review: ascorbic acid and cancer 25 years later. Integr. Cancer Therapy **4**(1), 32–44 (2005)
4. Kamel, D., Gray, C., Walia, J.S., Kumar, V.: PARP inhibitor drugs in the treatment of breast, ovarian, prostate and pancreatic cancers: an update of clinical trials. Curr. Drug. Targets **19**(1), 21–37 (2018)
5. Flaherty, R.L., Intabli, H., Falcinelli, M., Bucca, G., Hesketh, A., Patel, B.A., Allen, M.C., Smith, C.P., Flint, M.S.: Stress hormone-mediated acceleration of breast cancer metastasis is halted by inhibition of nitric oxide synthase. Cancer Lett. (2019)
6. Vilalta, M., Rafat, M., Graves, E.E.: Effects of radiation on metastasis and tumor cell migration. Cellul. Molec. Life Sci. **73**(16) (2016)
7. Gallagher, J.: Health and science correspondent: immune discovery 'may treat all cancer'. January,20, 2020
8. Jain, M.K: Numerical Solution of Differential Equations, Second Edition, Halsted Press (1978)
9. Breastcancer.org
10. Cancer Research UK
11. National Cancer Institute. www.cancer.gov
12. Snow, H.L.: Cancer and the cancer process. J&A. Churchill, London (1893); Apud Bleiker, E.M., van der Ploeg, H.M.: Psychosocial factors in the etiology of breast cancer: review of a popular link. Patient Educ. Couns. 37(3), 201–214 (1999)
13. Memorial Sloan Kettering Cancer Center. www.mskcc.org, Sept. 2019
14. Dey, S.K: Computational modeling of the breast cancer treatment by immunotherapy, Rad. Estrogen Inhib. Sci. Mathematicae Japonicae **8** (2003)
15. Cheng, R., Ma, J.-X.: Angiogenesis in diabetes and obesity. Rev. Endocr. Metab. Disord. **16**(1) (2015)
16. Chatterjee, A., Magee, J.L., Dey, S.K.: The role of homogeneous reactions in the radiolysis of water. Rad. Res. **96** (1983)
17. Dey, S.K., Dey, C.: An explicit predictor-corrector solver with applications to Burgers' equation. NASA Tech. Memo. 84402 (1983)
18. Dey, S.K: Numerical modeling of d-mapping with applications to chemical kinetics. NASA Tech. Memo. 84332 (1984)
19. Dey, S.K: Analysis of convergence of parallelized pfi for coupled large-scale nonlinear systems. Int. J. Comput. Math. **75** (2000)
20. Dey, S.K: A Massively Parallel Algorithm for Large-Scale Nonlinear Computations with Application to Nonlinear Parabolic PDEs. Computational Acoustics. North-Holland (1991)
21. Dey, S.K: Numerical solution of Euler's equations by perturbed functionals. Lect. Appl. Math. Amer. Math. Soc. **22** (1985)
22. Dey, S.K., Bharadwaj, D.: Parallelized PFI for Large-Scale Nonlinear Systems in a Distributed Computing Environment. Applicable Mathematics (Edited by J.C.Misra). Narosa Publication (2001)
23. Spencer, D.S., Puranik, A.S., Peppas, N.A.: Intelligent nanoparticles for advanced drug delivery in cancer treatment. Curr. Opin. Chem. Eng. **7**, 84–92 (2015)
24. Jain, R.K., Stylianopoulos, T.: Delivering nanomedicine to solid tumors. Nat. Rev. Clin. Oncol. **7**(11), 653–664 (2010)
25. Karsch-Bluman, A. etal.: Tissue necrosis and its role in cancer progression. Oncogene **38** (2018)
26. Michor, F., Beal, K.: Improving cancer treatment via mathematical modeling: surmounting the challenges is worth the effort. Cell (2016)

27. Katouli, A.A., Komarova, N.L.: The worst drug rule revisited: mathematical modeling of cyclic cancer treatments. Bull. Math. Biol. (2011)
28. Coldman, A.J., Murray, J.M.: Optimal control for a stochastic model of cancer chemotherapy. Math. Biosci. 168 (2000)
29. Goldie, J.H., Coldman, A.J.: A mathematical model for relating the drug sensitivity of tumors to their spontaneous mutation rate. Cancer Treat. Reports 63 (1979)
30. Altrock, P.M., Liu, L.L., Michor, F.: The mathematics of cancer: integrating quantitative models. Nat. Rev. Cancer **15**, 730–745 (2015)

Chapter 7
Gene Therapy

Abstract In this chapter, mathematical modeling on gene therapy has been discussed. Our emphasis is how poisonous proteins released by mRNAs due to mutations in genes could be corrected.

7.1 Rationale

Gene is the unit of heredity. Body functions following the codes implanted in genes. They are inside the nucleus of the cells, located on the chromosomes. They are in our deoxyribonucleic acid (DNA) in a region which encodes proteins. "When genes are expressed, the genetic information (base sequence) on DNA is first copied to a molecule of mRNA (transcription). The mRNA molecules then leave the cell nucleus and enter the cytoplasm, where they participate in protein synthesis by specifying the particular amino acids that make up individual proteins (translation)." [1]

None can say with any precision how many genes are in our DNA. An estimation is between 25,000 and 35,000. However, scientists have identified functions of many genes.

According to the National Cancer Institute, cancer is a genetic disease. Mutations of genes cause mutations of cells, and then, cells behave abnormally violating all the essential rules of homeostasis. They grow abnormally. They keep on invading tissues around them and change normal cells into abnormal cells. They employ some drivers to make their move. So cancer is a major global disease of the body affecting every organ. Right after it starts, it affects almost all the organs of the body directly and sometimes indirectly. By the time doctors try to resolve the problem in one area, it might have already evaded another area which could be very remote from the primary site. This invasion could be so subtle that expert oncoradiologists may fail to recognize.

Since cancer is a genetic disease, we need to study how genetic therapy could be accomplished by applying mathematical models.

© Springer Nature Singapore Pte Ltd. 2021

S. Dey and C. Dey, *Mathematical and Computational Studies on Progress, Prognosis, Prevention & Panacea of Breast Cancer*, Forum for Interdisciplinary Mathematics,
https://doi.org/10.1007/978-981-16-6077-1_7

Let us first review the process and progress of cancer. The National Institute of Cancer depicted a very comprehensive picture of this topic in https://www.cancer.gov/about-cancer/understanding/what-is-cancer. Some of these are as follows.

"The genetic changes that contribute to cancer tend to affect three main types of genes—proto-oncogenes, tumor suppressor genes, and DNA repair genes. These changes are sometimes called "drivers" of cancer.

Proto-oncogenes are involved in normal cell growth and division. However, when these genes are altered in certain ways or are more active than normal, they may become cancer-causing genes (or oncogenes), allowing cells to grow and survive when they should not.

Tumor suppressor genes are also involved in controlling cell growth and division. Cells with certain alterations in tumor suppressor genes may divide in an uncontrolled manner.

DNA repair genes are involved in fixing damaged DNA. Cells with mutations in these genes tend to develop additional mutations in other genes. Together, these mutations may cause the cells to become cancerous.

As scientists have learned more about the molecular changes that lead to cancer, they have found certain mutations commonly occur in many types of cancer. Because of this, cancers are sometimes characterized by the types of genetic alterations that are believed to be driving them, not just by where they develop in the body and how the cancer cells look under the microscope."

A germline mutation, or germinal mutation, may be found only in the germ cells, cells that become sperm or ovum. Then the offspring inherits them. Somatic mutations are very general. They can occur in any cell but germ cells. They are not passed on to children.

It is true that malignant cells move from one organ to another through the flow of blood and lymph. But except for leukemias, they form tumors in the tissues and invade tissues in a three-dimensional manner arbitrarily as time changes while moving almost aimlessly through tissues one after another. A weaker body makes the immune system weaker giving cancer cells a golden opportunity to spread.

Gene transfer or gene modification is a very serious and subtle issue. When a gene mutates, in general, DNA repairs that gene and corrects it or it goes to apoptosis. But sometimes it mutates multiple times and cannot be corrected. Then it behaves abnormally. That leads to cancer. To correct this scenario, researchers then think of gene transfer [2]. "First, affected cells are removed from the body. In the laboratory, functional genetic material is introduced into the cells and is then delivered back into the patient's body (i.e., CAR-T therapy)" [2]. This functional gene inserted into the cell is expected to work normally and keep the mutated gene suppressed.

"Gene therapy implies an approach that aims to modify, delete, or replace abnormal gene(s) at a target cell. Such target cells may be malignant primary or metastatic nodules, circulating tumor cells or dormant stem cells, and specific cells such as T-cell lymphocytes or dendritic cells. With the presence of over 20,000 active genes in human cells, exposed to numerous factors whether hereditary, environmental, infectious or spontaneous, unlimited possibilities for gene mutation, aberration, dysfunction, or deletion have been expected, leading to clinical presentation of various medical disorders, including cancer" [3].

How this may be achieved through mathematical modeling is our topic here. We will look into the pathway of genetic information mathematically. Through gene expression, the information encoded in a gene is decoded in a particular protein. Some recent studies on Gene therapy may be found in [8–13].

Symbolically, gene expression is

$$DNA \rightarrow mRNA \rightarrow Protein. \tag{7.1}$$

Messenger ribonucleic acid (mRNA) is a nucleic acid present in all living cells. RNAs are molecules carrying codes generated by DNA to the protein synthesis sites in the ribosomes (cytoplasm). Eventually, these molecules are the mRNA in our body.

This process is certainly not as simple as we have described. So in case there are mistakes, it may require another set or sets of proteins to correct them.

Each cell has over 35,000 genes. More than 100 are cancer-causing genes or oncogenes. There are also many tumor suppressing or anti-oncogenes. When tumor suppressing genes mutate and become oncogenes, malignancy ensues. For example, p53, BRCA1, and BRCA2 are all tumor suppressing genes. When they mutate and improper amounts of proteins are released, cancer starts.

So we are thinking of replacing the defective proteins transcribed by mRNA, by a set of new proteins which will bring back the desired state of homeostasis.

Let the set of oncogenes overly/underly expressed be $X = \{x_1, x_2, \ldots, x_m\}$. Body can produce over 100,000 proteins from about 20,000 genes, and each is transcribed into mRNA and then translated into a protein. Typically, one mRNA may produce 1000 proteins. "Imbalance of protein homeostasis (proteostasis) is known to cause cellular malfunction, cell death, and diseases. Elaborate regulation of protein synthesis and degradation is one of the important processes in maintaining normal cellular functions." [4] "De Bruin carries out chemical-biological research on proteasomes. They are a kind of degradation factory that cuts proteins in cells into smaller fragments. It's a highly useful activity: it's a way for cells to clean up proteins that are redundant or that have been damaged….With particular types of cancer, protein production in the cell is much higher than normal. If you inhibit the proteasomes, this stops degradation and the proteins pile up, which eventually kills the cancer cell" [1].

We will attempt to simulate this mathematically.

The most commonly mutated gene in people with cancer is p53 or TP53. More than 50% of cancers involve a missing or damaged p53 gene. Most p53 gene mutations are acquired. Germline p53 mutations are rare, but patients who carry them are at a higher risk of developing many different types of cancer.

We are not thinking of turning off the mutated genes by conducting direct gene modifications, rather correcting mRNAs with cRNAs which will correct redundant or damaged proteins and become a part of the newly formed mRNA family. This will certainly eliminate the scary aspects of "genetic modification by gene transfer."

Furthermore, they get dissolved, and the dosages are repeated until the desired goal is achieved. Certainly, one set of cRNA will not get the job done in general. It could take a sequence of cRNAs to achieve the goal. This is similar to taking daily dosages of medications.

By breaking down the proteins released by the mRNAs, it is possible to modify gene expressions by applying modified mRNAs which we will call cRNAs or corrected RNAs.

Furthermore, it has been found that a hormone (which could be a signal molecule) released by one cell binding to a receptor protein on a target cell initiates a sequence of biochemical changes resulting in changes in target cells (here these are cancer cells). These changes may include increased or decreased transcription finally modifying the proteins. The signal cell is the cell which sends a signal molecule carrying the good protein to the infected (mutated) target cell. The target cell's receptor receives that information and relays that protein to an activated transcription factor in the target cell. This goes into the nucleus of the target cell of the mutated gene, and it sends this information to the cRNA, which translates the information into the new corrected protein. In this way, hundreds of proteins may be corrected.

Let us denote the defective mRNA by dRNA. We will develop mathematical models on how to eliminate them.

7.1.1 Mathematical Preliminaries [3]

For a sequence of positive scalars $a_0, a_1, \ldots a_k \ldots$, if the product: $P = \prod_{k=0}^{\infty} a_k = 0$, then each a_k is called a D-element or a decaying element. A sufficient condition that a_k is a D-element is

$$\max a_k \leq \alpha < 1 \text{ for } \forall k > K.$$

That implies that initially $a_k's$ may increase and exceed 1 significantly, but after some point they all must be less than 1, so that their product will go to zero.

So if $\Theta(t) = \{\theta(t_n)\}$ is a time-dependent sequence of positive numbers, then if $\forall t_n > T, \theta(t_n) \leq \alpha < 1$, then $\Theta(t)$ is a decreasing sequence $\forall t_n > T$ and the product of its elements goes to zero as $n \to \infty$.

7.1.2 An Application

Let x and y be the quantifications of two interacting biochemicals such that under a chemical process, $x(t)$ changes with time and should go to zero while y remains a constant. Let $g : D \subset R \to D \subset R$ be an operator which simulates this chemical operation, where R represents the points of the real number axis representing the values of the chemical x.

Let $x_{n+1} = g(x_n, y) > 0 \forall x_n$ and $x^* = g(x^*, y)$. Let $g(x_n, y)$ satisfy

$$x_{n+1} - x^* = g(x_n, y) - g(x^*, y) < \alpha_n(x_n - x^*). \tag{7.2}$$

where for $\forall n > N$, $g(x_n, y)$ is a contracting function of x_n and $x^* < x_n \forall n$. Also $\forall n > N$, $\alpha_{n+1} < \alpha_n$, and $\alpha_n < 1$. It is now rather easy to prove that as $n \to \infty$, $x_n \to x^*$. If x^* is null, the biochemical interaction $g(x_n, y)$ will damp out the chemical x.

7.1.3 A Preliminary Model

Let x be a protein that should be eliminated by a protein y as x changes with time. This change is given by $dx/dt = f(x, y)$. Let us approximate it by a simple finite difference equation $x_{n+1} = x_n + hf(x_n, y)$, *where* $h = time\ step$. This may be represented by $x_{n+1} = g(x_n, y)$. Applying the concept of D-mapping, as done in the Sect. 7.3 it could be proved that, if $g(x_n, y)$ satisfies (7.2), then x_n will be damped out.

7.2 Introduction of D-Matrices [5–7]

A sequence of square matrices $A_0, A_1, \ldots A_k \ldots$ of the same type is called a sequences of D-matrices, if as $k \to \infty$, the product

$$A_k A_{k-1} \ldots A_0 = 0 \tag{7.3}$$

A necessary and sufficient condition is that if

$$\forall k > K, \max|a_{ij}^k| < \varepsilon \tag{7.4}$$

where $a_{ij}^k \in A_k$ and ε is positive and arbitrarily small, then A_k is a D-matrix and (7.2) is valid. The proof is trivial.

Also a sufficient condition is, for a given q norm,

$$||A_m||_q \leq \alpha < 1 \forall m > K \tag{7.5}$$

Proof is trivial.

The second condition (7.5) implies that all square matrices $A_m \forall m \geq K$ are convergent matrices.

However, a convergent matrix is not necessarily a D-matrix, and a D-matrix may not be a convergent matrix. For example:

Let $A_k = \text{diag}(1,1,.0.1,\alpha_{kk}, 1,.0.1)$, where $0 < \alpha_{kk} < 1$, then $\forall k$, A_k is a D-matrix, and none is a convergent matrix because the spectral radius of each of them is 1. Similarly, if

$$A_m = \begin{bmatrix} 0 & 0 & 1 \\ 0 & 0 & 0 \\ 0 & 0 & 0 \end{bmatrix}, \text{ and } A_{m+1} = A_m^T \quad \forall m \geq k, \text{ then each matrix is a conver-}$$

gent matrix, but the sequence does not represent D-matrices because it violates the condition (7.3).

7.3 A Model for Eliminating Multiple Proteins

Let $x_1, x_2, \ldots, x_I$ be proteins, released by mutated genes, which are malfunctioning. So these must be replaced by a set of proteins $y_1, y_2, \ldots, y_J$. These are the proteins released by corrector RNAs (cRNAs). If this is achieved, it means the defective gene is corrected. We assume that it is a time-dependent operation which reduces all $x_i's$ to null and the proteins $y_j's$ remain unchanged.

A simple mathematical model is with Y remaining a constant,

$$dX/dt = F(X, Y), \tag{7.6}$$

where

$$X = (x_1, x_2, \ldots, x_I)^T \in D \subset R^I \text{ and } Y = (y_1, y_2, \ldots, y_J)^T \in R^J \subseteq D \subset R^I$$
$$\text{and } F : D \subset R^I \to D \subset R^I$$

R^I is $I-dimensional\ Real\ space$ and D is an $I-dimensional$ Banach Space. A finite difference model is:

$$X^{n+1} = G(X^n, Y), \tag{7.7}$$

where $(X^n, Y) = X^n + \Delta t F(X^n, Y)$, where $\Delta t = the\ step\ size$. We want to see under what condition,
$\lim_{n \to \infty} X^n = X^*$, Y remaining unchanged.

7.3.1 Definition. D-Mapping [5–7]

If $X^n \in D$, and $Y \in R^J \subseteq R^I$ a parameter, and G is a mapping defined by $G : D \subset R^I \to D \subset R^I$, be such that

$$G(X^{n+1}, Y) - G(X^n, Y) = A_n(X^n, Y)(X^{n+1} - X^n) \tag{7.8}$$

where A_n is a D-matrix changing at each n, Y does not change. Then G is a D-mapping on $D \subset R^I$.

In fact $A_n(X^n, Y)$ is the Frechet derivative of G on D.

7.3.2   Theorem on Elimination of Defective mRNA

If G is a D-Mapping on $D \subset R^I$, D is a Banach Space, then the iterative method defined by

$$X^{n+1} = G(X^n, Y), Y \text{ is a fixed parameter} \tag{7.9}$$

X attains its point of attraction $X^* \in D$ as $n \to \infty$.

Proof: Let $X^{n+1} = X^* + \varepsilon^{n+1}$. Then from (7.8), $\varepsilon^{n+1} = G(X^n, Y) - G(X^*, Y) = G(X^n, Y) - G(X^n - \varepsilon^n, Y)$.

$= A_n(X^n, Y)\varepsilon^n$, where $A_n(X^n, Y) = G'(X^n, Y)$. G' is the Frechet derivative of G in D. Thus,

$\varepsilon^{n+1} = A_n A_{n-1}...A_0 \varepsilon^0$ since A_n is a D-matrix, as $n \to \infty$, $A_n A_{n-1}...A_0 \to 0$. It proves the theorem.

Since D is a Banach Space, we may write, $||G(X^n, Y) - G(X^*, Y)||$, here $A_n = A_n(X^n, Y)$. So, $||G(X^n, Y) - G(X^*, Y)|| \leq ||.||\varepsilon^n||$. Convergence is established since A_n is a D-matrix.

From this theorem, we get that as Y interacts with X in time, at the very outset X may not diminish, it may even increase. However, as time progresses, X ultimately merges with X^*. If X^* is null, at the end X vanishes and Y stays on. So X is damped out.

Let us consider an application. We will now consider solutions of two nonlinear systems like (7.6) using Charlie's convex method.

7.4   Two Examples

Ex.1. Solve:

$$dx_1/dt = x_1 - 0.75x_1y_1 - 0.5x_1y_2 + 0.005x_1x_2 \tag{7.10}$$

$$dx_2/dt = x_2 - 0.4602925x_2y_1 - 0.05x_2y_2 + 0.005x_1x_2 \tag{7.11}$$

subject to the following conditions, $x_1(0) = 2.3$, $x_2(0) = 1.3$, $y_1(t) = y_2(t) = 2 \ \forall t$.

First we should study how stiff this system is, because most models in biochemistry are stiff.

Considering $dx_1/dt = f_1(x_1, x_2, y_1, y_2)$ and $f_2(x_1, x_2, y_1, y_2)$ and noting that y_1, y_2 are constants, we considered the matrix: (expressed in a rowwise form)

$$[\partial f_1/\partial x_1 \quad \partial f_1/\partial x_2]$$
$$[\partial f_2/\partial x_1 \quad \partial f_2/\partial x_2]$$

where all the terms are evaluated at $x_1(0) = 2.3$, $x_2(0) = 1.3$, $y_1(t) = y_2(t) = 2$.

Then the eigenvalues are $\lambda_1 = -0.009035$ and $\lambda_2 = -1.4936$. Both eigenvalues being negative, solutions will go to zero as t increases. The stiffness ratio is $|\lambda_2|/|\lambda_1| = 165.3127$. So it is stiff. It is also evident from the graphs, that while x_1 is decaying very fast, x_2 is decaying very slowly. This is the very nature of stiff systems.

The system has been solved by Charlie's method. As we have seen before, this rather simple numerical method is quite powerful regarding solving stiff equations (Fig. 7.1).

It is very clear from the graph that the equations are quite stiff. While X_1 is moving toward zero very fast, X_2 is going to zero relatively slowly. Also these results reveal that from the point of view of numerical analysis, the system is very stiff and Charlie's method is very powerful.

Ex: 2.

Let us consider a set of equations:

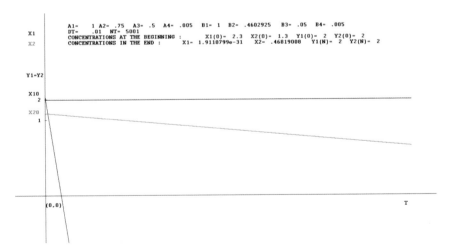

Fig. 7.1 Decadance of both X1 and X2 (In the equations these are x1 and x2)

$$dx_1/dt = a_1x_1 + a_2x_1x_2 + a_3x_1x_3 - a_4x_1y_1 - a_5x_1y_2 + a_6x_1x_2x_3 - a_7y_1y_2y_3$$
$$= f_1(x_1, x_2, x_3, y_1, y_2) \tag{7.12}$$

$$dx_2/dt = b_1x_2 + b_2x_1x_2 + b_3x_2x_3 - b_4x_2y_1 - b_5x_2y_2 + b_6x_1x_2x_3 - b_7y_1y_2y_3$$
$$= f_2(x_1, x_2, x_3, y_1, y_2) \tag{7.13}$$

$$dx_3/dt = c_1x_3 + c_2x_1x_3 + c_3x_2x_3 - c_4x_3y_1 - c_5x_3y_2 + c_6x_1x_2x_3 - c_7y_1y_2y_3$$
$$= f_2(x_1, x_2, x_3, y_1, y_2) \tag{7.14}$$

x_1, x_2, x_3 are the three defective proteins. The proteins y_1 and y_2 are the laboratory made corrective proteins. So y_1 and y_2 remain constant.

The following inputs were used:

Inputs for Fig. 7.2.

```
CHARLIE'S CONVEX CORRECTOR; INPUTS FOR FIG.7.2

A1=    .05   A2= 7.5e-10   A3= 7.5e-8   A4=  .015   A5=  .025   A6=  5.e-11   A7= .00006
B1=    .1    B2= 8.5e-8    B3= 6.5e-8   B4=  .031   B5=  .031   B6= .0000001  B7= .00005
C1=    .1    C2= 3.5e-10   C3= 3.5e-10  C4=  .025   C5= .0399   C6= 1.25e-9   C7= .0005
X10=   2.8   X20= 3.5      X30= 4.5     Y1= 1.95    Y2= 1.45    Y1 & Y2 ASSUMED TO REMAIN CONSTANTS.
DT=    .005  NUMBER OF TIME STEPS=   20000
```

Certainly this model is totally a hypothetical model. However, the last terms on the right side of the equations indicate that when the laboratory made proteins y_1, y_2 and y_3 interact among themselves, those interaction exert some efforts to reduce the defective proteins x_1, x_2 and x_3. This is an attempt to simulate placebo effects. We will now observe how effective the technique of solution is.

The Fig. 7.2 shows that the defective proteins $X1$, $X2$, $X3$ could be destroyed by the corrective proteins $Y1$ and $Y2$, respectively.

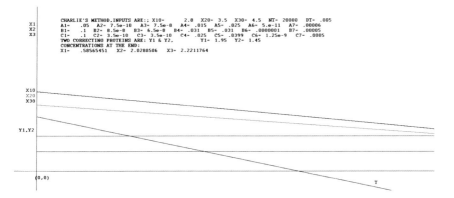

Fig. 7.2 Treatment worked.

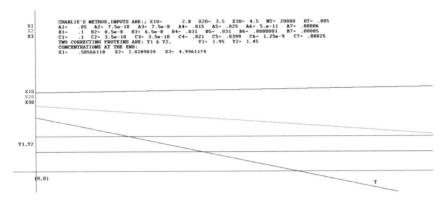

Fig. 7.3 Treatment failed. X1, X2 decreasing, but X3 increasing

The graphs of $X1$, $X2$, $X3$ show that the model is computationally stiff (because $X1$, presented by the redline is decreasing much faster than the other $X2$ and $X3$). The Jacobian matrix at $t = 0$ is :for the Fig. 7.2

$$[-1.5499659e-2 \qquad 2.73e-9 \qquad 2.1049e-7]$$
$$[1.8725e-6 \qquad -5.3982095e-3 \qquad 1.2075e-6]$$
$$[2.12625e-8 \qquad 1.7325e-8 \qquad -6.6049855e-3]$$

The eigenvalues are $1.6248e{-}12$, $-6.605e{-}3$, and $-5.3982e{-}3$.

Stiffness $= 6.605e{-}3/\ 1.6248e{-}12 = (6.605e{-}3/1.6248e{-}3) \times 10^9 = 4.07 \times 10^9$

It is a stiff system.

Next we have made a small change, replacing $c_4 = 0.025$ by $c_4 = 0.021$. It has a very strong impact on the results. The protein x_3 slowly started rising. This may be noticed in Fig. 7.3.

Inputs for Fig. 7.3.

```
CHARLIE'S CONVEX CORRECTOR; INPUTS FOR FIG.7.3

A1=    .05   A2= 7.5e-10   A3= 7.5e-8   A4= .015   A5= .025   A6= 5.e-11   A7= .00006
B1=    .1    B2= 8.5e-8    B3= 6.5e-8   B4= .031   B5= .031   B6= .0000001  B7= .00005
C1=    .1    C2= 3.5e-10   C3= 3.5e-10  C4= .021   C5= .0399  C6= 1.25e-9   C7= .00025
X10=   2.8   X20= 3.5      X30= 4.5     Y1= 1.95   Y2= 1.45   Y1 & Y2 ARE CONSTANTS.
DT=    .005  NUMBER OF TIME STEPS= 20000
```

Here the Jacobian matrix is as follows

$$\begin{pmatrix} -1.5499659e-2 & 2.73e-9 & 2.1049e-7 \\ 1.8725e-6 & -5.3982095e-3 & 1.2075e-6 \\ 2.12625e-8 & 1.7325e-8 & 1.195014e-3 \end{pmatrix}$$

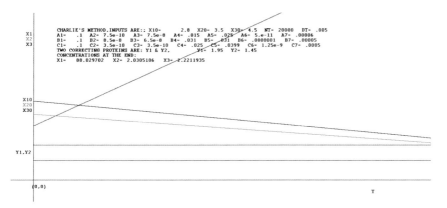

Fig. 7.4 Treatment failed while X2, X3 all decreasing, X1 is sharpely increasing

Eigenvalues are: $-0.0155, -5.3982e-3, 1.195e-3$.

Stiffness $= 1.55e-2/1.195e-3 = 7.9487$. So the system is very mildly stiff. The slopes of the graphs reveal that property in Fig. 7.3.

Finally, we doubled the value of a_1. x_1 started rising sharply.

Inputs for Fig. 7.4

```
CHARLIE'S CONVEX CORRECTOR; INPUTS FOR FIG.7.4

A1=   .1   A2= 7.5e-10   A3= 7.5e-8   A4=  .015   A5=  .025   A6= 5.e-11   A7= .00006
B1=   .1   B2= 8.5e-8    B3= 6.5e-8   B4=  .031   B5=  .031   B6= .0000001 B7= .00005
C1=   .1   C2= 3.5e-10   C3= 3.5e-10  C4=  .025   C5=  .0399  C6= 1.25e-9  C7= .0005
X10=  2.8  X20= 3.5      X30= 4.5     Y1= 1.95    Y2= 1.45    Y1 & Y2 ARE CONSTANTS
DT=   .005 NUMBER OF TIME STEPS=      20000
```

The Jacobian matrix is:

$$\begin{pmatrix} 3.4500341e-2 & 2.73e-9 & 2.1049e-7 \\ 1.8725e-6 & -5.3982095e-3 & 1.2075e-6 \\ 2.12625e-8 & 1.7325e-8 & -6.6049855e-3 \end{pmatrix}$$

The eigenvalues are: $0.0345, -6.61e-3$, and $-5.3982e-3$.

Stiffness $= 0.0345/5.3982e-3 = 6.391$. It is stiff.

The graphs show that $X2$ and $X3$ have basically the same slopes. So they do not contribute to higher stiffness. In fact, they bring the ratio down.

7.4.1 The Generalized Form

Following the equations (7.12), (7.13) and (7.14), we may construct a more generalized model, where the laboratory made correcting proteins do not change or interact among themselves.

$$dx_i/dt = \sigma_i(x_1, x_2, ..., x_{i-1}, x_{i+1}, ..., x_I, y_1, y_2, ..., y_J)x_i - \beta_i \prod_{j=1}^{J} y_j \quad i = 1, 2, ..., I$$

$$(7.15)$$

where

$$\sigma_i = a_1 x_1 + a_2 x_2 + ... + a_{i-1} x_{i-1} + a_i + a_{i+1} x_{i+1} + ... + a_I x_I - a_{I+1} y_1 - a_{I+2} y_2$$

$$- ... - a_{I+J} y_J + a_{I+J+1} \prod_{s=1, s \neq i}^{I} x_i$$

$$(7.16)$$

If $\beta_i = 0$, then if $\forall i$, $\sigma_i < 0$, *as* $t \rightarrow \infty$, $x_i \rightarrow 0$. (A fundamental property of the system).

We notice the last term in the Eq. (7.15) has no x_i. Yet it affects the rate of change of x_i. This is possible because once the proteins y_i 's are infused in the body and they start acting and reacting with each other, the biochemicals x_i 's become aware of their reactions and act accordingly. Since the very presence of y_j 's is a threat to them, so their joint reaction should reduce the rate of change of x_i 's.

7.4.2 Error Analysis for a Coupled System

Let us now assume that the laboratory designed proteins interact with each other as they try to dissolve the incorrect proteins generated by the defective mRNAs. The mathematical model will be

$$dX/dt = F(X, Y) \qquad\qquad (7.17)$$

$$dY/dt = G(X, Y) \qquad\qquad (7.18)$$

$F : R^I \times R^J \rightarrow R^I$ and $G : R^I \times R^J \rightarrow R^J$, R^I and R^J are, respectively, I and J dimensional real space.

$X = (x_1, x_2, \ldots, x_I)^T \in R^I$, $Y = (y_1, y_2, \ldots, y_J)^T \in R^J$. Here x_i 's are the defective proteins and y_j 's are the laboratory made corrective proteins which are now assumed to change as they will be infused in the body at the steady state.

By an iterative scheme, this nonlinear coupled system could be solved. So we should do an error analysis.

$$dX^n/dt = F(X^n, Y^n) \tag{7.19}$$

$$dY^n/dt = G(X^n, Y^n) \tag{7.20}$$

Please note that X^n is not X raised to a power n. It is a superscript denoting the value of X at the nth iteration (time step). The same is true for Y. X^0 and Y^0 are initial values which are known.

Then,

$$X^{n+1} = X^n + hF(X^n, Y^n) \tag{7.21}$$

And

$$Y^{n+1} = Y^n + hG(X^n, Y^n), \text{ where } h = \Delta t \tag{7.22}$$

If X^* and Y^* are the steady-state solutions, then

$$X^* = X^* + hF(X^*, Y^*)$$

and

$$Y^* = Y^* + hG(X^*, Y^*)$$

Let $E_1^{n+1} = X^{n+1} - X^*$ and $E_2^{n+1} = Y^{n+1} - Y^*$ be the error vectors at t_{n+1}. Also,

$$F(X^n, Y^n) - F(X^*, Y^*) = F(X^n, Y^n) - F(X^*, Y^n) + F(X^*, Y^n) - F(X^*, Y^*)$$
$$= F_x'(X^n, Y^n)E_1^n + F_y'(X^*, Y^n)E_2^n$$

Similarly,

$$G(X^n, Y^n) - G(X^*, Y^*) = G_x'(X^n, Y^n)E_1^n + G_y'(X^*, Y^n)E_2^n,$$

where $F_x'(X^n, Y^n)$ is the Jacobian matrix with respect to X evaluated at t_n, and $F_y'(X^*, Y^n)$ is the Jacobian matrix with respect to Y evaluated at t_n. The same is true for G_x' and G_y'

Hence,

$$E_1^{n+1} = X^{n+1} - X^* = X^n + hF(X^n, Y^n) - X^* - hF(X^*, Y^*)$$
$$= E_1^n + h(F_x'(X^n, Y^n)E_1^n + F_y'(X^*, Y^n)E_2^n)$$
$$= (I + hF_x'(X^n, Y^n))E_1^n + hF_y'(X^*, Y^n)E_2^n$$

Similarly,

$$E_2^{n+1} = hG_x'(X^n, Y^n)E_1^n + (I + hG_y'(X^*, Y^n))E_2^n$$

Thus,

$$(E_1^{n+1} \quad E_2^{n+1})^T = Z^n.(E_1^n \quad E_2^n)^T,$$

where Z^n is the block matrix given by

$$Z^n = \begin{bmatrix} I + hF_x'(X^n, Y^n) & hF_y'(X^*, Y^n) \\ hG_x'(X^n, Y^n) & (I + hG_y'(X^*, Y^n)) \end{bmatrix}$$

If this block matrix is a D-matrix, errors will be damped out, and steady-state solutions will be found.

At the present time, the rate constants are not known, so we could not do a computational study. What we have found is a mathematical model which shows that a set of $x_i's$, a set of chemicals could be removed in a chemical mixture and replaced by another set of chemicals which are y_i 's.

7.5 Solution of Interacting Proteins

Let $x_1, x_2, ..., I$ be the proteins which must be corrected by $y_1, y_2, ..., y_J$.
Then the model is, as given by (7.17) and (7.18)

$$dX/dt = F(X, Y)$$

$$dY/dt = G(X, Y)$$

In the element form,

$$dx_i/dt = a_1(Y, t)x_i - \left(\sum_{k=1}^{J} b_{1k} y_k\right)x_i + \prod_{k=1}^{I} m_k x_k - \prod_{k=1}^{J} n_k y_k \tag{7.23}$$

$$dy_j/dt = a_2(t)y_j - \left(\sum_{k=1,}^{I} b_{2k} x_k\right)y_j + \prod_{k=1}^{J} p_k y_k - \prod_{k=1}^{J} q_k x_k \tag{7.24}$$

The constants are rate constants to be determined by the chemists and biochemists. It must be noted that fourth term of the equation contains no x_i.

7.5.1 An Example

Let us consider the system with two defective proteins, x_1, and x_2 being corrected by y_1 and y_2 two laboratory made proteins. The model is

$$dx_1/dt = a_1x_1 + a_2x_1x_2 - a_3x_1y_1 - a_4x_1y_2 - a_5y_1y_2 = f_1(x_1, x_2, y_1, y_2) \quad (7.25)$$

$$dx_2/dt = b_1x_2 + b_2x_1x_2 - b_3x_2y_1 - b_4x_2y_2 - b_5y_1y_2 = f_2(x_1, x_2, y_1, y_2) \quad (7.26)$$

$$dy_1/dt = c_1y_1 - c_2x_1y_1 - c_3x_2y_1 - c_4x_1x_2 + c_5y_1y_2 = f_3(x_1, x_2, x_3, y_1, y_2) \quad (7.27)$$

$$dy_2/dt = d_1y_1 - d_2x_1y_1 - d_3x_2y_2 - d_4x_1x_2 - 4 + d_5y_1y_2 = f_4(x_1, x_2, x_3, y_1, y_2) \quad (7.28)$$

We need $x_i \to 0$ and y_j's $\to$ constants which are non zeros as $t \to \infty$
We have considered the following set of input data first.

Inputs for Fig. 7.5

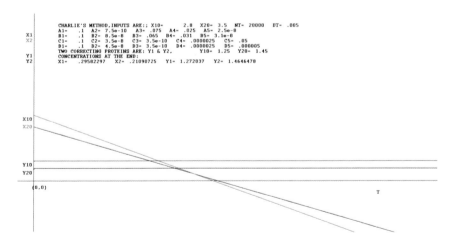

Fig. 7.5 Treatment worked

```
CHARLIE'S CONVEX CORRECTOR: INPUTS FOR FIG.7.5

A1=    .1   A2= 7.5e-10   A3=  .075   A4=  .025   A5= 2.5e-8   A6= 5.e-11
B1=    .1   B2= 8.5e-8    B3=  .065   B4=  .031   B5= 3.1e-8   B6=  .00001
C1=    .1   C2= 3.5e-8    C3= 3.5e-10 C4= .0000025   C5= .05
D1=    .1   D2= 4.5e-8    D3= 3.5e-10 D4= .0000025   D5= .000005
X10=   2.8  X20= 3.5   Y10= 1.25   Y20= 1.45
DT=   .005  NUMBER OF TIME STEPS=   20000
```

The Jacobian matrix is:

$$\begin{pmatrix} -2.9999997e-2 & 2.1e-9 & -.21000004 & -7.0000031e-2 \\ 2.975e-7 & -2.6199762e-2 & -.22750004 & -.10850004 \\ -8.79375e-6 & -7.0004375e-6 & .1724999 & .0625 \\ -8.81525e-6 & -7.0005075e-6 & .00000725 & .10000612 \end{pmatrix}$$

The eigenvalues are: 0.17253, −0.030009, −0.02621, and 0.09988.

Stiffness = 0.17253/0.02621 = 6.6. This is not very stiff.

Now we decided to change just the value for a_3 from 0.075 to 0.055, and the treatment failed as may be seen in Fig. 7.6.

Inputs for Fig. 7.6.

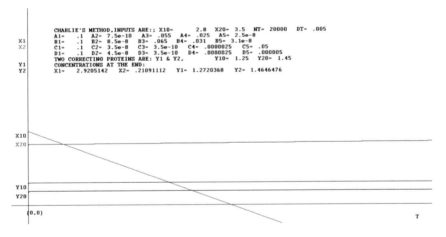

Fig. 7.6 Treatment barely failed

```
CHARLIE'S CONVEX CORRECTOR; INPUTS FOR FIG.7.6

A1=   .1   A2= 7.5e-10   A3= .055   A4= .025   A5= 2.5e-8   A6= 5.e-11
B1=   .1   B2= 8.5e-8    B3= .065   B4= .031   B5= 3.1e-8   B6= .00001
C1=   .1   C2= 3.5e-8    C3= 3.5e-10   C4= .0000025   C5= .05
D1=   .1   D2= 4.5e-8    D3= 3.5e-10   D4= .0000025   D5= .000005
X10=  2.8  X20= 3.5   Y10= 1.25   Y20= 1.45
DT=   .005  NUMBER OF TIME STEPS=   20000
```

$X1$ slowly increased from 2.8 to 2.92 in 20,000 time steps.

7.6 Conclusion

In this chapter, we did not discuss gene transfer where doctors try to make use of new genes to cause apoptosis of cancer cells of surrounding tissues to inhibit the progress of cancer. We did not do that, rather we tried to model how gene expressions could be modified so that cancer cells will be unable to do any damage. Ours is a mathematical approach. Bioscientists will judge its applicability.

References

1. https://www.ncbi.nlm.nih.gov/probe/docs/applexpression/
2. https://www.hemdifferently.com/what-is-gene-therapy/
3. Magid, H.: Gene therapy for cancer: present status and future perspective. Am. Mol. Cell. Ther. **2**, 27 (2014). http://www.molcelltherapies.com/content/2/1/27
4. Watson, J.D., Crick, F.H.: Molecular structure of nucleic acids; a structure for deoxyribose nucleic acid. Nature **171** (1953)
5. Dey, S.K.: Lectures on applied mathematics. Am. Math. Soc. (1985)
6. Dey, S.K.: D-mapping analysis for numerical solution of nonlinear systems with applications to differential equations. Int. J. Model. Simul. **6** (1996)
7. Dey, S.K.: Numerical modeling of D-mapping with applications to chemical kinetics. NASA Technical Memorandum 84332 (March, 1984)
8. Cross, D., Burmester, J.K.: Gene therapy for cancer treatment: past, present and future. Clin. Med. Res. **4**, 218–227 (2006)
9. Walther, W., Schlag, P.M.: Current status of gene therapy for cancer. Curr. Opin. Oncol. **25**, 659–664 (2013)
10. Ay, A., Arnosti, D.N.: Mathematical modeling of gene expression: a guide for the perplexed biologist. Published in final edited form as Critical Reviews in Biochemistry and Molecular Biology (April 2011)
11. Struhl, K.: Fundamentally different logic of gene regulation in eukaryotes and prokaryotes. Cell **98** (1999)
12. Tsygvintsev, A., Marino, S., Kirschner, D.E.: A mathematical model of gene therapy for the treatment of cancer. Medicine (2013). https://doi.org/10.1007/978-1-4614-4178-6_13
13. Struhl, K.: Fundamentally different logic of gene regulation in eukaryotes and prokaryotes. Cell **98**, 1–4 (1999)

Chapter 8
The Smartest Fighters

Abstract The smartest fighters are combat ready and extremely resourceful. When fighting gets very intense and standard drugs are failing, then some smart combat drugs (SCD) and skilled killer drugs (SKD) have been applied. The basic differences between them and standardized medicines are their art of fighting. While SCDs fight along with the other fighting agents, SKDs fight independent of other fighters. But they have one objective: Victory must be achieved, and patients must survive. Some medical professionals told the first author that replications of cancer cells should be prevented first. Although this is true, even if this is done, the existing cells could cause huge damage while circulating in the body. So they must be destroyed as well. For instance, if we consider a model like: $du/dt = a(t)u$, where $a(t) = 1/t$, at $t = t_0$, $u = u_0$, then the replication is given by $a(t)$, which is decreasing in time, but the solution is $u(t) = (u_0/t_0)t$. So u is continuously increasing as t increases. This may be an exceptional case, but when it comes to cancer what is exceptional and what is normal is very difficult to ascertain. Even the experts get confused sometimes. In such cases what we may do is using a drug that will get attached to every cancer cell and kill it. So cancer must be attacked in two ways: (1) Replications must be inhibited, and (2) each malignant cell should be destroyed regardless where it exists. These concepts have been applied in this chapter.

8.1 Rationale

Let us reiterate some most salient points on cancer treatments. The very first point is that we must clarify that a dead cancer cell may not be really dead. Necrotic areas of a malignant tumor keep on creating new cells which are often more dangerous.

Cancer is a very complex disease, most notoriously complex. It is a genetic disease. It requires several mutations of a gene to become an oncogene which causes cancer. According to the National Institute of Health, "So far, 291 cancer genes have been reported, more than 1% of all the genes in the human genome. 90% of cancer genes show somatic mutations in cancer, 20% show germline mutations and 10% show both." Somatic mutations and germline mutations are two different kinds of gene

© Springer Nature Singapore Pte Ltd. 2021
S. Dey and C. Dey, *Mathematical and Computational Studies on Progress, Prognosis, Prevention & Panacea of Breast Cancer*, Forum for Interdisciplinary Mathematics, https://doi.org/10.1007/978-981-16-6077-1_8

mutations. The first kind may happen in any of the cells in the body and cannot be passed on to the offspring, whereas the second kind could occur in sperm or ovum and be passed on to the children. These mutations are highly subtle, and it is rather impossible to detect when they happen. As cancer progresses, tumors become highly heterogeneous creating a mixed population of cells characterized by different molecular features, and diverse responsivity to therapies [1].

In Chap. 1, we talked about two Super Combat Drugs (SCD1 and SCD2). They are like monoclonal drugs working together. While SCD1 reduces the rate of replication of cancer cells and destroys them wherever they may be found in the body, SCD2 surrounds the cancer cells and destroys them by increasing the rate constant for destruction. Concepts and some discussion about these drugs are available in [2]. In this section, we will discuss two different forms of treatments.

Breast cancer starts at the breast, and like any other cancer it attacks various organs, both near and far. The immune system attacks back making the entire body a battlefield. Cancer cells apply various tactics, similar to those used by the attackers in a real war, to overcome and overpower the defensive forces of the body, and unfortunately, sometimes they win. Their primary weapons are poisonous proteins that they release to confuse the defense, to convoy together, and to camp at different locations in the body. In fact, cancer cells are some of the smartest cells in nature surviving through the most grueling, most arduously severe bioenvironmental situations where life becomes totally unbearable, even without oxygen (hypoxic). They send out friendly signals to make friends with several members of the immune system who not only stop attacking them, but build blood vessels so that cancer cells could get fresh supplies of blood and nutrients. So thinking of a silver bullet to treat cancer could be considered to be a utopian idea. Yet, our mathematical studies imply that someday this concept could be made factual. Scientists will be able to find the medications mentioned here as Skilled Killer Drugs (SKD) that will save lives. So the most significant question that the whole world is eager to know is: Is there a silver bullet to win the war on cancer? Our mathematical answer is YES. Although we are fully aware of the fact that cancer is an extremely subtle and complicated process, yet we state firmly that whatever process we have taken mathematically will be put into a real-world scenario some day, and the war on cancer will be won. In this chapter, we will look into that. First we must reiterate that targeted treatment is not just injecting some medications into the tumors but blocking and/or destroying all the possible tumor pathways. These pathways consist of a series of reactions which certain enzymes control, catalyze, and fully coordinate with other biomolecules, finally converting a normal cell into a malignant cell.

Targeted medications attempt to block and destroy these pathways. However, so long even one cancer cell is present in the blood, it has the ability to generate these pathways. Tumors can generate hundreds of such pathways. This is truly a life-threatening event happening inside a cancer patient. That has to be prevented. So pathways should be nipped in the bud. How that could be done mathematically is the topic of this chapter.

Some targeted antibodies are proteins produced by the immune system that can be customized to target specific markers, which are antigens on cancer cells. They inhibit cancerous activities. The antibody drug conjugates (ADCs) are able to carry anti-cancer drugs and deliver them to malignant cells. The antibodies (BiTEs) bind both cancer cells with T-cells and are very effective for better immunotherapy. Regarding breast cancer clinical trials, antibodies could target: (1) angiopoietin, a protein that promotes blood vessel formation in tumors; (2) DLL/notch: a pathway that induces faster cell growth and EGFR (often mutated in cancer), another pathway for cell growth, (3) HER2 that controls cell growth, often overexpressed in breast cancer and cause metastasis, and mesothelin, another protein that also does the same. (4)TROP2: a protein, commonly overexpressed in cancer aiding cancer cell reconstruction, proliferation, spread, and survival of malignant cells, and (4) VEGF/VEGF-R: a pathway helping tumors to build new blood vessels (angiogenesis).

So this battle must be fought just like any other battle. The strategy of the fight should be at four stages: (i) search for the enemy, (ii) denigrate their convoys wherever and whenever these could be found, (iii) clamp them down, and (iv) destroy their camps. In short, the strategy is SDCD. This SDCD scheme has been mathematically employed in this chapter. While doing so, we have to keep in mind how to minimize the losses of the defense and side effects of the medications to be used. The difference between this war and the war between two rival groups of people are that inside the body, defense has strict limitations, whereas in any regular war, defense could have unlimited power. Some recent studies on cancer treatment have been given in [4–7].

8.2 The Irresistible Fighters Against Cancer

In this monograph, we have tried to study various mathematical models with one motivation: How to conquer cancer, especially breast cancer? We have found that success is never an easy way to achieve. Cancer cells have every skill to fight most meticulously, masking their appearances by hiding their antigens. They skillfully hide under the skin and inside adipose cells and at places where they expect the least resistance. They often hide for years looking for the right opportunity to attack the body.

So in order to destroy them, the war tactics should be (1) to stop their growths wherever and whenever even one such cell is found, (2) surround them and destroy them causing none or the least amount of harm to the body, and (3) promote the strength of the immune system. These medications are the irresistible fighters. When they fight together, cancer has practically no place to survive. We have mathematicized this very concept in this chapter.

8.2.1 The Mathematical Modeling and Computational Studies: Two Smart Combat Drugs SCD1(w) and SCD2(r)

Let w and r represent two Smart Combat Drugs (SCD1 and SCD2, respectively). While SCD1 reduces the rate of growth of cancer, SCD2 increases the strength of the immune system. The chemical war is modeled by the following nonlinear reaction–diffusion equations:

$$\partial u/\partial t = a_1(t)u - a_2(t)uv - a_3uw - a_4ur + \kappa_1\nabla^2 u \qquad (8.1)$$

$$\partial v/\partial t = b_1(t)u - b_2uv + b_3uw + b_4vr + \kappa_2\nabla^2 v \qquad (8.2)$$

$$\partial w/\partial t = c_1u - c_2uw - c_3vw + c_4wr + \kappa_3\nabla^2 w \qquad (8.3)$$

$$\partial r/\partial t = d_1u - d_2ur - d_3rw + \kappa_4\nabla^2 r \qquad (8.4)$$

where $a_1(t)$ = time-dependent growth of u, to be controlled by w, $a_2(t)$ = time-dependent rate constant for chemical interaction between u and v, the immune response, controlled by the drug SCD1 (w), a_3 and a_4 are rate constants for the interactions between u, w and u, r. Mathematically, mobility of u could be controlled if κ_1 is controlled as time changes. A smaller value of κ_1 will cause slower dispersion. $b_1(t)$ is the rate of chemotaxis. If it increases, that means more rate of assemblage of fighters of the defense against the antigen u.

At the outset, no fixed amounts of SCD1 and SCD2 have been used. It could be noticed from the Eqs. (8.3) and (8.4) that the rates of growths of w and r, the two Smart Combat Drugs are triggered by the rates of growths of u. The plan is to reduce the rate of replication of u. So $a_1(t)$ will be reduced by w. The second drug r will strengthen the immune system by increasing $a_2(t)$ or sometimes increasing $b_1(t)$ and reducing the dispersion of u. However, $a_1(t)$ could be reduced to zero, $a_2(t)$ and $b_1(t)$ cannot be increased more than what body permits. Every drug has a strict limitation, and the number of v is restricted physiologically. So the two drugs SCD1 and SCD2 are playing two different roles to fight cancer.

Considering the properties of w and how it interacts with $a_1(t)$, we need to formulate the relation between them. The formula that we have used is

$$da_1/dt = -\lambda a_1 W. \qquad (8.5)$$

where $W = maxw(t_n) = maxw(i, j, k)$ at t_n.

Let us explain the formulation of (8.5). It has been assumed that if $dW/dt > 0$, then we should have $da_1/dW < 0$. Because when W increases with t, a_1 decreases.

And, if in case, $dW/dt < 0$, then $da_1/dW > 0$. So, if we apply these concepts, then writing, $da_1/dt = (da_1/dW).(dW/dt)$ we note if $da_1/dW = \xi a_1$ and $dW/dt = \eta W$, then if $\xi > 0$, $\eta < 0$ and vice versa. So in (8.5), $\lambda > 0$. Now we understand the formula (8.5). Integrating (8.5) numerically by trapezoidal rule, we find

$$a_n = a_{n-1} exp \left(-\lambda \int_{t_{n-1}}^{t_n} W dt \right) = a_{n-1} exp\left(-\lambda (W_{n-1} + W_n)(\Delta t/2) \right), \Delta t = t_n - t_{n-1}$$

$$(8.6)$$

$n = 2,3,...,$ a_n = value of a at time $t = t_n$ and $a_1 = a_1(t_1)$ is given at the outset equals the rate of growth of the antigen u. How λ should be chosen depends on how the delivery of the drug w will be monitored.

Here we have used the trapezoidal rule to approximate the integral. We have also chosen $\lambda = 0.01$. The information of a_0 will be given by a pathologist or oncologist.

The differential Eqs. (8.1), (8.2), (8.3) and (8.4) were approximated by explicit finite difference schemes as before. There are $105 \times 105 \times 105$ grid points in the computational field including the boundary. So there are in all 1,157,625 number of points in the entire computational field. We have also considered mesh sizes as $\Delta x = \Delta y = \Delta z = 0.02$ and the time step, $\Delta t = 0.02$. For most explicit finite difference equations, $\Delta t << \Delta x$ the mesh size in order to maintain numerical stability. That is certainly not applicable to the CDey-Simpson formula. As before, this algorithm has been used in this monograph for numerical solutions of all our models. The boundaries are the free boundaries in the sense that at these points, values of all the variables have been extrapolated by three point extrapolation formulas as shown in the Chap. 4. In short, the formula is

$$f(x_1, y, z) = 2.5f(x_4, y, z) - 6f(x_3, y, z) + 4.5f(x_2, y, z)$$

$$f(x_{I+1}, y, z) = 2.5f(x_{I-2}, y, z) - 6f(x_{I-1}, y, z) + 4.5f(x_I, y, z)$$

The same is done for $f(x, y_1, z)$, $f(x, y_{J+1}, z)$, $f(x, y, z_1)$, $f(x, y, z_{K+1})$. x_1 and x_{I+1}, y_1 and y_{J+1}, and, z_1 and z_{K+1} are points on the boundary. f could be any dependent variable like u, v, w, r. Since I, J, K should be large and mesh sizes should relatively small, we may assume in some cases that $f(x_1, y_1, z_1) = f(x_1, y_2, z_2)$, etc. We have discussed this in Chap. 4.

Here we have considered metastatic breast cancer which is multicentric, meaning that there are more than one tumor in the breast. These are three cubical tumors, each with 64 grid points. At the first tumor, ranging from (13, 13, 13) to (16, 16, 16), there are 15,000 cancer cells at each grid point, the second tumor ranging from (26, 26, 26) to (29, 29, 29) having 10,000 cancer cells at each grid point, and at the third tumor ranging from (47, 47, 47) to (50, 50, 50) having 7500 cancer cells at each grid point, respectively. However at all these tumors, there are only ten leukocytes

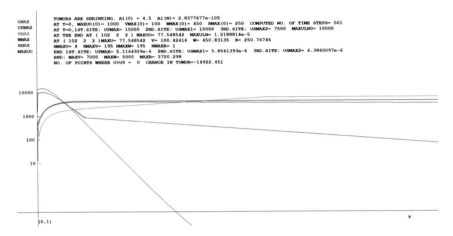

Fig. 8.1 Use of SCD1 (W) and SCD2 (R) drugs. A success

at each grid point. There are also three affected lymph nodes at three grid points at (3, 3, 3), (20, 20, 20), and (55, 55, 55) and at each point in the field there are only 25 leukocytes, whereas $u_0(i, j, k) = 1000$ at each lymph node. $a_1(0) = 4.5$. So cancer is aggressive, and the immune system is very poor. We have assumed that the immune response denoted by $v(t)$ is very poor. The inputs are as follows.

Inputs for Fig. 8.1

```
*****************************************************************************************************
! MODEL STRATEGIC TREATMENT FOR METASTATIC:  TREATMENT WITH TARGETTED DRUGS SCD1 AND SCD2 MAY 22,2020
!
! DU/DT  = A1 (N) *U-A2*U*V - A3*U*W - A4*U*R + NU1*DELSQ(U)  ! ANTIGEN ; FIG.8.1
! DV/DT  = B1*U-B2*V*U + B3*V*W + B4*V*R + NU2*DELSQ(V)  !IMMUNE RESPONSE
! DW/DT  = C1*U -C2*W*U - C3*W*V +C5*W*R+ NU3*DELSQ(W)  !  SCD1 DRUG
! DR/DT  = D1*U - D2*U*R - D3*R*W + NU4 * DELSQ(R)   ! SCD2 DRUG, DOES NOT AFFECT V NEGATIVELY
! DX= .02  DY= .02  DZ= .02  DT= .02  NX= 101  NY= 101  NZ = 101  NT= 500
! A1(1)= 4.5  A2= .00001  A3= .00075  A4= .00095
! B1= .5  B2= .00005  B3= .00005  B4= .00005
! C1= 1  C2= .00001  C3= .00001  C5= .00005
! D1= 1  D2= .00001  D3= .00001
! MAXUO(0)= 1000  MAXU IN THE FIELD= 15000  VMAX(0)= 100  W(0)= 450  R(0)= 250
! NU1= .0000001  NU2= .0000001  NU3= .0000001  NU4= .0000001
! THREE TUMORS LOCATED AT ( 13 , 13 , 13 ) AND ( 26 , 26 , 26 ) WHERE U= 15000
! AT THE LYMPH NODES( 3  3  3 )( 20  20  20 )( 55  55  55 ) AT T=0, ULNO= 10000
*****************************************************************************************************
```

Here $a_1 = 4.5$. That indicates that the rate of growth of cancer is 350%. Max $v(i, j, k)$, cells of the immune system, over the entire field is only 100 at each grid point, and their rate of growth of the immune response is $b_1 = 0.5$. So chemotaxis is relatively poor. The strength of the response of v is $a_2 = 0.00001$, which is poor too. However, $a_3 = 0.00075$ and $a_4 = 0.00095$. So the drugs are relatively powerful. At the outset, $t = 0$, two Smart Combat Drugs, namely w (SCD1) and r (SCD2) were injected intravenously. $w_0(i, j, k) = 450$ units at each grid point and $r_0(i, j, k) = 250$ units at each grid point in the computational field. Furthermore, we have assumed that whereas the growth of $u(i, j, k)$ is totally unrestricted, at all time and at all points in the computational field, $v(i, j, k) \leq 7000$, $w(i, j, k) \leq 5000$ and $r(i, j, k) \leq 4000$ for all t. These restrictions have been imposed in the code. These variables always

Table 8.1   A#1

```
|**********************************************************************************************
 SYSTEM TIME - 10:13:58  THREE LYMPH NODES AFFECTED: MAX CONCENTRATION OVER THE FIELD AT TIME -     .04  N- 2  NT- 500  A1(N)- 4.5
 AT  55 , 55 , 55  MAXUO(N)-  10394.296    V- 61.401112    W- 502.87393    R- 352.93361
 TOTAL NO. OF POINTS U- MAXUO-     3
 AT  15 , 15 , 15  UMAX(N)-  15580.273    V- 176.27267    W- 554.19373
 R- 404.31415
 AT   3 , 3 , 1  VMAX(N)-   367.62676    U- 1040.6912    W- 0  R- 0
 R- 0
 AT  14 , 14 , 14  WMAX(N)-   554.19373    U- 15580.273    V- 176.27267    R- 404.31415
 R- 404.31415
 AT  14 , 14 , 14  RMAX(N)-   404.31415    U- 15580.273    V- 176.27267
 W- 554.19373
 TOTAL NO.OF POINTS WHERE U IS MAX-        1  WHERE V IS MAX-     3  WHERE W IS MAX-     1  WHERE R IS MAX-     1
 CHANGE OF UMAX-     580.2727   CHANGE OF VMAX-    367.62676   CHANGE OF WMAX-   154.19373   CHANGE OF RMAX-   154.31415
 AT THE PRIMARY SITE:   UMAX- 15580.273    VMAX-  176.27267    WMAX-   554.19373    RMAX-   404.31415
 AT THE SECONDARY SITE:   UMAX-  10394.345   VMAX-  150.97417    WMAX-  502.87111    RMAX-  352.93512
 AT THE THIRD SITE:    MAXU-  7798.5712   VMAX-  138.28109    WMAX-  477.18732    RMAX-  327.22224
 USTAR1-  15580.161   USTAR2-  10394.273    USTAR3 -   7798.5193
 UNODE1-  10394.296   UNODE1-1-  1040.7852   UNODE2-  10394.296   UNODE2+1-  1040.7852   U(NX+4,NY+4,NZ+4)-   1040.7852
 UNODE3-  10394.296   UNODE3-1-  1040.7852   UNODE3+1-  1040.7852
 VNODE1-  61.401112   VNODE2-  61.401112   VNODE3-  61.401112   V(NX+4,NY+4,NZ+4)-      100
 WNODE1-  502.87393   WNODE2-  502.87393   WNODE3-  502.87393   W(NX+4,NY+4,NZ+4)-      400
 RNODE1-  352.93361   RNODE2-  352.93361   RNODE3-  352.93361   R(NX+4,NY+4,NZ+4)-      250
 NO.OF POINTS WHERE U(N)>U(N-1)-       1157625   MAX.CHANGE FROM THE START - UMAX - UOMAX-   580.2727
```

have some limitations in a real situation. Because based upon the blood reports, doctors always check the amounts of drugs given to a patient. The amounts were often restricted by the state of health of the patient. The list of all the inputs are in inputs for Fig. 8.1 (Table 8.1).

At the very beginning, when $N = 2$, this is really the very first time step, the values of U_{ijk} at the three lymph nodes increased first from 10,000 to 10,394, and at the second tumor it did the same. At the primary site, it jumped from 15,000 to 15,558. Outside the tumors and lymph nodes, U_{ijk} at $t = 0$ was 1000 in the field. It jumped up to 1040. So cancer is increasing fast as expected. And this is monitored by the number 1157625 where $U^n > U^{n-1}$. This is the total number of points in the entire computational field. So at each and every point, cancer is growing (Table 8.2).

When $N = 56$, a_1 is reduced from 4.5 to 1.379026E-08. There is not a single point where cancer is on the rise (number of points where $U^n > U^{n-1}$ is zero.) At $(NX + 4, NY + 4, NZ + 4), NX = NY = NZ = 101$, the most remote point, $U = 789.54$ and at the remotest tumor site (50,50,50), *not mentioned in the input list*, $U = 1389.86$, is the largest, showing that the cancer cells are on the run moving far away from the primary site, and at all lymph nodes $U = 1212.64$, the largest. This is the largest value of U outside the tumor sites. At the primary site $U = 881.32$. It has shrunk significantly. Interestingly, it could be observed, U has been surrounded by the forces of defense, and U is moving closer to the boundary (102, 2, 2) where defense is maintaining its strong presence. Technically in this code, cancer has no place to hide. So they will lose the

Table 8.2   A#2

```
|**********************************************************************************************
 SYSTEM TIME - 10:55:48  THREE LYMPH NODES AFFECTED: MAX CONCENTRATION OVER THE FIELD AT TIME -     1.12  N- 56  NT- 500  A1(N)-
 1.3790261e-8
 AT  55 , 55 , 55  MAXUO(N)-  1212.6415    V- 1835.4627    W- 4085.3898    R- 3688.1178
 TOTAL NO. OF POINTS U- MAXUO-     3
 AT  50 , 50 , 50  UMAX(N)-  1389.858    V- 1606.6722    W- 3448.6883
 R- 3130.8961
 AT  15 , 15 , 15  VMAX(N)-  2457.5835    U- 881.12636    W- 5000    R- 4581.2501
 R- 4581.2501
 AT  16 , 16 , 16  WMAX(N)-  5000    U- 881.31586    V- 2456.8041    R- 4579.9335
 R- 4579.9335
 AT  15 , 15 , 15  RMAX(N)-  4581.2501    U- 881.12636    V- 2457.5835
 W- 5000
 TOTAL NO.OF POINTS WHERE U IS MAX-        1  WHERE V IS MAX-     1  WHERE W IS MAX-    64  WHERE R IS MAX-     1
 CHANGE OF UMAX-    -78.075997   CHANGE OF VMAX-   15.139057   CHANGE OF WMAX-    0   CHANGE OF RMAX-   6.1974253
 AT THE PRIMARY SITE:   UMAX-  881.31586    VMAX-  2457.5835    WMAX-  5000    RMAX-  4581.2501
 AT THE SECONDARY SITE:   UMAX-  1212.4208   VMAX-  1923.7128    WMAX-  4085.959    RMAX-  3689.8157
 AT THE THIRD SITE:    MAXU-  1389.858   VMAX-  1607.128    WMAX-  3449.6006    RMAX-  3131.7241
 USTAR1-  881.31586   USTAR2-  1212.4208   USTAR3 -   1389.858
 UNODE1-  1212.6414   UNODE1-1-  789.49893   UNODE2-  1212.6415   UNODE2+1-  789.49342   U(NX+4,NY+4,NZ+4)-   789.53999
 UNODE3-  1212.6415   UNODE3-1-  789.49342   UNODE3+1-  789.49342
 VNODE1-  1835.4624   VNODE2-  1835.4627   VNODE3-  1835.4627   V(NX+4,NY+4,NZ+4)-      100
 WNODE1-  4085.3893   WNODE2-  4085.3898   WNODE3-  4085.3898   W(NX+4,NY+4,NZ+4)-      400
 RNODE1-  3688.1173   RNODE2-  3688.1178   RNODE3-  3688.1178   R(NX+4,NY+4,NZ+4)-      250
 NO.OF POINTS WHERE U(N)>U(N-1)-       0   MAX.CHANGE FROM THE START - UMAX - UOMAX-   -13610.142
```

Table 8.3 A#3

```
******************************************************************************************
SYSTEM TIME  = 11:33:40  THREE LYMPH NODES AFFECTED: MAX CONCENTRATION OVER THE FIELD AT TIME =          2  N= 100  NT= 500  A1(N)=
3.8467492e-18
AT  102 .  2 .  2  MAXUO(N)=  799.36247     V= 103.99863    W= 458.02326    R= 257.9737
TOTAL NO. OF POINTS U= MAXUO=     11025
AT  102 .  2 .  2  UMAX(N)=   799.36247     V= 103.99863    W= 458.02326
R= 257.9737
AT  15 . 15 .  15  VMAX(N)=   3082.9285     U= 26.988641    W= 5000      R= 4579.5722
R= 4579.5722
AT  16 . 16 .  16  WMAX(N)=   5000      U= 27.046518     V= 3081.0214    R= 4577.1501
R= 4577.1501
AT  15 . 15 .  15  RMAX(N)=   4579.5722     U= 26.988641    V= 3082.9285
W= 5000
TOTAL NO.OF POINTS WHERE U IS MAX=       1    WHERE V IS MAX=      4    WHERE W IS MAX=     64    WHERE R IS MAX=       1
CHANGE OF UMAX= -4.7023949   CHANGE OF VMAX=  14.862496    CHANGE OF WMAX=       0   CHANGE OF RMAX= -2.0336701
AT THE PRIMARY SITE:   UMAX=  27.046518    VMAX=  3082.9285   WMAX=  5000      RMAX=  4579.5722
AT THE SECONDARY SITE:  UMAX=  63.393809   VMAX=  2378.0394   WMAX=  4568.776   RMAX=  3780.4614
AT THE THIRD SITE:   MAXU=  109.62011   VMAX=  1975.5199   WMAX=  3898.79    RMAX=  3291.9662
USTAR1=  27.046518    USTAR2=  63.393809   USTAR3 =  109.62011
UNODE1=  63.467289    UNODE1-1=  359.77747   UNODE2=  63.467268   UNODE2+1=  359.7646    U(NX+4,NY+4,NZ+4)=    359.89177
UNODE3=  63.467268    UNODE3-1=  359.7646    UNODE3+1=  359.7646
VNODE1=  2272.1091    VNODE2=  2272.1101   VNODE3=  2272.1101   V(NX+4,NY+4,NZ+4)=    100
WNODE1=  4567.7687    WNODE2=  4567.7706   WNODE3=  4567.7706   W(NX+4,NY+4,NZ+4)=    400
RNODE1=  3777.1122    RNODE2=  3777.1137   RNODE3=  3777.1137   R(NX+4,NY+4,NZ+4)=    250
NO.OF POINTS WHERE U(N)>U(N-1)=       0   MAX.CHANGE FROM THE START - UMAX - UOMAX=     -14200.638
```

war. At $N = 100$, the value of $a_1(t)$ is practically zero. Cancer is not replicating at all. So cancer is contained; however, the malignant cells in circulation are still fighting with full strength.

From its initial value, UMAX has been reduced to almost 800, the difference being $15{,}000 - 800 = 14{,}200$ as may be seen in the computer output given in Table 8.3.

At $N = 271$, $a_1(t) = 2.8374 \, E - 55$, there is absolutely no growth of cancer, and they are fully surrounded by their enemies, yet the remaining cells are giving a reasonably strong fight (Table 8.4).

From all the above tables, it is quite evident that wherever the tumors were, the defensive forces are most strongly present at those locations. And, wherever cancer cells are stronger, the defense is stronger too. This is a realistic frame of treatment revealed by the model. We saw very similar pictures all through our computer solutions.

Finally, we see the victory of the defense in Fig. 8.1. All tumors are practically gone as revealed by the black bottom line. The redline above it shows that in the field, UMAX = MAXUO = 77.55, at the point (102, 2, 2). Searching the Internet, we found that in the year 2005 a similar drug was found.

Dr. Gordon Mills from the Univ. of Texas at Houston (M.D.Anderson Center) worked on a drug Sphingomab, which could work very similarly to SCD1. He found

Table 8.4 A#4

```
******************************************************************************************
SYSTEM TIME  = 17:15:57  THREE LYMPH NODES AFFECTED: MAX CONCENTRATION OVER THE FIELD AT TIME =          5.42  N= 271  NT= 500  A1(N)=
2.837368e-55
AT  102 .  2 .  2  MAXUO(N)=  295.22619    V= 101.50164    W= 452.99985    R= 252.94105
TOTAL NO. OF POINTS U= MAXUO=     11025
AT  102 .  2 .  2  UMAX(N)=   295.22619    V= 101.50164    W= 452.99985
R= 252.94105
AT  15 . 15 .  15  VMAX(N)=   6896.6043    U= 4.7647438e-5   W= 5000      R= 4206.7646
R= 4206.7646
AT  55 . 55 .  55  WMAX(N)=   5000      U= 2.9619652e-3   V= 4726.8137    R= 3473.4753
R= 3473.4753
AT  15 . 15 .  15  RMAX(N)=   4206.7646    U= 4.7647438e-5   V= 6896.6043
W= 5000
TOTAL NO.OF POINTS WHERE U IS MAX=       4    WHERE V IS MAX=      1    WHERE W IS MAX=    131   WHERE R IS MAX=       1
CHANGE OF UMAX= -1.7160882   CHANGE OF VMAX=  31.744521    CHANGE OF WMAX=       0   CHANGE OF RMAX= -2.1060576
AT THE PRIMARY SITE:   UMAX=  1.1625146e-3   VMAX=  6896.6043   WMAX=  5000      RMAX=  4206.7646
AT THE SECONDARY SITE:  UMAX=  1.6779829e-3   VMAX=  4959.6942   WMAX=  5000      RMAX=  3482.461
AT THE THIRD SITE:   MAXU=  4.0944626e-3   VMAX=  3799.5429   WMAX=  4904.9685   RMAX=  3068.9442
USTAR1=  1.1625141e-3   USTAR2=  1.6779829e-3   USTAR3 =  4.0944626e-3
UNODE1=  2.9879975e-3   UNODE1-1=  8.2676957   UNODE2=  2.9619652e-3   UNODE2+1=  8.2637874   U(NX+4,NY+4,NZ+4)=    8.3614598
UNODE3=  2.9619652e-3   UNODE3-1=  8.2637874   UNODE3+1=  8.2637874
VNODE1=  4726.7998    VNODE2=  4726.8137   VNODE3=  4726.8137   V(NX+4,NY+4,NZ+4)=    100
WNODE1=  5000      WNODE2=  5000      WNODE3=  5000      W(NX+4,NY+4,NZ+4)=    400
RNODE1=  3473.4662    RNODE2=  3473.4753   RNODE3=  3473.4753   R(NX+4,NY+4,NZ+4)=    250
NO.OF POINTS WHERE U(N)>U(N-1)=       0   MAX.CHANGE FROM THE START - UMAX - UOMAX=     -14704.774
```

that the medication Sphingomab "zeroes in on a molecule that stimulates cancer growth known as sphingosine-1-phosphate or SIP."

Dr. Roger Sabbadini, from Lpath Therapeutics Inc., San Diego, found that high levels of SIP are associated with aggressive cancers, resistant to traditional chemotherapy www.webmd.com.

8.2.2 Total Immunotherapy by SCD1 (w) and SCD2 (r)

We now look into another major step to defeat cancer by increasing the role of $r(i, j, k)$. It will be a total immunotherapy. While SCD1 will reduce the rate of growth of cancer cells, SCD2 will increase the strength of the immune system by increasing its ability to fight back and by assembling more soldiers at every point in the body affected by cancer. So the more the cancer cells, the more the soldiers of defense to fight back. This approach should lead to better results much faster, provided it could be applicable to a patient.

Mathematically while SCD1 will reduce $a_1(t)$, SCD2 will increase the strengths of a_2 and b_1 the response of the immune system, as time goes. With those in view, we introduce the following equations:

$a_2 = a_2(t, max(r_n))$ and $b_1 = b_1(t, max(r_n))$.

$da_2/dt = (da_2/d(max(r_n))) \cdot (d(max(r_n))/dt)$. Now as t increases, max (r_n) increases, and as max (r_n) increases, a_2 increases.

That gives, $da_2/dt = \alpha(a_2 RMAX(n))$. where $RMAX(n) = max_{ijk}(r_{ijk})$ at t_n.

If we choose $\alpha = 0.01$ (which must be determined by the pathologists and oncologists), and $a_2(n) = a_2(t = t_n)$ then by Trapezoidal rule, we get

$$a_2(n) = a_2(n - 1) \cdot exp((0.01)(\Delta t/2)(RMAX(n) + RMAX(n - 1))) \quad (8.7)$$

In the code, $a_2(n)$ has a restriction. That is if, $a_2(t) > 0.0075$, then $a_2(t) = 0.0075$. This is done because the strength of each and every drug has a limitation. Here we have made an educated guess on the basis of our experience. Another strategy to strengthen immunotherapy is to increase $b_1(t)$ which increases the rate of growth and transport of the immune response wherever cancer cells are gathering. This means an enhancement of chemotaxis. The formula for $b_1(t)$ is different. We have assumed that $db_1/dt = \beta \max r(i, j, k)$. That gives

$$b_1(n) = b_1(n - 1) + \beta(\Delta t/2)((RMAX(n) + RMAX(n - 1)) \quad (8.8)$$

We have applied the following inputs. Results are given in Fig. 8.2.

Inputs for Fig. 8.2.

```
:•••••••••••••••••••••••••••••••••••••••••••••••••••••••••••••••••••••••••••••••••••••
! MODEL STRATEGIC TREATMENT FOR METASTATIC CANCER : ADVANCED IMMUNOTHERAPY. MAY30,2020
!
! DU/DT = A1(N)•U - A2(N)•U•V- A3•U•W - A4•U•R + NU1•DELSQ(U) ! ANTIGEN ; FIG.8.2
!
! DV/DT = B1(N)•U-B2•V•U + B3•V•W + B4•V•R + NU2•DELSQ(V) !IMMUNE RESPONSE
! DW/DT = C1•U -C2•W•U +C5•W•R+ NU3•DELSQ(W) ! SCD1 DRUG
! DR/DT = D1•U - D2•U•R - D3•R•W + NU4 • DELSQ(R)    ! SCD2 DRUG REDUCES MOBILITY OF CANCER CELLS
! A1(N) = A1(N-1)• EXP(-0.01•(DT/2)•(WMAX(N) + WMAX(N-1)))
! A2(N) = A2(N-1)•EXP(0.01•(DT/2)•(RMAX(N) + RMAX(N-1)))
! IF A2(N) > 0.0075 THEN A2(N) = 0.0075
! B1(N) = LAMBDA•(RMAX(N)+RMAX(N-1))(DT/2) +B1(N-1), LAMBDA = 0.05.
! IF B1(N) >3 THEN B1(N) = 3
! DX=  .02   DY=  .02   DZ=  .02   DT=  .02   NX= 101   NY= 101   NZ = 101  NT= 350
! A1(1)=    4.5  A2(1)=  .00001   A3= .00005   A4= .00005
! B1(1)=    .25  B2= .00005  B3= .00005  B4= .00005
! C1=  1  C2= .00001   C3= 0  C5= .00001
! D1=  1  D2= .0001  D3= .0001
! MAXUO(0)=    1000   MAXU IN THE FIELD=   15000   VO= 100   W(0)= 250   R(0)= 250
! NU1=   .0000001   NU2=  .0000001    NU3=  .0000001   NU4=  .0000001
! THREE TUMORS LOCATED AT      ( 13 , 13 , 13 ) ( 26 , 26 , 26 )
! AND   ( 47 , 47 , 47 ) WHERE UO=  15000   10000  AND 7500  RESPECTIVELY.
! AT THE LYMPH NODES    ( 3  3   3 )( 20   20   20 )( 55   55   55 ) AT T=0, ULNO=   10000
:•••••••••••••••••••••••••••••••••••••••••••••••••••••••••••••••••••••••••••••••••••••
```

From the inputs, it is evident that the patient was medically in a better physical condition with the inputs for Fig. 8.1 than for the treatment in Fig. 8.2. Previously, $b_1(0) = 0.5$. Now it is 0.25.

Previously, $a_3 = 0.00075$, now it is 0.00005, and previously $a_4 = 0.00095$, now it is 0.00005. Yet previously, more medications were used at $t = 0$, $w(i, j, k) = 450$. Now it is 250. Here the strength of r has been increased 10 times. The results in Fig. 8.2 showed the results are better running the same code for 351 time steps than what we have found in 501 time steps previously in Fig. 8.1.

The power of the two drugs is prominent. Yet the runaways are not fully controlled and confined and reduced to null as the drugs did to the tumors. This is given by the slope of the redline. Cancer is that dangerous. However, the slope is negative, meaning that they are not able to increase.

Fig. 8.2 shows that mathematically the system is very stiff.

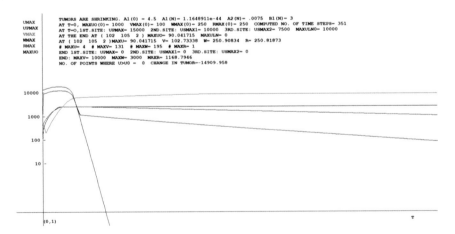

Fig. 8.2 SCD1 (W) reduces growth of cancer and SCD2 (R) strengthens the immune system

In the next figure, Fig. 8.3, fixed values of SCD1 and SCD2 were used. Only one equation is used.

$$\partial u / \partial t = a_1(t)u - a_2 u V - a_3 u (SCD1) - a_4 u (SCD2) + \kappa_1 \nabla^2 u \qquad (8.9)$$

The inputs are as follows.

Inputs for Fig. 8.3.

```
! CHAPTER 8, FIG 8.3, MARCH.15, 2020
! THERE ARE THREE TUMORS , TWO LYMPH NODES ARE AFFECTED.
! DU/DT - A1(N)*U-A2*U*V0-A3*U*SCD1-A4*U*SCD2+ (NU1)*DELSQ(U)
!
! AT T=0, U0 IS - 1000   V0-FIXED -   100   SCD1-FIXED= 500   SCD2=FIXED= 500
! DX- .02   DY- .02   DZ- .02   DT- .02   NX- 60   NY- 60   NZ - 60   NT- 45
! A1(1)- 5   A2- .00001            A3- .00075   A4- .00095
! A1(N) - A1(N-1)* EXP(-0.01*(DT)*(W0))
! NU1- .0000001
! GAMMA1- .015
! THREE LYMPH NODES ARE LOCATED AT: ( 3 , 3 , 3 ) ( 20 , 20 , 20 )( 20 , 20 , 20 )
! AT THE LYMPH NODE1 U0- 10000   AT THE LYMPH NODE2 U0- 10000   AT THE LYMPH NODE2 U0- 10000
! PRIMARY TUMOR IS LOCATED AT:   ( 13 , 13 , 13 ) UPRIMARY- 15000
! SECOND TUMOR IS LOCATED AT:   ( 26 , 26 , 26 ) US1- 10000
! THIRD TUMOR IS LOCATED AT:   ( 47 , 47 , 47 ) US2- 7500
!
```

Here the Eq. (8.6) has become:

$$a_n = a_{n-1} exp \left(-\lambda \int_{t_{n-1}}^{t_n} W_0 dt \right) = a_{n-1} exp(-\lambda W_0 \Delta t) \qquad (8.10)$$

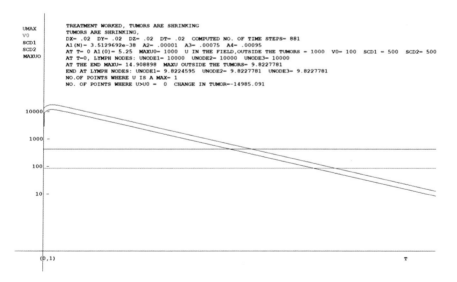

```
                TREATMENT WORKED, TUMORS ARE SHRINKING
UMAX            TUMORS ARE SHRINKING,
V0              DX- .02   DY- .02   DZ- .02   DT- .02   COMPUTED NO. OF TIME STEPS= 881
SCD1            A1(N)- 3.5129692e-38   A2- .00001   A3- .00075   A4- .00095
SCD2            AT T- 0 A1(0)- 5.25   MAXU0- 1000   U IN THE FIELD,OUTSIDE THE TUMORS - 1000   V0- 100   SCD1 - 500   SCD2- 500
MAXU0           AT T-0, LYMPH NODES: UNODE1- 10000   UNODE2- 10000   UNODE3- 10000
                AT THE END MAXU- 14.908898   MAXU OUTSIDE THE TUMORS- 9.8227781
                END AT LYMPH NODES: UNODE1- 9.8224595   UNODE2- 9.8227781   UNODE3- 9.8227781
                NO.OF POINTS WHERE U IS A MAX- 1
                NO. OF POINTS WHERE U>U0 -   0   CHANGE IN TUMOR--14985.091
```

Fig. 8.3 With fixed SCD1 and SCD2. Treatment is a success

Here, $SCD1 = W = Fixed. = W_0$, and we just validated that if we keep SCD1 and SCD2 fixed and start diminishing the rate of growth of cancer cells, the treatment shall work. We notice that in Fig. 8.3.

8.3   Enhancement of the Immune Response

Our next enquiry is to study how SCD1 (w) and SCD2 (r) will work, if the accelerated chemotaxis is turned off and a poor immune response $a_2(t)$ is enhanced only by the drug SCD2 or r in the code as done in Fig. 8.2, whereas SCD1 keeps on lowering the rate of growth of cancer $a_1(t)$. Furthermore, the strength of SCD2 has also been increased. It worked quite well as expected.

Inputs for Fig. 8.4

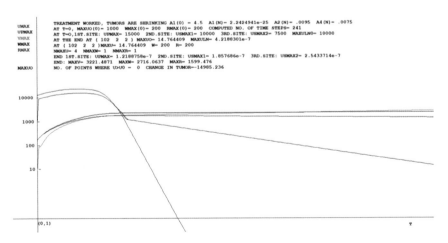

Fig. 8.4 SCD1 (W) reduces A1 growths of cancer cells, and SCD2 (R) increases both A2 and A4 to destroy cancer cells

We notice that the $b_1 = 0.25$, meaning that chemotaxis is very weak. In fact in the model v is not giving any fight against cancer (which could also mean that the standard medication used to boost the immune system is not effective). So, $a_2(t)$ is being increased by the drug SCD2 (r). The treatment appeared to be a grand success.

We could observe this in Fig. 8.4 (MAXULN0 = max value at a lymph node = 10,000).

The black line showing the graph of UPMAX(n), the value of u at the primary tumor, taking a strong nose dive (having the largest negative slope) shows shrinking of the tumor at the primary site. The redline above this with negative slope indicates the largest accumulation of cancer cells in the entire field (UMAX(n)). It is evident from the graph that at some point, this value is the same as the largest value of cancer cells outside the tumor sites represented by MAXUO(n). Other lines show that v, w (SCD1) and r (SCD2) have reached steady states. (Note: steady-state values mean the amounts of drugs to be used to maintain the pattern of the graphs of cancer cells.) That means these medications with these dosages should continue. Only oncologists should make an assessment regarding at which point they will determine a change of the course of treatments.

8.4 Fixed Values of SCD1 and SCD2 for Cancer Cells Dispersing Faster in Tissues

As we have mentioned earlier, the nonlinear partial differential equations are not really reaction–diffusion equations. Because unless all boundary conditions show that both cancer cells and all the medications are zero or some constant values on the boundary, there is dispersion. In the codes, we have used those boundary conditions which could be appropriate for the kind of treatments under consideration. When cancer cells float through blood or lymph, they may not just float, they may swim, because there are many threats to their lives. In fact, they move in groups and still most of them die before they reach any destination to metastasize. The rate of dispersion could be much larger than the value of κ, which often is represented by the parameter NU in the input list, used for blood circulation. This is investigated in this section.

Here we took $\kappa = 0.001$, $a_1(t) = 5.25$, SCD1 $= w = 1000$ (fixed), and SCD2 $= r$ $= 500$(fixed) and considered that the immune system is so weak that it is practically ineffective. The drug SCD1 is continuously reducing the rate of growth of cancer cells.

The equation is

$$\partial u / \partial t = a_1(t)u - a_3 u(SCD1) - a_4 u(SCD2) + \kappa_1 \nabla^2 u \qquad (8.11)$$

Since $SCD1 = w = a\ constant$, the formula (8.10) has been used to decrease $a_1(t)$.

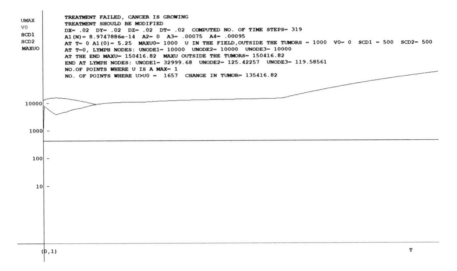

Fig. 8.5 With fixed SCD1 and SCD2 moving faster in blood and no immune activities. Two (not three) lymph nodes are affected

Inputs for Fig. 8.5.

```
*************************************************************************************
! CHAPTER 8, FIG 8.5, JULY 2ND. 2020
! THERE ARE THREE TUMORS , TWO LYMPH NODES ARE AFFECTED.
! DU/DT - A1(N)*U-A2*U*V0-A3*U*SCD1-A4*U*SCD2+ (NU1)*DELSQ(U)
!
! AT T-0, U0 IS - 1000  V0-FIXED -   0  SCD1-FIXED- 500  SCD2-FIXED- 500
! DX- .02  DY- .02  DZ- .02  DT- .02  NX- 60  NY- 60  NZ - 60  NT- 5
! A1(1)- 5.25  A2- 0              A3- .00075  A4- .00095
! A1(N) - A1(N-1)* EXP(-0.01*(DT)*(W0))
! NU1- .001
! GAMMA1- .015
! THREE LYMPH NODES ARE LOCATED AT: ( 3 , 3 , 3 ) ( 20 , 20 , 20 )( 20 , 20 , 20 )
! AT THE LYMPH NODE1 U0- 10000  AT THE LYMPH NODE2 U0- 10000  AT THE LYMPH NODE2 U0- 10000
! PRIMARY TUMOR IS LOCATED AT:  ( 13 , 13 , 13 ) UPRIMARY- 15000
! SECOND TUMOR IS LOCATED AT:   ( 26 , 26 , 26 ) US1- 10000
! THIRD TUMOR IS LOCATED AT:    ( 47 , 47 , 47 ) US2- 7500
!
*************************************************************************************
```

It should be noted that here NU1 $= \kappa_1 = 0.001$ (two not three lymph nodes are affected) much larger than what we have generally used before. Cancer cells are dispersing (for metastasization) much faster as revealed by the increasing slope of the redline. The method of treatment, at first, seemed to be working. But later it failed miserably.

This is not an instability caused by the numerical method. We will study this in the Appendix A.

The drug SCD1 has reduced the rate of growth of cancer from 5.25 to practically zero. Yet cancer is growing. Why? As we have mentioned before, these cells are coming out via dispersion from the dying three tumors. Dead cancer cells are really not dead. Live tumors are not the only source of their growths.

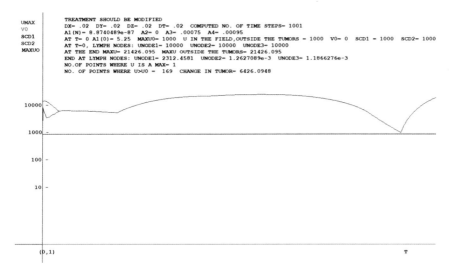

Fig. 8.6 With fixed SCD1 & SCD2 moving faster in blood and no immune activities

As evidenced from Fig. 8.5, the treatment method failed ultimately, even though it appeared to be having some success. Unfortunately, the first author noticed such cases time and again.

So our next attempt was to observe what could happen if we start with low dosage of SCD2, and if and when cancer cells start increasing, we will slowly increase the dosage and bring the number of cells down. Then, as the rate of growth becomes sufficiently low, the dosage will be reduced again, so that cancer will face death. But surprisingly that did not happen.

Inputs for Fig. 8.6.

```
*********************************************************************************
! CHAPTER 8, FIG 8.6, JULY 2ND. 2020
! THERE ARE THREE TUMORS , TWO LYMPH NODES ARE AFFECTED.
! DU/DT = A1(N)*U-A2*U*V0-A3*U*SCD1-A4*U*SCD2+ (NU1)*DELSQ(U)
!
! AT T=0, U0 IS = 1000   V0-FIXED =   0   SCD1=FIXED= 1000  SCD2=FIXED= 500
! DX= .02  DY= .02  DZ= .02  DT= .02  NX= 101  NY= 101  NZ = 101  NT= 1000
! A1(1)= 5.25  A2= 0           A3= .00075  A4= .00095
! A1(N) = A1(N-1)* EXP(-0.01*(DT)*(W0))
! NU1= .001
! GAMMA1= .015
! THREE LYMPH NODES ARE LOCATED AT: ( 3 , 3 , 3 ) ( 20 , 20 , 20 )( 20 , 20 , 20 )
! AT THE LYMPH NODE1 U0= 10000  AT THE LYMPH NODE2 U0= 10000  AT THE LYMPH NODE2 U0= 10000
! PRIMARY TUMOR IS LOCATED AT:   ( 13 , 13 , 13 ) UPRIMARY= 15000
! SECOND TUMOR IS LOCATED AT:    ( 26 , 26 , 26 ) US1= 10000
! THIRD TUMOR IS LOCATED AT:     ( 47 , 47 , 47 ) US2= 7500
!
*********************************************************************************
```

After about 800 time steps (Fig. 8.6) when the rate of growth of cancer was practically zero, and SCD2 was about its initial value, the number of cells started rising again! That indicates that this method of treatment must go on forever. So the question comes back: Will the patient be able to tolerate such a treatment for a great length of time.

Here we have considered that on the boundary, the variables are not zeros. Because if we assume that, u must go to zero as $t \to \infty$ that is the property of a diffusion model.

Here cancer is returning in a cyclic pattern.

In order to explain to the readers that this is not a computational problem, we considered a simpler model where all boundary conditions are zero. The inputs are as follows.

Inputs for Fig. 8.7.

```
************************************************************************************************
! CHAPTER 8, FIG 8.7 , JULY 2ND. 2020
! THERE ARE THREE TUMORS , TWO LYMPH NODES ARE AFFECTED. **** BOUNDARY CONDITIONS ARE SET TO ZERO *****
! DU/DT - A1(N)*U-A3*U*SCD1+ (NU1)*DELSQ(U)                     ! ANTIGEN
!
! AT T=0, U0 IN THE FIELD IS = 1000   SCD1=FIXED= 1000   SCD2= 0
! DX= .02  DY= .02  DZ= .02  DT= .02  NX= 60  NY= 60  NZ = 60  NT= 500
! A1(1)= 5.25  A3= .00075
! A1(N) = A1(N-1)* EXP(-0.01*(DT)*(W(N)))
! NU1= .001
! GAMMA1= .015
! THREE LYMPH NODES ARE LOCATED AT: ( 3 , 3 , 3 ) ( 20 , 20 , 20 )( 20 , 20 , 20 )
! AT THE LYMPH NODE1 U0= 10000  AT THE LYMPH NODE2 U0= 10000  AT THE LYMPH NODE2 U0= 10000
! PRIMARY TUMOR IS LOCATED AT:  ( 13 , 13 , 13 ) UPRIMARY= 15000
! SECOND TUMOR IS LOCATED AT:  ( 26 , 26 , 26 ) US1= 10000
! THIRD TUMOR IS LOCATED AT:  ( 47 , 47 , 47 ) US2= 7500
!
************************************************************************************************
```

We note that $\kappa_1 = \text{NU1} = 0.001$ (two lymph nodes are affected).

The results are given in Fig. 8.7.

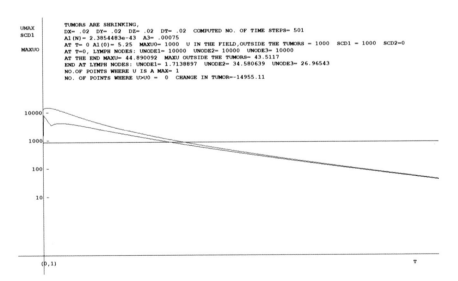

Fig. 8.7 With fixed SCD1 and no SCD2, bounday conditions $= 0$

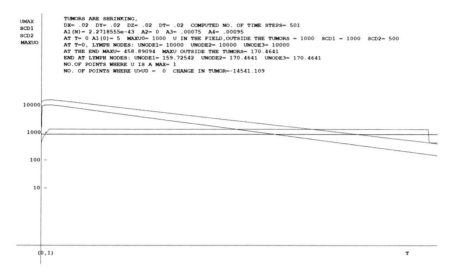

UMAX
SCD1
SCD2
MAXUO

```
TUMORS ARE SHRINKING,
DX= .02  DY= .02  DZ= .02  DT= .02  COMPUTED NO. OF TIME STEPS= 501
A1(N)= 2.2718555e-43  A2= 0  A3= .00075  A4= .00095
AT T= 0 A1(0)= 5  MAXU0= 1000  U IN THE FIELD,OUTSIDE THE TUMORS = 1000  SCD1 = 1000  SCD2= 500
AT T=0, LYMPH NODES: UNODE1= 10000  UNODE2= 10000  UNODE3= 10000
AT THE END MAXU= 458.89094  MAXU OUTSIDE THE TUMORS= 170.4641
END AT LYMPH NODES: UNODE1= 159.72542  UNODE2= 170.4641  UNODE3= 170.4641
NO.OF POINTS WHERE U IS A MAX= 1
NO. OF POINTS WHERE U>U0 =  0  CHANGE IN TUMOR=-14541.109
```

Fig. 8.8 With fixed SCD1. SCD goes up if u goes up

Our next attempt is to slow down the rate of dispersion $NU1 = \kappa_1 = 0.00001$ and use the same inputs as shown in Fig. 8.6. A much better result has been found in Fig. 8.8.

Inputs for Fig. 8.8.

```
**********************************************************************
! CHAPTER 8, FIG 8.8, JULY 2ND. 2020
! THERE ARE THREE TUMORS , TWO LYMPH NODES ARE AFFECTED.
! DU/DT = A1(N)*U-A3*U*SCD1-A4*U*SCD2+ (NU1)*DELSQ(U)                    ! ANTIGEN
!
! AT T=0, U0 IS = 1000  SCD1(1)=FIXED= 1000  SCD2(1)= 500
! DX= .02  DY= .02  DZ= .02  DT= .02  NX= 101  NY= 101  NZ = 101  NT= 1000
! A1(1)= 5  A3= .00075  A4= .00095
! A1(N) = A1(N-1)* EXP(-0.01*(DT)*(W(N)))
! NU1= .00001
! GAMMA1= .015
! THREE LYMPH NODES ARE LOCATED AT: ( 3 , 3 , 3 ) ( 20 , 20 , 20 )( 20 , 20 , 20 )
!! AT THE LYMPH NODE1 U0= 10000  AT THE LYMPH NODE2 U0= 10000  AT THE LYMPH NODE2 U0= 10000
! PRIMARY TUMOR IS LOCATED AT:  ( 13 , 13 , 13 ) UPRIMARY= 15000
! SECOND TUMOR IS LOCATED AT:  ( 26 , 26 , 26 ) US1= 10000
! THIRD TUMOR IS LOCATED AT:  ( 47 , 47 , 47 ) US2= 7500
! IF MAXUF(N)>MAXUF(N-1) THEN  R1(N)=R1(N-1) + 100, IF R1(N) > 1500 THEN R1(N) = 1500 ELSE IF MAXUF(N) < 500 THEN  R1(N) = R1(1)
! ELSE R1(N) = R1(N-1)
!
**********************************************************************
```

8.5 Introduction of the Skilled Killer Drug (SKD)

SKD is a drug that kills cancer cells and reduces their rate of growth at the same time. The model is

$$\partial u / \partial t = a_1(t)u - a_3 u (SKD) + \kappa_1 \nabla^2 u$$

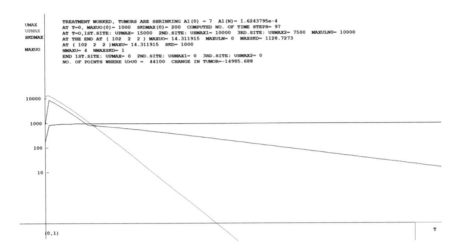

Fig. 8.9 The silver bullet (SKD). The skilled killer drug

$$\partial(SKD)/\partial t = c_1 u + \kappa_3 \nabla^2 (SKD)$$

Together with

$$da_1/dt = -0.01 a_1 \times (max(SKD(t)))$$

Integrating from t_{n-1} to t_n by trapezoidal rule, we get: $a_1(n) = a_1(n-1)exp(-0.01(\Delta t/2)(max(SKD(t_n)) + max(SKD(t_{n-1}))))$.

Inputs for Fig. 8.9.

```
•••••••••••••••••••••••••••••••••••••••••••••••••••••••••••••••••••••••••••••••••••
I THE SKILLED KILLER DRUG (SKD) FOR METASTATIC CANCER MAY 31,2020
I
I DU/DT - A1(N)•U - A3•U•(SKD) + NU1•DELSQ(U)
I
I D(SKD)/DT - C1•U + NU3•DELSQ(SKD); (SKD) DOES NOT DIRECTLY KILL U. IT REDUCES THE RATE OF GROWTH OF CANCER
I & GETS ATTACHED TO CANCER CELLS TO DESTROY THEM.
I U- U-A5•U•(SKD);A1(N) - A1(N-1)• EXP(-0.01•(DT/2)•(SKDMAX(N) + SKDMAX(N-1)))
I RESTRICTIONS: IF SKD(I,J,K) > 3000 THEN SKD(I,J,K) - 3000, IF SKD(I,J,K) <1000 THEN SKD(I,J,K) - 1000
I DX-    .02   DY-  .02   DZ- .02   DT- .02   NX- 101   NY- 101   NZ - 101   NT- 200
I A1(1)-    7   A3-  .00005   A5- .00025
I C1-   .25   C2- 0
I MAXUO(0)-    1000   MAXU IN THE FIELD-    15000   SKD(0)- 200
I NU1-   .0000001   NU3-   .0000001
I THREE TUMORS LOCATED AT        ( 13 , 13 , 13 ) ( 26 , 26 , 26 )
I AND    ( 47 , 47 , 47 ) WHERE U0- 15000    10000   AND  7500   RESPECTIVELY.
I AT THE LYMPH NODES       ( 3  3  3 )( 20  20  20 )( 55  55  55 ) AT T-0, ULNO- 10000
•••••••••••••••••••••••••••••••••••••••••••••••••••••••••••••••••••••••••••••••••••
```

There is a restriction on the drug (SKD), $namely$ $1000 \leq (SKD) \leq 3000$. Again, in the code $NU1 = \kappa_1$ and $NU2 = \kappa_2$. The inputs show that $a_1(t = 0) = 7$. It means initially cancer cells are replicating at the rate of 600%. It is a very fast-growing tumor. The defense with nutrition and other medications are trying to slow down some and that is given by $a_3 = 0.00005$ (significantly small) rate of reduction of the antigen. However, the drug SKD simply gets attached to each cancer cell and starts killing it. So, $u = (1 - a_5 \times (SKD))u$ at each point of the computational field where $a_5 = 0.00025$. The rate of growth of SKD is given by $c_2 = 0.25u$. The dispersion rates of u and SKD are the same, because wherever u is present SKD is also present. The results in Fig. 8.9 show that as SKD goes to a steady state, the largest tumor (green graph) is drastically reduced and the largest collection of cancer cells outside this, $UMAX$ and $MAXUO$ both are the same (black graph) and going down. First they jumped up, being attacked, then they took a nose drive (large negative slope) and then started decreasing slowly although rate of growth is $a_1(n) = 1.6243795 \times 10^{-4}$ (extremely small).

From a computational standpoint, these three graphs show that the mathematical model is quite stiff. It should be noted that there are strict limitations imposed upon SKD. First its rate of growth is $c_1 = 0.25$.

8.6 SKD: The Intelligent Nanoparticle

Here we thought of the applications of intelligent nanoparticles. Cancer researchers have already started conducting research on this. We have designed it such that cancer cells could move faster through blood, lymph, and tissues with $\kappa = 0.005 = NU1$, and the nanoparticles are sticking to each cancer cell and destroying its power of growth and its destructive capabilities all over the body where even one cancer cell is present. If $SKD = 0.65$, then $u = u * (SKD)$ means u has become 65% of what it was before. So it has been reduced by 35%. There were three tumors and three lymph nodes.

$a_1 = 20$. Cancer is growing at the rate of 1900% very fast.

Inputs for Fig. 8.10.

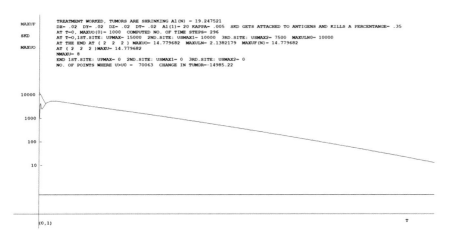

Fig. 8.10 The skilled killer drug (SKD) reduces growths of cancer cells, gets attached to them to kill them

```
••••••••••••••••••••••••••••••••••••••••••••••••••••••••••••••••••••••••••••••••••••••••••••••••••••••••
! THE SILVER BULLET,SKILLED KILLER DRUG(SKD)DELIVERED BY INTELLIGENT NANOPARTICLES FOR METASTATIC CANCER, FIG 8.10
!
! DU/DT - A1(N)*U  + NU1*DELSQ(U), U=U*SKD, WHERE SKD < 1 SO THE PERCENTAGE OF REDUCTION = (1 -SKD)
!
! DX- .02   DY- .02   DZ- .02   DT- .02   NX- 101   NY- 101   NZ - 101   NT- 500
! A1(1)- 20   KAPPA- .005 SKD(1) =  .65
! MAXUO(0)- 1000  MAXU IN THE FIELD- 15000
! THREE TUMORS LOCATED AT ( 13 , 13 , 13 )  ( 26 ,  26 ,  26 )
! AND ( 47 , 47 , 47 )  WHERE U0- 15000  10000  AND 7500  RESPECTIVELY.
! AT THE LYMPH NODES( 3   3   3 )( 20   20   20 )( 55   55   55 ) AT T=0, ULNO= 10000
••••••••••••••••••••••••••••••••••••••••••••••••••••••••••••••••••••••••••••••••••••••••••••••••••••••••
```

Figure 8.10 vividly shows the success of the treatments.

While SKD remains the same, max value of u is going down significantly. At all cancer sites, tumors shrank to zero. However, all is not good. There are 70,063 points in the field, where $U_{ijk}^n > U_{ijk}^{n-1}$. So the cancer cells have moved around, all over the field, and many are taking shelters at various locations of the body. They are planning to hide. So applications of intelligent nanoparticles are not enough to complete full destruction of cancer cells. Possibly at this point, some form of adjuvant therapy should be continued as being directed and monitored by oncologists.

8.7   Faster Dispersion of SKD

The next is basically the same inputs with 25% reduction of cancer cells and a much faster dispersion of the drug SKD. We also assume that there are no other defending elements that are destroying cancer. It is given by the Inputs for Fig. 8.11

Everything else is the same.

Inputs for Fig. 8.11.

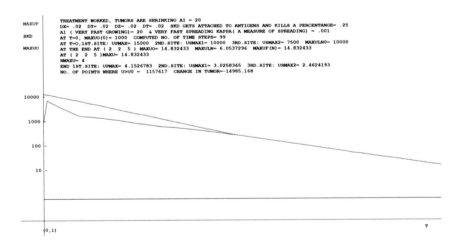

Fig. 8.11 A silver bullet for a very fast growing and very fast spreading cancer

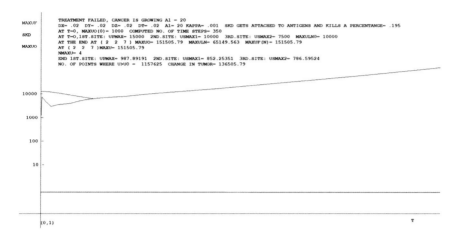

Fig. 8.12 A silver bullet (SKD) destroying fast growing fast spreading cancer

```
***************************************************************************************************
! THE SILVER BULLET,SKILLED KILLER DRUG(SKD)DELIVERED BY INTELLIGENT NANOPARTICLES FOR METASTATIC CANCER, FIG 8.11
!
! DU/DT = A1*U  + NU1*DELSQ(U), U=U*SKD, WHERE SKD < 1 SO THE PERCENTAGE OF REDUCTION = (1 -SKD)
!
! DX= .02  DY= .02  DZ= .02  DT= .02  NX= 101  NY= 101  NZ = 101  NT= 300
! A1= 20  KAPPA= .001
! MAXUO(0)= 1000  MAXU IN THE FIELD= 15000
! IF A1<15.5 THEN SKD= 0.85 =MEANS 15% REDUCTION OF CANCER IF A1 = 20 SKD=0.75 =25% REDUCTION OF CANCER AT EACH T(N)
! THREE TUMORS LOCATED AT ( 13 , 13 , 13 )  ( 26 , 26 , 26 )
! AND ( 47 , 47 , 47 ) WHERE UO= 15000  10000  AND 7500  RESPECTIVELY.
! AT THE LYMPH NODES( 3  3  3 )( 20  20  20 )( 55  55  55 ) AT T=0, ULNO= 10000
***************************************************************************************************
```

Figure 8.11 reveals some very interesting scenarios. All three tumors are practically gone after 99 time steps. However, cancer spreads all over the computational field with a pattern to replicate at every point ($U_{ijk}^{99} > U_{ijk}^{98}$ at 1,157,617).

In the next computer run, we wanted to see if we decrease the mass accumulation of the cancer cells (u) by 20% and assume that cancer cells may disperse faster, $\kappa = 0.001 = $ NU1, then what happens. It did not work.

Inputs for Fig. 8.12.

```
***************************************************************************************************
! THE SILVER BULLET,SKILLED KILLER DRUG(SKD)DELIVERED BY INTELLIGENT NANOPARTICLES FOR METASTATIC CANCER, FIG 8.12
!
! DU/DT = A1*U  + NU1*DELSQ(U), U=U*SD, WHERE SKD < 1 SO THE PERCENTAGE OF REDUCTION = SKD= (1 -SD)
!
! DX= .02  DY= .02  DZ= .02  DT= .02  NX= 101  NY= 101  NZ = 101  NT= 400
! A1= 20  KAPPA= .001 SD= .805
! MAXUO(0)= 1000  MAXU IN THE FIELD= 15000
! THREE TUMORS LOCATED AT ( 13 , 13 , 13 )  ( 26 , 26 , 26 )
! AND ( 47 , 47 , 47 ) WHERE UO= 15000  10000  AND 7500  RESPECTIVELY.
! AT THE LYMPH NODES( 3  3  3 )( 20  20  20 )( 55  55  55 ) AT T=0, ULNO= 10000
***************************************************************************************************
```

From Fig. 8.12, we can notice that cancer has spread all around, everywhere in the field, started leaving the three tumor sites and moved to a new location at (2, 2, 7) with a mass of 151,505. Initially, at the largest tumor site, $u = 15000$. We ran the code for 350 time steps.

This treatment did not work as may be evidenced by the redline. So we must remodel this treatment and did another attempt to modify the treatment in the figure

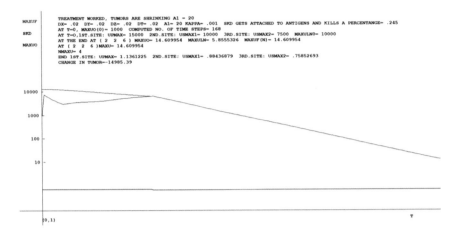

MAXUF
SKD
MAXUO

TREATMENT WORKED, TUMORS ARE SHRINKING A1 = 20
DX= .02 DY= .02 DZ= .02 DT= .02 A1= 20 KAPPA= .001 SKD GETS ATTACHED TO ANTIGENS AND KILLS A PERCENTANGE= .245
AT T=0, MAXUO(0)= 1000 COMPUTED NO. OF TIME STEPS= 168
AT T=0,1ST.SITE: UFMAX= 15000 2ND.SITE: USMAX1= 10000 3RD.SITE: USMAX2= 7500 MAXULNO= 10000
AT THE END AT (2   2   6) MAXUO= 14.609954 MAXULN= 5.8555326 MAXUF(N)= 14.609954
AT (2   2   6)MAXU= 14.609954
NMAXU= 4
END 1ST.SITE: UFMAX= 1.1361225 2ND.SITE: USMAX1= .88436879 3RD.SITE: USMAX2= .75852693
CHANGE IN TUMOR=-14985.39

Fig. 8.13 The silver bullet (SKD) destroying fast growing fast spreading cancer

(after Fig. 8.13) to show that even with $a_1 = 22$, treatment shall work. It is in our second Fig. 8.12. This is the only one example which shows that just killing the mass of cancer cells wherever they start and wherever they move is not enough. We have to do more. Either in the rate of growth, a_1 must be diminished more vigorously or the strength of the drug SKD must be increased or both. So using the same inputs with an exception, here we are reducing the cancer mass by 24.5%, the medication worked as shown in Fig. 8.13.

It is obvious that here in this monograph, these concepts are only in theory. But the models could be very realistic in the future, especially, every day there is progress done by a number of scientists around the globe working untiringly to find a cure of cancer or at the very least to keep it under control.

These mathematical graphs could be useful in that regard. This is indeed our primary objective while writing this monograph.

Just out of curiosity, when we increased a_1 from $a_1 = 20$ to $a_1 = 22$, the treatment still worked. Figure 8.14 (second figure correcting previous Fig. 8.14).

8.8 Conclusion

Applications of nanotechnology to treat are not a new concept here. We have only done a mathematical model regarding its use. More details on applications are available on the Internet. We just mention one below: "Nanotechnology and Nanoparticles in Drug Delivery–Cancer Radiation Therapy: Researchers are investigating the use of bismuth nanoparticles to concentrate radiation used in radiation therapy to treat cancer tumors. Initial results indicate that the bismuth nanoparticles would increase the radiation dose to the tumor by 90%. A method to make radiation therapy more effective in fighting prostate cancer is using radioactive gold nanoparticles attached

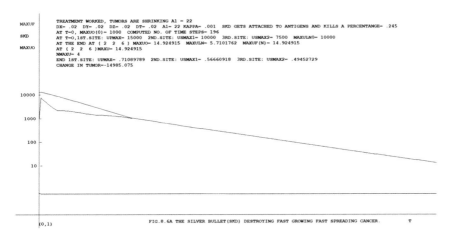

TREATMENT WORKED, TUMORS ARE SHRINKING A1 = 22
DX= .02 DY= .02 DZ= .02 DT= .02 A1= 22 KAPPA= .001 SKD GETS ATTACHED TO ANTIGENS AND KILLS A PERCENTAGE= .245
AT T=0, MAXUO(0)= 1000 COMPUTED NO. OF TIME STEPS= 196
AT T=0,1ST.SITE: UPMAX= 15000 2ND.SITE: USMAX1= 10000 3RD.SITE: USMAX2= 7500 MAXULNO= 10000
AT THE END AT (2 2 6) MAXUO= 14.924915 MAXULN= 5.7101762 MAXUF(N)= 14.924915
AT (2 2 6)MAXU= 14.924915
NMAXU= 4
END 1ST.SITE: UPMAX= .71089789 2ND.SITE: USMAX1= .56660918 3RD.SITE: USMAX2= .49452729
CHANGE IN TUMOR=-14985.075

FIG.8.6A THE SILVER BULLET (SKD) DESTROYING FAST GROWING FAST SPREADING CANCER.

Fig. 8.14 The silver bullet (SKD) destroying fast growing fast spreading cancer

to a molecule that is attracted to prostate tumor cells. Researchers believe that this method will help to concentrate the radioactive nanoparticles at the cancer tumors, allowing treatment of the tumors with minimal damage to healthy tissue.

Nanotechnology and Nanoparticles in Drug Delivery–Cancer Heat Therapy: Another technique delivers chemotherapy drugs to cancer cells and also applies heat to the cell. Researchers are using gold nanorods to which DNA strands are attached. The DNA strands act as a scaffold, holding together the nanorod and the chemotherapy drug. When Infrared light illuminates the cancer tumor the gold nanorod absorbs the infrared light, turning it into heat. The heat both releases the chemotherapy drug and helps destroy the cancer cells. Targeted heat therapy is being developed to destroy breast cancer tumors. In this method antibodies that are strongly attracted to proteins produced in one type of breast cancer cell are attached to nanotubes, causing the nanotubes to accumulate at the tumor. Infrared light from a laser is absorbed by the nanotubes and produces heat that incinerates the tumor."

Reference: https://www.understandingnano.com/nanotechnology-drug-delivery.html.

One major problem could be the amount of toxicity that could be caused by such treatments. There are some ongoing research works in this regard [3]. The authors have mentioned: "Besides the potential beneficial use also attention is drawn to the questions how we should proceed with the safety evaluation of the nanoparticle formulations for drug delivery. For such testing the lessons learned from particle toxicity as applied in inhalation toxicology may be of use. Although for pharmaceutical use the current requirements seem to be adequate to detect most of the adverse effects of nanoparticle formulations, it can not be expected that all aspects of nanoparticle toxicology will be detected. So, probably additional more specific testing would be needed."

References

1. Hossena, S., Hossain, M.K., Basherb, M.K., Mia, M.N.H., Rahman, M.T., Jalal Uddin, M.: Smart nanocarrier-based drug delivery systems for cancer therapy and toxicity studies: a review Author links open overlay panel. J. Adv. Res. **15**, 1–18 (January 2019)
2. Masoud, V., Pagès, G.: Targeted therapies in breast cancer: new challenges to fight against resistance. World J. Clin. Oncol. **8**(2), 120–134 (2017 Apr 10)
3. De Jong, W.H., Borm, P.J.A.: Drug delivery and nanoparticles: applications and hazards. Int. J. Nanomed. **3**(2), 133–149 (2008 Jun)
4. https://www.webmd.com/cancer/news/20080610/treatment-can-make-cancer-stronger#1
5. Vilalta, M., Rafat, M., Giaccia, A.J., Graves, E.E.: Recruitment of circulating breast cancer cells is stimulated by Radiotherapy. Cell Rep. (2014). [PMC free article] [PubMed] [Google Scholar]
6. Pucci, C. et al.: Innovative approaches for cancer treatment : current perspectives and new challenges. Ecancermedicalscience **13**, 961 (2019)
7. Lombardo, D., Kiselev, M.A., Caccamo, M.T.: Smart nanoparticles for drug delivery application: development of nanocarrier platforms in biotechnology and nanomedicine. J. Nanomat. (2019 Feb)

Chapter 9
Nutritional Therapy

Abstract Food sustains life. From time immemorial, all over the world, men and women used nutrition to fight against all kinds of ailments, including cancer and certainly it did work. Why? Because it kept them alive, and we are their lineage. Here a mathematical study has been done in this regard. Since fat fuels growth of cancer, we have added a section of this topic with a statistical analysis.

9.1 Rationale

From the very beginning of human civilization, it was nutritional therapy that helped men to conquer various ailments for ages. It is a therapy for both body and mind. Only good food is not enough. A good state of mind is absolutely necessary for better health. But what we eat affects our mind and what we think affects our choice of food. So nutritional therapy is truly a lifelong practice. It is a holistic approach to enjoy a better state of health. For cancer patients, on a short-term basis, its effectiveness could be very limited. So it should be done along with other standard procedures to treat this disease.

According to our experience, most of the time, this is done after cancer is already detected. Doctors, nurses, and other medical professionals often recommend nutritional therapy along with regular treatments. In our opinion, better food habits should be adopted throughout our life.

With the progress of time, with ever growing demands for better housing, better transportation, better food, better communication, and unquenchable thirst for more wealth and power rampant deforestation, urbanization, and industrialization brought us today on the arenas of hundreds and thousands of diseases, causing often incurable genetic complications of which cancer is possibly the number one. To these persons, all advancement of science is more a curse than a bliss.

Now, many scientists think cancer is caused mostly by lifestyles, environmental pollution, and food. So nutritional therapy is a primary topic of study by many researchers. We have attempted here to mathematize this concept. Obviously, this is indeed a preliminary attempt.

© Springer Nature Singapore Pte Ltd. 2021   281
S. Dey and C. Dey, *Mathematical and Computational Studies on Progress, Prognosis, Prevention & Panacea of Breast Cancer*, Forum for Interdisciplinary Mathematics, https://doi.org/10.1007/978-981-16-6077-1_9

Food is our primary source of energy. If we do not eat, we get hungry and cannot work. As food gets digested, it helps every cell in the body to do their respective jobs in an environment of dynamic equilibrium (homeostasis). Food helps all cells, including cancer cells. However, there are some foods which help primarily the good cells that could strengthen the law of homeostasis and prevent growths of cancer cells who destroy that law and grow and move around the body invading other good cells.

From: https://www.webmd.com/cancer/features/seven-easy-to-find-foods-that-may-help-fight-cancer#1, we found: "A comprehensive review of thousands of studies on diet, physical activity, and weight conducted for the World Cancer Research Fund and the American Institute for Cancer Research pointed to the benefits of eating mostly foods of plant origin. Foods such as broccoli, berries, and garlic showed some of the strongest links to cancer prevention". Dr. Wendy Demark-Wahnefried, PhD, RD, associate director for Cancer Prevention and Control at the University of Alabama's Comprehensive Cancer Center thinks that a "diet that could ward off cancer really doesn't look that different from the healthy foods you should be eating anyway. That means plenty of fruits and vegetables, as well as whole grains and lean meat or fish."

"Diet is one the most important factors for the formation and prevention of cancer. Thus, in part to reduce cancer risk, two dietary goals have to be achieved for the year 2000: (i) the caloric contribution of fat should not be more than 30% of total caloric intake which can be achieved by reducing present fat intake by 18%; (ii) The consumption of carbohydrate and fiber-containing foods should be doubled by increasing fruit and vegetable consumption to five servings per day. If alcoholic beverages are consumed anyway, according to NCI guidelines, it should be taken in moderate dose (less than 40 g per day). Retinoids, carotenoids, vitamin C, and vitamin E should be taken in optimum amounts." [1]

Many scientists think that a diet high in whole foods like fruits, vegetables, whole grains, healthy fats, and lean protein may prevent cancer. Conversely, processed meats, refined carbs, salt, and alcohol may increase your risk. Though no diet has been proven to cure cancer. In the article "The Diet That Could Stop Cancer from Spreading", Jill Waldbieser stated in the following reference: Oct. 17, 2018 [https://www.thehealthy.com/cancer/this-diet-could-stop-cancer-from-spreading/].

Power of Nutrition therapy [https://healthprep.com/cancer/5-strange-cures-for-cancer/] must not be underestimated: "Soursop is a fruit present in the Caribbean, Mexico, as well as Central and South America. The effect of this plant in the body is reportedly ten thousand times more powerful than chemotherapy, though the soursop acts on the damaged cells without affecting healthy ones. Artemisinin, a substance present in this fruit, is responsible for inhibiting the membrane of cancer cells, stopping their growth and eventually causing death (apoptosis)." There are many more.

In a 2002 study, researchers found that an estimated one-third of all cancer deaths in the USA can be attributed to diet. "Good nutrition may reduce the incidence of breast cancer and the risk of breast cancer progression or recurrence," wrote Natalie Ledesma, RD, in Women's Health Matters, a publication from the University of California, San Francisco. Research published in Nutrition Reviews found that the foods you eat—and, importantly, the things you avoid—can dramatically reduce the risk of cancer returning. In this review of studies on diet and cancer recurrence, the researchers found that a generally healthy diet could lower future cancer risk by about 25%; following a high-sugar, high-fat Western-style diet nearly doubled the risk of return. Alcohol did not help either: The more people drank, the higher their risk of recurrence is.

Following all these article and more, we attempted to set up a mathematical model as follows.

9.2 A Mathematical Model

This first model is a model very similar to those we have discussed before.

Let u = the foreign antigen (cancer). We are considering foods that will do the following: (1) lower the growth of cancer, (2) help the defensive forces of the body, and (3) inhibit the spread of cancer. Certainly, food does not work like chemo or radiation which work much faster. Here the time scale is large. Food feeds the entire body and is not targeted. Cancer cells use the same food. But those foods which are anti-inflammatory, they fight against cancer. Because cancer causes inflammation even in the microstates. Mediterranean diets like plenty of colorful fruits and vegetables and fresh seafood strengthen cells such that cancer cells could not easily use their microscissors to cut the barriers between the cells and attack the neighboring cells. So while food strengthens both normal cells and cancer cells, since the normal cells, initially, outnumber cancel cells, they are more empowered to stop invasion of cancer cells. So if a_1 = the rate of growth of cancer cells, proper food can reduce that. Food can also shrink cancer cells and strengthen the immune system. When any cell consumes healthy food, we may assume this makes the cell a newer stronger cell. That means if r_1 is the food that shrinks the rate of growth of each unit of u, r_2 is at which each unit of u is destroyed and r_3 is the rate at which each unit of v, the immune response is strengthened, then the model is

$$\partial u / \partial t = a_1(r_1, t)(ur_2) - a_2(vr_3) + \kappa \nabla^2 u \qquad (9.1)$$

where r_1, r_2, and r_3 are all functions of time. They should be chosen in such a way so that r_1 and r_2 will reduce $a_1(t)$ and u, respectively, and r_3 will increase the strength of the immune response v.

Following our discussions in previous chapters, we chose

$$\text{(i)} \quad dr_1/dt = -\eta_1 r_1 \qquad (9.2)$$

$$\text{(ii)} \quad dr_2/dt = -\eta_2 r_2 \qquad (9.3)$$

$$\text{(iii)} \quad dr_3/dt = \eta_3 r_3 \tag{9.4}$$

where η_1, η_2, and η_3 are all positive.

At present, there are medications which can do all of these. So these mathematical formulas are not really all hypothetical. Only they could serve as giving some directions regarding administration of the medications.

Integrating these equations from t_{n-1} to t_n and noting that $\Delta t = t_n - t_{n-1}$, we get (assuming that r_1, r_2, r_3 each is 1 at the beginning. The reason is after digestion process is over, body absorbs all the nutrients and changes to body happen very slowly which are hard to detect from day to day. So, virtually there is no noticeable change. So, the values of these parameters at the previous time step should be taken as 1) [15, 16]:

$$r_1 = exp(-\eta_1 \Delta t), r_2 = exp(-\eta_2 \Delta t) \text{ and } r_3 = exp(\eta_3 \Delta t). \tag{9.5}$$

We chose $\eta_1 = 1.25$, $\eta_2 = 1.25$, and $\eta_3 = 0.1$, respectively. These are some educated guesses after analyzing various data.

Inputs for the first computer run are given below. If the rate of growth of cancer $a_1 = 3.5$ is high for just nutritional therapy alone, a chemo or poison is routinely used by doctors in such cases. However, we decided to make an attempt to see if such a therapy could make some sense. So we have made this attempt. Figure 9.1 shows that there may exist a case where such a therapy could be fruitful, of course, over a long period of time.

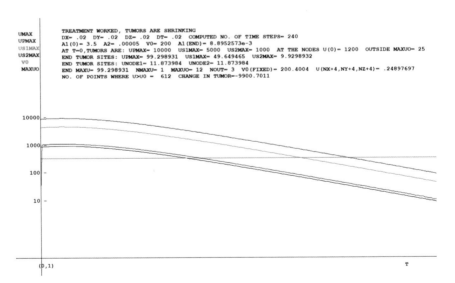

Fig. 9.1 Nutrition therapy reduces accumulation of cancer cells and strengthens immunity

The parameters r_1 and r_2 are being used at each time step like medications used at a given interval. Although these are not medications, these are the results of consuming certain therapeutic foods that make the body lower the growth of cancer. On the contrary, r_3 has been used only once to boost the immune system. The logic behind this is whatever lowers the growth of cancer continuously will implicitly strengthen the power of resistance continuously. So there may be no need to give the patients more food to be used as drugs. According to the experience of the first author's observations, most patients lose their appetite during any therapy. So we kept v as a constant during computations.

Inputs for Fig. 9.1

```
*****************************************************************************************
! CHAPTER 9 NUTRITION THERAPY FIG 9.1
! THERE ARE THREE TUMORS , THREE LYMPH NODES ARE AFFECTED
! DU/DT = A1(N)*U-A2*U*VO*R2+(NU1)*DELSQ(U)
! AT T=0,  UO = 10000  AT THE PRIMARY SITE, U = 10000  SECOND TUMOR= 5000  3RD = 1000
! DX= .02  DY= .02  DZ= .02  DT= .02  NX= 101  NY= 101  NZ = 101  NT= 450
! A1(0)= 3.5  A2= .00005  UO(NX+4,NY+4,NZ+4)= 25  VO= 200
! NU1= .0000001  GAMMA1= .015  ETA1= 1.25  ETA2= 1.25  ETA3= .1  R1= .97530991  R2= .97530991  R3= 1.002002
! NUTRITION THERAPY DOES THE FOLLOWING :
! R1 = EXP(-ETA1*DT), R2= EXP(-ETA2*DT);R3= EXP(ETA3*DT); A(N)=R1*A1(N-1), U=R2*U, VO=R3*VO.
! THREE AFFECTED LYMPH NODES ARE LOCATED AT: ( 3 , 3 , 3 ) ( 20 , 20 , 20 )
! AND ( 35 , 35 , 35 ) AT ALL THE LYMPH NODES UO= 1200
! PRIMARY TUMOR IS LOCATED AT:    ( 13 , 13 , 13 ) UPRIMARY= 10000
! SECOND TUMOR IS LOCATED AT:     ( 26 , 26 , 26 ) US1= 5000
! THIRD TUMOR IS LOCATED AT:      ( 47 , 47 , 47 ) US2= 1000
!
*****************************************************************************************
```

The model worked well. It may not look very practical when we look into the outputs in a shorter time frame. As we have mentioned earlier, this treatment is a long-term treatment meaning that it should be a part of the lifestyle. The diet should be used by cancer patients as well as non-patients.

In our next case, we have considered cancer growing at the rate of 50% over a long period of time. These are relatively slow growing tumors. So a nutritional treatment could work. However, it should be done following the discretions of an oncologist.

Figure 9.2 shows a much better output than what we have noticed in Fig. 9.1. All tumors are regressing uniformly. Again this is a long-term effect. Here $a_1 = 1.5$. Growth is slower than before. Nutritions have been expected to work.

Inputs for Fig. 9.2

```
*****************************************************************************************
! CHAPTER 9 NUTRITION THERAPY FIG 9.2
! THERE ARE THREE TUMORS , THREE LYMPH NODES ARE AFFECTED
! DU/DT = A1(N)*U-A2*U*VO*R2+(NU1)*DELSQ(U)
! AT T=0,  UO = 10000  AT THE PRIMARY SITE, U = 10000  SECOND TUMOR= 5000  3RD = 1000
! DX= .02  DY= .02  DZ= .02  DT= .02  NX= 101  NY= 101  NZ = 101  NT= 450
! A1(0)= 1.5  A2= .00005  UO(NX+4,NY+4,NZ+4)= 25  VO= 200
! NU1= .0000001  GAMMA1= .015  ETA1= 1.05  ETA2= 1.05  ETA3= .025  R1= .97921896  R2= .97921896  R3= 1.0005001
! NUTRITION THERAPY DOES THE FOLLOWING :
! R1 = EXP(-ETA1*DT), R2= EXP(-ETA2*DT);R3= EXP(ETA3*DT); A(N)=R1*A1(N-1), U=R2*U, VO=R3*VO.
! THREE AFFECTED LYMPH NODES ARE LOCATED AT: ( 3 , 3 , 3 ) ( 20 , 20 , 20 )
! AND ( 35 , 35 , 35 ) AT ALL THE LYMPH NODES UO= 1200
! PRIMARY TUMOR IS LOCATED AT:    ( 13 , 13 , 13 ) UPRIMARY= 10000
! SECOND TUMOR IS LOCATED AT:     ( 26 , 26 , 26 ) US1= 5000
! THIRD TUMOR IS LOCATED AT:      ( 47 , 47 , 47 ) US2= 1000
!
*****************************************************************************************
```

It should bring it down much further. r_1, r_2, r_3 are all close to one, meaning that nutrition are working slowly. In Fig. 9.2, we have tried to see the results in a compact form. They were appropriately scaled, and graphs were drawn.

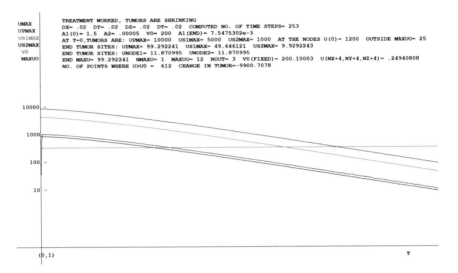

UMAX
UPMAX
US1MAX
US2MAX
V0
MAXUO

TREATMENT WORKED, TUMORS ARE SHRINKING
DX= .02 DY= .02 DZ= .02 DT= .02 COMPUTED NO. OF TIME STEPS= 253
A1(0)= 1.5 A2= .00005 V0= 200 A1(END)= 7.5475302e-3
AT T=0,TUMORS ARE: UPMAX= 10000 US1MAX= 5000 US2MAX= 1000 AT THE NODES U(0)= 1200 OUTSIDE MAXUO= 25
END TUMOR SITES: UPMAX= 99.292241 US1MAX= 49.646121 US2MAX= 9.9292243
END TUMOR SITES: UNODE1= 11.870995 UNODE2= 11.870995
END MAXU= 99.292241 NMAXU= 1 MAXUO= 12 NOUT= 3 V0(FIXED)= 200.10003 U(NX+4,NY+4,NZ+4)= .24940808
NO. OF POINTS WHERE U>U0 = 612 CHANGE IN TUMOR=-9900.7078

Fig. 9.2 Nutrition slowly but certainly works

All tumors together with concentration of cancer cells outside the tumors are steadily shrinking as time goes by. We should remember that taking nutritional foods is a primary part of a lifestyle. It does not work overnight. Or in other words, it is not a fast track solution like chemo, radiation, or surgery. It also has a lot less or possibly none at all side effects like chemo, radiation, or surgery.

Our last computer experiment is a bit more realistic. There is no tumor, yet cancer cells have been slowly growing all around the body. The defense is also working very slowly. Sometimes it does that when the body is attacked by several different pathogens. "Studies of the immune system have been at the forefront of characterizing how different gene programmes function. Immune cells read the environment through their receptors and then modify how they use the genes encoded by their DNA. Some groups of genes are switched on, and others are switched off. This gives the different cell types a great deal of flexibility in how they handle an infection. Sometimes these genes programme changes of the cytokines that cells secrete, sometimes they change the pattern of receptors on the surface and sometimes they change how resistant the cell is to infections. Information in the environment can label a specific location, keeping immune cells from moving away…. The adaptations that we make in response to infection are measured over many time scales. They may occur rapidly in minutes and resolve just as fast; they may continue for days until a viral infection is cleared or they may be long-lasting and change the local anatomy of a tissue…" [2] https://www.ncbi.nlm.nih.gov/pmc/articles/PMC5091071/.

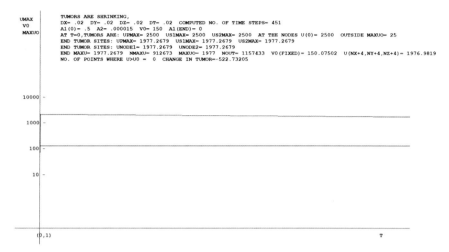

Fig. 9.3 No tumor, yet cancer cells are present and nutrition slowly worked

Inputs for Fig. 9.3

```
**********************************************************************************************
! CHAPTER 9 NUTRITION THERAPY FIG 9.3
! THERE ARE THREE TUMORS , THREE LYMPH NODES ARE AFFECTED
! DU/DT = A1(N)*U-A2*U*VO*R2+(NU1)*DELSQ(U)
! AT T=0,   U0 = 2500  AT THE PRIMARY SITE, U = 2500  SECOND TUMOR= 2500  3RD = 2500
! DX= .02  DY= .02  DZ= .02  DT= .02  NX= 101  NY= 101  NZ = 101  NT= 450
! A1(0)= .5  A2= .000015  U0(NX+4,NY+4,NZ+4)= 2500  VO= 150
! NU1= .0000001   GAMMA1= .015  ETA1= .025  ETA2= .025  ETA3= .025  R1= .99950012  R2= .99950012  R3= 1.0005001
! NUTRITION THERAPY DOES THE FOLLOWING :
! R1 = EXP(-ETA1*DT), R2= EXP(-ETA2*DT);R3= EXP(ETA3*DT) ; A(N)=R1*A1(N-1), U=R2*U, VO=R3*VO.
! THREE AFFECTED LYMPH NODES ARE LOCATED AT: ( 3 , 3 , 3 ) ( 20 , 20 , 20 )
! AND ( 35 , 35 , 35 ) AT ALL THE LYMPH NODES U0= 2500
! PRIMARY TUMOR IS LOCATED AT:  ( 13 , 13 , 13 ) UPRIMARY= 2500
! SECOND TUMOR IS LOCATED AT:   ( 26 , 26 , 26 ) US1= 2500
! THIRD TUMOR IS LOCATED AT:    ( 47 , 47 , 47 ) US2= 2500
!
**********************************************************************************************
```

The rate of growth of cancer is quite slow. The response of the immune system is also very feeble. Actually, cancer cells are everywhere. With nutrition therapy, all are dying very slowly. The graph displays that trend. From UPMAX = 2500, it has come down to 1977.2649 in 451 time steps, each time step = Δt = 0.02. In the log scale, it is not quite noticeable.

9.3 If a Tumor has Attained a Steady State

A malignant tumor hardly goes and stays in a steady state. Nutritional therapy however is a part of living.

It is a significant part of treatment once cancer is found and other treatments begin.

"All India Institute of Ayurveda and the Central Council for Research in Ayurvedic Sciences to delve deeper into the probability of treating cancer with Ayurvedic drugs, it brings a sense of hope to many who have been turning to Ayurveda for an alternative treatment…various health institutions are including Ayurveda in their treatment, along with radiotherapy and chemotherapy to reduce the side effects. A pilot study done by AIIMS (All India Institute of Medical Science) also found that Ayurvedic drugs significantly reduced side effects in breast cancer patients" [3]. In [4] Stanford Medical Center has some guidelines in this regard. According to a study at Mayo Clinic: "It's true plant-based foods, such as fruits, vegetables, legumes, nuts and whole grains, are packed with nutrition. And research has also shown that eating lots of them is linked with lower cancer rates." That indicates that plant-based foods have some power to resist the onset of cancer. At present, no scientific studies show that these foods could cure cancer. But the first author has noticed several breast cancer patients regularly taking plant-based foods, especially Ayurvedic treatments, during adjuvant therapy. These treatments have many positive benefits both on body and mind [14]. So, it gave us some incentives to develop the models given below.

For the first part, we assume a logistic model:

$$du/dt = au(1 - u/Q), at \ t = 0, u = u_0 \tag{9.6}$$

$u(t)$ = concentration of cancer cells in the blood.

a = rate of growth of cancer cells.

Q = the capacity of carrying cancer cells. (We need to remember that cancer cells move in a group in the bloodstream and lymphatic vessels before and after forming a tumor). There are few cases which are known as "miraculous healing" where cancer heals all by itself [Spontaneous Regression of Cancer: Explanations and Predictions, https://www.cancertherapyadvisor. com › general-oncology]. These are very rare.

At the steady state, $du/dt = 0$, and $u = Q$. This is just a speculative information. In the real world, cancer may never stop growing. (It must be noted that a cancerous tumor hardly goes to a steady state. Cancer is dynamic. It grows, metastasizes, damaging the tissues every moment, even when it seems to be stationary. Yet some assume that it is reaching a steady state).

The analytical solution of (9.6) is

$$u = (u_0 Q)/\{u_0 + (Q - u_0)exp(-at)\} \tag{9.7}$$

From (9.6), assuming $a > 0$ as $t \to \infty$, $u \to Q$. (If $a < 0$, u will tend to 0. If Q happens to be the same as the initial value of u, the tumor will be like a benign growth for all times). This is the steady-state solution. Also from (9.7), the steady-state solution is $u = Q$. (Q is obviously positive).

If we now assume that there are three kinds of nutrition, namely ayurvedic medicines [3, 9, 14, 17] given to a patient, then, (9.6) becomes

$$du/dt = au(1 - u/Q) - b_1 u V_1^0 - b_2 u V_2^0 - b_3 u V_3^0 \ at \ t = 0, u = u_0 \tag{9.8}$$

where $b_i = $ *the rate at which u is destroyed by each unit of* $V_i^0 (i = 1, 2, 3)$. These are the rate constants as discussed earlier. It is assumed that V_i^0's $(i = 1, 2, 3)$ are just fixed doses of nutrition used at a given interval.

Equation (9.8) may be expressed as follows:

$$du/dt = (a - b_1 V_1^0 - b_2 V_2^0 - b_3 V_3^0)u - u^2/Q \qquad (9.9)$$

$$du/dt = \alpha u - u^2/Q \qquad (9.10)$$

where $\alpha = (a - b_1 V_1^0 - b_2 V_2^0 - b_3 V_3^0)$

From (9.9), if $\alpha \to 0$, du/dt will be negative and u will be decreasing continuously reaching $u = 0$ in a steady state. That will mark the success of the treatment. In general, nutrition alone may not do this.

Inputs for Fig. 9.4

```
*********************************************************************************
! CHAPTER 9 NUTRITION THERAPY WITH LOGISTIC EQUATION FIG 9.4
! ONE TUMOR & THREE LYMPH NODES ARE AFFECTED
! DU/DT = A1*U*(1-U/Q(N))-A2*U*VO*R2+(NU1)*DELSQ(U)
! AT T=0,   U0 = 10000   AT THE PRIMARY SITE, U = 10000   INITIAL VO= 100
! DX= .02  DY= .02  DZ= .02  DT= .02  NX= 101  NY= 101  NZ = 101  NT= 450
! A1= 2.5  A2= .00001  U0(NX+4,NY+4,NZ+4)= 250  Q(0) = 25000
! NU1= .0000001    GAMMA1= .015  ETA1= .98  ETA2= 1.25  ETA3= .125
! R1= .98  R2= EXP(-ETA2*DT)= .97530991  R3= EXP(ETA3*DT)= 1.0025031  Q(N) = R1*Q(N-1)  U=R2*U  VO=VO*R3= 100.25031
! THREE AFFECTED LYMPH NODES ARE LOCATED AT: ( 3 , 3 , 3 ) ( 20 , 20 , 20 )
! AND ( 35 , 35 , 35 ) AT ALL THE LYMPH NODES U0= 1200
! PRIMARY TUMOR IS LOCATED AT:  ( 13 , 13 , 13 ) UPRIMARY= 10000
! NUTRITION THERAPY DOES THE FOLLOWING :
!
*********************************************************************************
```

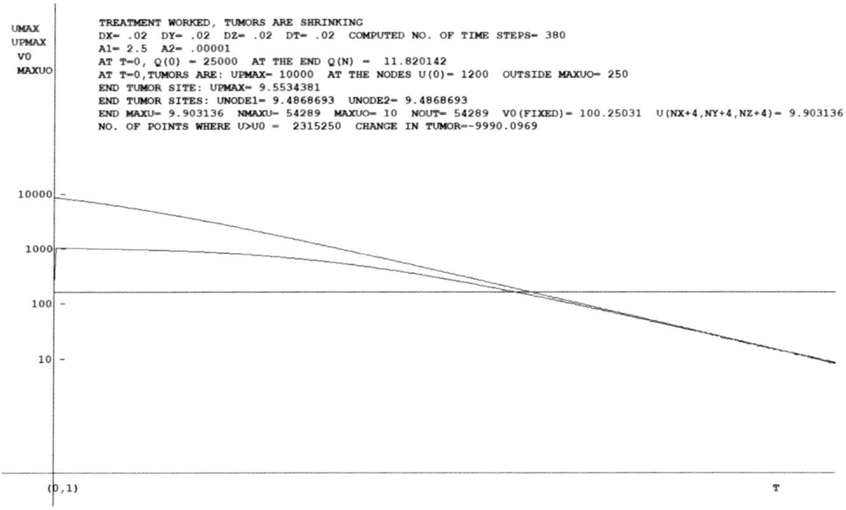

```
UMAX      TREATMENT WORKED, TUMORS ARE SHRINKING
UPMAX     DX= .02  DY= .02  DZ= .02  DT= .02  COMPUTED NO. OF TIME STEPS= 380
VO        A1= 2.5  A2= .00001
MAXUO     AT T=0, Q(0) = 25000  AT THE END Q(N) =  11.820142
          AT T=0,TUMORS ARE: UPMAX= 10000  AT THE NODES U(0)= 1200  OUTSIDE MAXUO= 250
          END TUMOR SITE: UPMAX= 9.5534381
          END TUMOR SITES: UNODE1= 9.4868693  UNODE2= 9.4868693
          END MAXU= 9.903136  NMAXU= 54289  MAXUO= 10  NOUT= 54289  VO(FIXED)= 100.25031  U(NX+4,NY+4,NZ+4)= 9.903136
          NO. OF POINTS WHERE U>U0 =  2315250  CHANGE IN TUMOR=-9990.0969
```

Fig. 9.4 Nutrition therapy applying logistic equation

However, this is only time dependent. So its validity in reality is questionable.

The pattern of lifestyles determines how effective nutrition therapy will be. In this regard in the Ref. [5], the authors have mentioned: "This year, more than 1 million Americans and more than 10 million people worldwide are expected to be diagnosed with cancer, a disease commonly believed to be preventable. Only 5–10% of all cancer cases can be attributed to genetic defects, whereas the remaining 90–95% have their roots in the environment and lifestyle."

Nutritional therapy lasts all life and should be in continuum. And it is indeed a lifelong therapy. It can combat all different kinds of diseases to keep the body fit. It is a true deterrent of both physical and psychological ailments.

Another aspect of eating healthy food is that food is a driver of mind. A sound mental health means reduction of stress. It means less glucocorticoids in blood. It means strengthening the immune system. A rule of thumb is in our body; the number of good normal cells really far outnumbers the bad harmful pathogens by trillions. Although not all cells fight, if they are strong enough, the bad cells like cancer may not be able to affect their genetic structures.

9.3.1 A Sheer Myth

The above discussion is almost a sheer myth, even if we consider only one tumor. It is hardly applicable to malignant tumors. By the time, a tumor stops growing and starts becoming necrotic, other similar ones start growing in abundance elsewhere in the body. Cancerous tumors hardly go into a steady state. So an assumption of the existence of Q is a myth.

So another simplistic model is as follows:

$$du/dt = (a(t) - b(t) - c(t))u, \qquad (9.11)$$

$a(t) = \alpha t, \quad b(t) = \beta t^2, \beta = 1 \ if \ 0 \leq t \leq T \ , \ \beta = 0.01, \ if \ t > T, c(t) = \gamma/t,$
$\gamma = 0.01 \ if \ 0 \leq t < T, \gamma = K, if \ t \geq T.$

Value of K is adjusted as needed. In fact, value of β will also be adjusted periodically. The objective is how to minimize the coefficient of u on the right side of (9.11) or make it negative.

9.4   Statistical Studies on Fat Intake and Breast Cancer [6]

Cancer cells require a lot of energy to reproduce rapidly and grow uncontrolled. "Pound for pound fat has more energy than any other nutrient. So it's perhaps not surprising that when cancer cells find themselves in fat tissue, they make quick use of these resources." [Memorial Sloan Kettering Cancer Center report, June 14, 2018]. Fat in the diet should be reduced.

Any nutritional guide book will be helpful. In this regard, a research has been conducted by Kline [6]. She found out the statistics regarding fat intakes and deaths due to breast cancer for a number of countries.

$x =$ estimated dietary fat intake, grams per day, per person.
$y =$ age-adjusted breast cancer deaths per 10,000 population.

Following this table, we have computed the correlation between fat intake and number of deaths caused by breast cancer in different countries in the world. In our computations, $\bar{x} = average\ fat\ intake = 90.6$; $\bar{y} = average\ number\ of\ deaths\ caused\ by\ breast\ cancer = 12.8$; $\sigma_x = standard\ deviation\ of\ x_i\text{'s} = 37.7$; $\sigma_y = standard\ deviation\ of\ y_i\text{'s} = 7.2$

Formula for correlation coefficient
$r = \left\{\sum_{i=1}^{n}(x_i - \bar{x})(y_i - \bar{y})\right\}/((n-1)\sigma_x\sigma_y) = 0.908$. Equation of the regression lines $y = 0.175x - 3.03$.

$n = 39$. When r $is\ very\ close\ to\ 1, correlation\ is\ very\ strong.$

It means there exists a strong correlation between fat intake and number of deaths due to breast cancer as may be observed from Table 9.1.

The US National Cancer Institute indicates that one of the main ways in which obesity can cause cancer is by promoting chronic low-level inflammation, which can, over time, cause DNA damage which leads to cancer. The Internet is literally flooded with this information.

According to a report in June 2017, from MD Anderson Cancer Center, Houston, Texas: "Visceral fat cells are large, and there are a lot of them. This excess fat doesn't have much room for oxygen. And that low-oxygen environment triggers inflammation." These inflammations cause insulin resistance, finally causing more estrogen which causes faster cell growth. Mathematically it means the following:

If $u=$ a cell of the body, a simple model for cell growth is $du/dt = \lambda u, \lambda > 0$. So, $u = u_0\ exp(\lambda t)$. Larger the value of λ, faster the rate of growth is. It leads to cancer.

This also shows indirectly how excessive sugar intake could lead to cancer. In [7], authors showed an association between dietary fat and breast cancer for women in the USA between the ages 50 and 71.

Table 9.1 Correlation between breast cancer and dietary fat intake

Nation	X	Y	Nation	X	Y
Thailand	24	0.5	Czech	90	14.2
Philippines	31	4.2	Greece	90	7.3
El Salvador	38	0.8	Hungary	95	12.9
Japan	40	3.3	Spain	95	8.1
Taiwan	44	4.0	Finland	112	13.1
Columbia	47	4.6	Austria	115	16.7
Ceylon	47	2.1	Australia	123	18.8
Chile	53	8.5	Norway	125	16.7
Panama	55	7.3	Sweden	125	17.9
Venezuela	60	8.1	Germany	128	17.1
Mexico	60	3.8	Iceland	130	20.8
Bulgaria	65	8.3	Switzerland	132	21.0
Romania	67	7.5	Belgium	135	20.2
Yugoslavia	68	6.7	United Kingdom	137	23.3
Hong Kong	68	9.6	Canada	138	22.3
Portugal	70	12.1	U.S.A	143	20.2
South Africa	75	21.7	Netherlands	147	24.6
Italy	83	15.0	New Zealand	148	22.3
Poland	87	10.0	Denmark	152	22.9
Israel	90	19.6			

x: Estimated dietary fat intake, grams per day per person
y: Age-adjusted breast cancer deaths per 10,000 population
N = 39

$\bar{x} = 90.6$ $\bar{y} = 12.8$ $r = 0.908$
$\sigma_x = 37.7$ $\sigma_y = 7.5$ $y \approx 0.175x - 3.03$

Correlation coefficient = 0.908
With   the   data   randomly   perturbed   by   a   maximum   of $\pm$ 5%:

$\bar{x} = 90.3$ $\bar{y} = 12.8$ $r = 0.906$
$\sigma_x = 37.4$ $\sigma_y = 7.3$ $y \approx 0.177x - 3.14$

9.5 A Mathematical Model on Statistical Studies

Another dangerous chemical is acrylamide. When food is overly exposed to heat, like grilled meat, deep fried chicken and potatoes, acrylamide is the result that goes into the food. "…antioxidant compounds like myrcitrin, genistein, resveratrol, and curcumin have been reported to be effective in the reduction of acrylamide toxicity in cell lines and/or animal studies" [8]. According to the Cancer Treatment Centers of America, October 26, 2017 report, the following food and drinks should be taken by strict moderations:

- Alcohol: When the body metabolizes alcohol, it produces acetaldehyde, a DNA damaging chemical compound, as an output that could cause cancer. Breast cancer and liver cancer are two of them.
- Processed meats: These are bacon, sausage, hot dogs, pepperoni, salami, beef jerky, where too much salt and preservatives are added.
- Meats cooked at high temperatures and/or burnt or charred have carcinogenic chemicals which damage DNA that may lead to cancer.
- Consumption of red meat may lead to cancer. Red meat is dipped in saturated fat. Also, most animals often eat a lot of carcinogenic foods. Finally, some of these go into the bloodstream of people eating such foods.
- Hot beverages, hotter than 149° Fahrenheit.

To look into these facts mathematically, we may look into the following equation:

$$\partial u / \partial t = (r_1 + r_{12} D_0) u - r_{13} V_0 u + \kappa \nabla^2 u \tag{9.12}$$

where u = cancer, D_0 = saturated fat and other carcinogenic foods helping u, and V_0 = the defense.

How strong are the effects of the fat and carcinogenesis are measured by r_1 and r_{12}. So defense must be strong enough to control u. How much strength defense needs will be measured by the rate constant r_{13}.

To inhibit cancer, one necessary requirement is $r_{13} V_0 > (r_1 + r_{12} D_0)$, or in other words, coefficient of u should be negative. In [9], we find a support of our observation. Donoldson stated "It has been estimated that 30–40% of all cancers can be prevented by lifestyle and dietary measures alone. Obesity, nutrient sparse foods such as concentrated sugars and refined flour products that contribute to impaired glucose metabolism (which leads to diabetes), low fiber intake, consumption of red meat, and imbalance of omega 3 and omega 6 fats all contribute to excess cancer risk. Intake of flax seed, especially its lignan fraction, and abundant portions of fruits and vegetables will lower cancer risk. Allium and cruciferous vegetables are especially beneficial, with broccoli sprouts being the densest source of sulforaphane. Protective elements in a cancer prevention diet include selenium, folic acid, vitamin B-12, vitamin D, chlorophyll, and antioxidants such as the carotenoids (α-carotene, β-carotene, lycopene, lutein, cryptoxanthin). Ascorbic acid has limited benefits orally, but could be very beneficial intravenously. Supplementary use of oral digestive enzymes and probiotics also has merit as anticancer dietary measures. When a diet is compiled according to the guidelines here it is likely that there would be at least a 60–70% decrease in breast, colorectal, and prostate cancers, and even a 40–50% decrease in lung cancer, along with similar reductions in cancers at other sites. Such a diet would be conducive to preventing cancer and would favor recovery from cancer as well." At the end the author has stated: "As reviewed above, reductions of 60% in breast cancer rates have already been seen in human diet studies, and a 71% reduction in colon cancer for men without the known modifiable risk factors. These reductions are without taking into account many of the other factors considered in this review, such as markedly increased fruit and vegetable intake, balanced omega 3 and 6 fats, vitamin D, reduced sugar intake, probiotics, and enzymes—factors which all are likely to have an impact on cancer. Certainly cancer prevention would be possible, and cancer reversal in some cases is quite likely."

9.6 Probabilistic Analysis of Diets

Working with several oncodieticians, the first author noticed the practical probabilistic tools that they often use. By conducting such studies, they avoid control groups. Because in dietetics studies, control groups are hard to control. Dieticians certainly do not use the same diet plan for all cancer patients. Their analyses are based upon statistical studies and applications in the real world. We will briefly mention one of these techniques which formed the background of our studies. We consider a very simplistic, yet a practical model.

Let us consider two diet plans: D_1 and D_2. They are disjoint. A dietician used these two plans to two ethnically different groups of patients G_1 and G_2 with similar physical and psychological states of health. These groups are mutually exclusive. They are kept geographically and socially apart.

It must be noted that each group being a distinct group, mutually exclusive, success or failure of a trial should be a unique property of the group.

The objective is to check whether ethnicity could make any significant difference regarding the effectiveness of these diets.

Let D^+ = a diet plan is a success, D^- = A diet plan is a failure. (We have considered two diet plans. These diet plans were used on both groups of patients).

Let G^+ = the group is a success and G^- = the group is a failure.

Please note: The symbol # denotes number of people or number of diet plans.

Each group has N number of patients for each diet plan. So, from Table 9.2, $n_{11}^+ + n_{11}^- = N$, and $n_{21}^+ + n_{21}^- = N$. Noting that those who were treated with both diet plans are mutually exclusive

Table 9.2 Simple model on probabilistic studies on the effectiveness of diet plans

#G_1 worked	#G_1 did not work	#G_2 worked	#G_2 did not work	# of Patients	# of Diet Plans
n_{11}^+	n_{11}^-	n_{21}^+	n_{21}^-	$2N$	$D_1 = Diet\#1$
n_{12}^+	n_{12}^-	n_{22}^+	n_{22}^-	$2N$	$D_2 = Diet\#2$
$n_{11}^+ + n_{12}^+$	$n_{11}^- + n_{12}^-$	$n_{21}^+ + n_{22}^+$	$n_{21}^- + n_{22}^-$	$4N$	Total

$$P\left(D_1^+\right) = P(\text{assuming } D_1 \text{ was used and the patient is a success}) = \left(n_{11}^+ + n_{21}^+\right)/2N$$
$$(9.13)$$

And, considering both plans together

$$P(G_2^+/D_1^+) = P(\text{assuming } D_1 \text{ worked for both groups, the patient is from } G_2)$$
$$= n_{21}^+/\left(n_{11}^+ + n_{21}^+\right) \qquad (9.14)$$

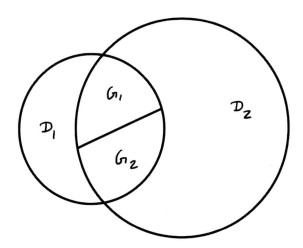

Note: Diet plans are disjoint. Users may use both and their groups are disjoint.

Now it should be noted that all members of the disjoint sets G_1 and G_2 used both D_1 and D_2. So,

$$P(D_1/G_1) = P(D_2/G_1) = P(D_1/G_2) = P(D_2/G_2) = 1 \qquad (9.15)$$

However, considering both groups together, $(j = 1, 2)$

$$P\left(\text{The Diet Plan } D_j \text{ is a Success}\right) = \left(n_{1j}^+ + n_{2j}^+\right)/2N \qquad (9.16)$$

Now let us assume that there are J number of diet plans and I number of groups of participants.

If in each group there are N number of participants, then there are $(I \times N)$ number of participants for each diet plan. So the total number of trials is $(I \times J \times N)$. Let $\mathbf{N} = (\mathbf{I} \times \mathbf{J} \times \mathbf{N})$.

If we chose a participant from the ith group arbitrarily, the probability that the jth diet plan worked for him is $(n_{ij}^+/\mathbf{N})$. The question is now: Find out for a given D_j^+, what is the probability that it has worked for a group G_i. So, we look for a given j:

$P\left(G_i^+/D_j^+\right) = n_{ij}^+/\left(\sum_{i=1}^{I} n_{ij}^+\right)$. Therefore, the best group on which D_j^+ is given

by $max_i\left(n_{ij}^+/\left(\sum_{i=1}^{I} n_{ij}^+\right)\right)$

By the same token, we need to find out what is the probability that for a given G_i^+ the diet plan D_j has worked best. So we need $max_j\left(P\left(D_j^+/G_i^+\right)\right)$ where $P\left(D_j^+/G_i^+\right) = n_{ij}^+/\left(\sum_{j=1}^{J} n_{ij}^+\right)$.

These should be long term studies. However, in such studies short term trials may be significant. Here we have not done any comparative study on various diet plans for a particular group of patients or comparative studies of groups for a particular diet plan.

9.7   Conclusion

Nutrition is the key to good health, which has been known since the beginning of civilization. I remember from a TV serial: "You are what you eat". All around the world, scientists are now pushing for better food habits as a key inhibitor of cancer. It is highly needed for cancer patients too. Let us see what the National Cancer Research Institute and the American Cancer Society state in this regard:

National Cancer Research Institute [10] guidelines state the following: (1) Good nutrition is important for cancer patients. (2) Healthy eating habits are important during and after cancer treatment. (3) A registered dietitian is an important part of the healthcare team. Cancer and cancer treatments may cause side effects that affect nutrition. (4) Cancer and cancer treatments may cause malnutrition. (5) Anorexia and cachexia are common causes of malnutrition in cancer patients.

The background of these guidelines is inevitably Statistics.

The American Cancer Society states [11].

"When you're healthy, eating enough food to get the nutrients and calories you need is not usually a problem. Most nutrition guidelines stress eating lots of vegetables, fruits, and whole-grain products; limiting the amount of red meat you eat, especially meats that are processed or high in fat; cutting back on fat, sugar, alcohol, and salt; and staying at a healthy weight. But when you're being treated for cancer, these things can be hard to do, especially if you have side effects or just don't feel well. Good nutrition is especially important if you have cancer because both the illness and its treatments can change the way you eat. They can also affect the way your body tolerates certain foods and uses nutrients. During cancer treatment you might need to change your diet to help build up your strength and withstand the effects of the cancer and its treatment. This may mean eating things that aren't normally recommended when you are in good health. For instance, you might need high-fat, high-calorie foods to keep up your weight, or thick, cool foods like ice cream or milk shakes because sores in your mouth and throat are making it hard to eat anything."

World Cancer Research [12] stated that: There is limited evidence that: (1) Consuming non-starchy vegetables might decrease the risk of oestrogen-receptor-negative (ER–) breast cancer (unspecified), (2) consuming foods containing carotenoids might decrease the risk of breast cancer (unspecified), (3) consuming diets high in calcium might decrease the risk of postmenopausal breast cancer.

All this information is based on statistical studies. Nutritional therapy [21] is the best when a person is not sick and applies that to his/her life.

References

1. Pal, D., Banerjee, S., Ghosh, A. K.: Dietary-induced cancer prevention: an expanding research arena of emerging diets related to the healthcare system. J. Adv. Pharm. Technol. Res. 3(#1) 2012
2. Nicholson, L.B.: The immune system. Essays Biochem. 60, 275–301 (2016)
3. https://food.ndtv.com/health/power-of-ayurveda-6-herbs-that-can-prevent-risk-of-cancer
4. https://stanfordhealthcare.org/medical-clinics/cancer-nutrition-services/during-cancer-treatment/nutrition-during-chemo.html
5. Anand, P. et.al.: Cancer is a preventable disease that requires major lifestyle changes. Pharm. Res. 25(9) (September 2008)
6. Kline, K.: Nutrition today (May/June 1986)
7. Khodarahmi, M., Azadbakht, L.: The association between different kinds of fat intake and breast cancer risk in women. Int. J. Prev. Med. 5(1), 6–15 (Jan 2014)
8. Kahkeshani, N., Saeidnia, S., Abdollahi, M.: Role of antioxidants and phytochemicals on acrylamide mitigation from food and reducing its toxicity. J. Food Sci. Technol. 52(6) (June 2015)
9. Donaldson, M.S.: Nutrition and cancer: a review of the evidence for an anti-cancer diet. Nutr. J. (October 2004)
10. https://www.cancer.gov/about-cancer/treatment/side-effects/appetite-loss/nutrition-pdq
11. https://www.cancer.org/content/dam/cancer-org/cancer-control/en/booklets-flyers/nutrition-for-the-patient-with-cancer-during-treatment.pdf
12. https://www.wcrf.org/dietandcancer/breast-cancer
13. Derossi, A., Husain, A., Caporizzi, R., Severini, C.: Manufacturing personalized food for people's uniqueness. An overview from traditional to emerging technologies. Crit. Rev. Food Sci. Nutr. 60(7) 2020
14. Smit, H.F., Woerdenbag, H.J., Singh, R.H., Meulenbeld, G.J., Labadie, R.P., Zwaving, J.H.: Ayurvedic herbal drugs with possible cytostatic activity. J. Ethnopharmacol.47, 75–84 (1995). [PubMed] [Google Scholar]
15. Patel, B., Das, S., Prakash, R., Yasir, M.: Natural bioactive compound with anticancer potential. Int. J. Adv. Pharmaceut. Sci. 1, 32–41 (2010). [Google Scholar]
16. Mathematical Models for Cancer Growth. https://chemoth.com/tumorgrowth
17. Key, T.J. et al.: Diet, nutrition and the prevention of cancer. Review Public Health Nutrition (Feb.2004)
18. National Cancer Institute. A Statistical Study on Fat Intake & Breast Cancer
19. Jang, H.H.: Regulation of protein degradation by proteasomes in cancer. J. Cancer Prev. 23(4) (December 2018)
20. Science News: Halting protein degradation may contribute to new cancer treatment (May 31, 2016)
21. Pence, B.C., Dunn, D.M.: Nutrition & women's cancer. CRC Press LLC. Boca Ratan, USA (1998)
22. MD Anderson Cancer center: How does obesity cause cancer? (June 2017)

Chapter 10
The Fateful Code and The Future Course

Abstract We will introduce a new model on growth, dispersion, and movements of cancer cells as they swim through blood, lymph, and interstitial fluid. We have derived one fourth-order, one second-order, and a first-order partial differential equations. They are linear and as such we have solved them analytically. We found that all these equations share the same solution revealing biophysics of cancer.

10.1 Rationale

Biological science and physical science are very different. Often we make mistakes by imposing the laws of physical science in biology and make predictions accordingly. In the previous chapters, we have imposed the reaction–dispersion law, Fick's law, and the viscosity law of fluid dynamics and derived certain conclusions which have many validities with regard to the experimental studies of the treatment of cancer.

A particle in physics or a chemical in chemistry is not the same as a cell in our body driven by DNA. It is very tiny, just a few microns in diameter. One micron is about 10^{-6} m, which is one thousandth of a millimeter or 1/25th of a thousandth of an inch. Yet it has a mind of its own and can perform many jobs defying the laws of physical science which still baffle many Nobel Laureates. Within exactly the same tumor, all the cancer cells do not follow the same law. Some decide to stick to the tumor. Some decide to leave. Some could grow differently than others. Being lawless is their only law. In this regard, we will look into some of our computational results first. Then, we will make an attempt to develop some mathematical models to review our previous studies.

An Observation: Erratic Behavior of Metastasization

"Cancer can occur anywhere in the body and cause "almost any" symptom, according to the American Cancer Society. Sometimes these symptoms are impossible to ignore. But sometimes they're vague and all too easy to brush aside or miss completely." [insider.com/subtle-signs-of-cancer-2018].

© Springer Nature Singapore Pte Ltd. 2021   299
S. Dey and C. Dey, *Mathematical and Computational Studies on Progress, Prognosis,*
Prevention & Panacea of Breast Cancer, Forum for Interdisciplinary Mathematics,
https://doi.org/10.1007/978-981-16-6077-1_10

Statistical studies on biological science are also incomplete because regardless how vast is the field of data, conclusions drawn by professional medical statisticians could be widely different. The reason is these data do not have the same homogeneous basis or pattern. Biological phenomena have some psychological backgrounds which are often so subtle, so inconspicuous, and so illusive that even experts could not comprehend or detect. So scientists get baffled.

"A cell accumulates mistakes in its DNA, which cause a gene or a set of genes to go awry. And because cells have safety mechanisms that work to stop cells growing and dividing more than they should, multiple faults need to appear before a cell tips over the edge and becomes cancerous" [11]. Then it could grow exponentially. Several researchers state this fact. However, an exponential growth does not always imply a very fast growing cell. Based upon what we have experienced working with groups of oncologists, sometimes the rate of growth curve is

$$f(t) = a \exp(t - k) \tag{10.1}$$

where k = a large number and a is a constant. They vary from patient to patient. At $t = 0, f(t)$ or growth could be practically undetectable, although it will increase.

The growth is exponential, but initially it is not fast at all. In fact, very slow, unnoticeable till, $t > k$. Unfortunately, k often varies with time, and it could even change from one group of cancer cells to the other.

So for several years, there could be absolutely no symptoms and suddenly cancer appears with a deadly blow. These are known as asymptotic cancers. They have no noticeable symptoms. The patient appears to be completely healthy. The first author observed several such cases. And that was the primary incentive to write this monograph.

So the danger of cancer is that it could stay totally invisible for months, maybe for years and appear all on a sudden with the most deadly strength, and becomes a killer within a few months. There are many cases like this. Some oncologists, headed by Dr. Charles Roper, from the Barnes Hospital at St. Louis told the first author that in a scar tissue in the lungs or in an adipose tissue, cancer cells can stay alive for twenty-five years showing no symptom, and later they could attack the body with a full force. It is that unpredictable.

So all precautionary steps must be taken like following a healthy lifestyle all life.

We will now study a case where cancer remained silently latent for years and suddenly took a sharp upturn. This chapter is enriched by information that we have gathered from each reference [1–14]. Especially references [10, 11, 13, 14] gave us many current discussions on various aspects of cancer.

10.2 The Fateful Code

Note: **This particular section describing detailed mathematical modeling relates vividly a true story of an unfortunate patient at every step of her stages of progression of cancer, her hopes and her frustrations and finally her sad demise.** Let us consider the model simulating the health condition of the patient who possibly developed a poor immune system during the course of treatment for about two years. Her cancer suddenly showed up and took a formidable turn. We simulated that very elaborately here.

We have seen such cases where the patient had absolutely no symptoms of any ailment for years. All medical tests were normal. There were no medically noticeable aberrations. None at all. Suddenly, almost like a bolt from the blue, cancer appeared when it had already metastasized at different parts of the body. At that time, all attempts to rescue the patient failed.

Let us consider the following model:

$$\partial u/\partial t = a_1(t)u - a_2 u V_0 + k\nabla^2 u \tag{10.2}$$

$a_1(t)$ = the variable rate of growth of cancer cells u. V_0 = the immune system attacking cancer per unit of time per unit u.

Let at $t = 0$, $a_1(0) = 0.05$, $a_2 = 0.00005$, $V_0 = 2000$. There is a primary tumor where, $u(t_0) = 5000$.

The value of a_2 indicates that the strength of defense is weak. We have assumed that three lymph nodes are also affected. At these three lymph nodes $u(t_0) = 200$, elsewhere in the body, $u(t_0) = 25$. We have assumed that the rate of growth of cancer is very slowly increasing with time and the defense becoming weaker with aging or other factors. These assumptions are mathematically described as follows.

$da_1/dt = \xi_1 a_1(t)$, and $V_0(n) = \xi_2 V_0(t_0)$, We chose $\xi_1 = 0.75$ and $\xi_2 = 0.99$. So V_0 is getting weaker and weaker very slowly. Here delta $t = 0.02$.

Integrating for $a_1(t)$ from t_{n-1} to t_n, we get $a_1(t_n) = \exp(\xi_1 \Delta t)a_1(t_{n-1})$. That gives $a_1(t_n) = 1.015 a_1(t_{n-1})$.

This is a slow increase in the rate of growth.

Inputs for Fig. 10.1.

```
!*****************************************************************
! CHAPTER 10 SLOW GROWING UNTREATED TUMOR & SAD END FIG 10.1
! THERE IS ONE PRIMARY TUMOR & THREE LYMPH NODES ARE AFFECTED
! DU/DT = A1(N)*U-A2*U*VO*R2+(NU1)*DELSQ(U)
! AT T=0,   UO = 5000   AT THE PRIMARY SITE, U = 5000
! DX= .02   DY= .02   DZ= .02   DT= .02   NX= 101   NY= 101   NZ = 101   NT= 300
! A1(0)= .05   A2= .00005   UO(NX+4,NY+4,NZ+4)= 25   VO= 2000
! NU1= .0000001   GAMMA1= .015
! THREE AFFECTED LYMPH NODES ARE LOCATED AT: ( 3 , 3 , 3 ) ( 20 , 20 , 20 )
! AND ( 35 , 35 , 35 ) AT ALL THE LYMPH NODES UO= 200
! PRIMARY TUMOR IS LOCATED AT:   ( 13 , 13 , 13 ) UPRIMARY= 5000
! TUMOR STARTED GROWING SILENTLY AND ITS RATE OF GROWTH INCREASES SLOWLY. :
!
!*****************************************************************
```

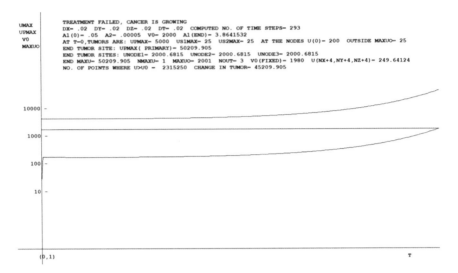

Fig. 10.1 Untreated tumor growth at the beginning

After 293 time steps, we saw Fig. 10.1. Cancer was increasing very slowly, hard to notice. Then around the time step 280, it jumped up very fast. This is certainly not an exceptional case. Cancer often goes for years completely unnoticed. The rate of growth has increased from $a_1(0) = 0.05$ to $a_1(n = 293) = 3.8642$. Outside the sites of the tumor, u has grown from $u = 100$ to u(outside, $at\ n = 293) = 249.64$, and cancer started increasing very sharply, both at the tumor site (redline) and outside the tumor (black line).

At first, it did not show any danger whatsoever and appeared to be very benign. But actually, it is life threatening. That is why doctors always try to keep track of any tumor for several years trying to notice any changes. Treatments are adjusted accordingly.

Even a simple cyst should be under observation.

Figure 10.1 bears the testimony that no abnormality could be accepted. Because at any moment, it could pose a threat to life. However, the body could correct it. Decisions should be left up to an expert, an experienced medical professional.

At this particular stage, we have looked into, if this was a small tumor and that it was caught by MRIs, what could have saved this precious life.

It is interesting to notice that at the very beginning, the spread of cancer was computed, and at every point of the computational field (105 X 105 X 105), cancer grew so that the total number of points was 1157625 (everywhere in the field). At this time, those points are again infected, meaning cancer grew at all those points again (causing the field to get twice bigger as it there are 2315250 number of points in the field of computation). The entire body is infected, and this overwhelming attack of cancer is gaining ground. It could become a reality if a cell has doubled its mass.

At this point, we decided to continue the code and see whether there is any way to halt the deadly motion of cancer. So we continued to the Sect. 10.2.

10.3   Cancer Subsides, Yet It Could be Life Threatening

When the growth of cancer suddenly gets its momentum to destroy the body, both patients and their families become totally bewildered and scared. Even at this point, it could happen that many standard tests to check the nature of the disease could be negative one after another. Even MRIs could be inconclusive and angiograms could be negative. The first author experienced all such cases where one of the most famous neurosurgeons did brain microsurgery because the patient had continuous seizures (for brain metastasis) and found nothing. Later when in a month a cancerous lesion was caught in the lungs and the brain surgery was done again, a golf ball size tumor was found hiding behind the cerebellum. At that point, chemos started to delay the growth rate and keep the cancer contained. Those chemos were not tolerated by the body. The sad end came slowly.

So we did not stop the code at $t_n = 293$ in the previous section. We continued to see states of the exponential growth of tumor. These time medications were added to reduce the deadly activities of cancer, and cancer was contained.

Several doctors from several hospitals cooperated and coordinated with each other conducting almost daily briefings about the health of the patient and the progress of the treatments. Families were duly informed.

So we ran the same code with a few modifications reversing the previous procedure and noticed what the outcomes were. With these ideas in view, we chose $a_1(t_0) = 3.8641532$, $u(t_0) = 50209.905$ and $V_0 = 2000$, and outside the regions of cancer, $u_0 = 249.6815$. These are the end values in the previous run as may be seen in Fig. 10.3.

Rate of growth of cancer is going down, yet cancer is on the rise.

```
****************************************************************************************************
! CHAPTER 10 INPUTS OF FIG.10.1 ARE REVERSED IN THE FIG 10.2A
! THERE IS ONE PRIMARY TUMOR & THREE LYMPH NODES ARE AFFECTED
! DU/DT  =  A1(N)*U-A2*U*VO*R2+(NU1)*DELSQ(U)
! AT T=0,   U0 = 50209.905  AT THE PRIMARY SITE,  U = 50209.905
! DX= .02   DY= .02   DZ= .02   DT= .02   NX= 101   NY= 101   NZ = 101   NT= 4000
! A1(0)= 3.8641532 -THE RATE OF GROWTH WHEN IT WAS DETECTED.
! A2= .00005   U0(NX+4,NY+4,NZ+4)= 249.6815   VO= 2000   NU1= .0000001   GAMMA1= .015
! THREE AFFECTED LYMPH NODES ARE LOCATED AT: ( 3 , 3 , 3 ) ( 20 , 20 , 20 )
! AND ( 35 , 35 , 35 ) AT ALL THE LYMPH NODES U0= 2000.6815  AT THE TIME OF DETECTION.
! PRIMARY TUMOR IS LOCATED AT:   ( 13 , 13 , 13 ) UPRIMARY= 50209.905  AT THE TIME OF DETECTION.
! TUMOR FIRST INCREASED THEN STARTED SHRINKING SHARPLY WITH A CHANGE OF LIFESTYLE AFTER IT WAS DETECTED.
!
****************************************************************************************************
```

We assumed that the patient has started living a more well-planned health conscious life. And a monoclonal drug has been used to reduce the rate of growth of cancer. So the rate of growth of cancer is reversed. Here $\xi_1 = -0.75$ and $da_1/dt = \xi_1 a_1(t)$. So $a_1(t_n) = (1/1.015) a_1(t_{n-1})$ giving

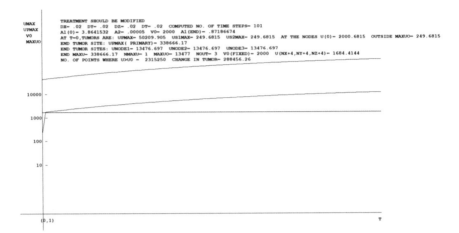

Fig. 10.2 Inputs are collected from the end results from Fig. 10.1. Cancer first started increasing sharply. (The computed value of A1 from Fig. 10.1 has been used at the start)

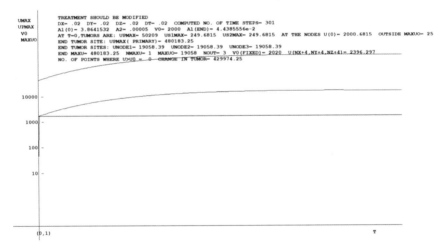

Fig. 10.3 Took a long computer run. NT $= 301$. (In Fig. 10.2, NT $= 101$). Results look promising

$a_1(t_n) = 0.985221675a_1(t_{n-1})$. So the rate of growth of cancer is slowly decreasing.

At this point, we decided to take a long computer run to find out what could be the longtime effect on the body.

At first, after the 101 time step run, we see cancer is still on the rise and is rising sharply with time. From $u(t_0) = 50209.905$, it has gone up to $u(t_{101}) = 338666.17$, a huge increase, although the rate of growth has gone down from $a_1(t_0) = 3.8641532$ to $a_1(t_{101}) = 0.87186674$ which is significantly lower than its initial value. Another quite disappointing result is that $u_{101} > u_{100}$ at 2315250 number of points in the entire computational field. Essentially when this code was started, $u_2 > u_1$ at 1157652 (=2315250/2) number of points, meaning in the entire field of computation, and then again it started increasing when the code was started again. So at each point in the computational field, cancer started increasing. Also all over the body, the $u_{101} = 1684.4144$ which means it has jumped up significantly from its initial value $u_0 = 249.6815$. This is certainly a very grim picture.

In the second run, (the second Fig. 10.2) after 301 time steps, $u(t)$ appears to be on a steady state (under control), with $u(t_{301}) = 480183.66$ and $a_1(t_{301}) = 0.044385556$, decrease of the rate of growth obtained by medications. One good point is that at no point in the field, $u_n > u_{n-1}$, (Fig. 10.3, No. of time steps $= 301$). So we decided to continue the code up to 4001 time steps and see the result (Fig. 10.4). The growth of cancer cells outside the tumor site went to a steady state, meaning cancer cells are not gaining any more strength.

Tumor is shrinking steadily. The value of cancer cells has gone down to 12278.642 from its initial value of 50209.905. The metastasization process has been successfully inhibited. Slopes of *UMAX* in the field and (*MAXUO*) outside the cancer sites are all strongly negative.

This shows that mathematically cancer is not only contained, in fact tumors are shrinking quite rapidly. However, the reality was that the patient did not survive. Body was too weak to deal with all the medications for such a long time.

So the glory of the theory ended in the gloom of reality. Science and mathematics failed in reality, although successful in theory. Such a case did happen as experienced by the first author.

This grim reality strongly suggests that even when the cancer is contained, survival of a patient may not happen.

At the outset, we have stated that we will attempt to keep cancer contained like asthma, diabetes, or hypertension. Here we see that cancer does not really fall into that category, although, mathematically, it does.

However, the stark truth is that no mathematical modeling in biological science is perfect or even complete. There could be a large set of unknown parameters which could make a significant difference when it comes to reality.

We are simply saying that in many cases like the one which we have tried to model mathematically in this section, there could be a person who may look completely normal, yet who may already be a victim of cruel cancer which could come into light at any time when it is practically incurable. In such a case, the very least thing that he could do to deter cancer is following a truly healthy lifestyle and following the guidelines of the doctors.

A Special Note

The following section has been taken out from the Chap. 5 and inserted here because it seems to be more meaningful here with regard to our discussions. Please consider this section as a section by itself. We did not change the figure numbers, etc. Because these cosmetic changes require a large amount of computer time.

10.4 Growth Factor Reduction is not Enough (A Continuation of the Treatment of the Same Patient)

Considering only a time-dependent model, it could be seen that cancer is contained and cured if the rate of growth is being continuously reduced. In the real world, this is not at all true. Cancer may not grow anywhere, still it could be life threatening.

10.4.1 An Example: Difference Between Only Time Dependent and Both Time and Space-Dependent Models

Let us consider a treatment model which is only time dependent.

$$du/dt = a_1 u - a_2 V u \tag{10.3}$$

where u = number of cancer cells, a_1 = rate of growth of cancer, and a_2 = rate of destruction of u by each unit of V. Obviously, $a_2 > 0$. Then, if $a_1 = 0$, $u \to 0$ as $t \to \infty$. This conclusion is completely invalid in a three-dimensional model. Let us look into that. Let us consider the model given by Eq. (10.2), which is both time and space dependent, where $a_1(t)$ is diminishing continuously as given by the formula:

$$a_1(t = t_n) = 0.25 \, a_1(t = t_{n-1}).$$

This formula has been chosen arbitrarily, to explain one possible aspect of cancer treatment by monoclonal drugs.

For simplicity, we have considered that the immune response is zero. This will not make it a very special case. Because the immune response for some patients could be negligibly small because of the state of health.

Inputs for Fig. 10.3

```
*****************************************************************************************
! TREATMENT WITH MONOCLONAL DRUG THAT JUST REDUCES THE RATE OF GROWTH OF CANCER. FIG.5.1H
! DU/DT = A1(N)*U-A2*U*V + NU1*DELSQ(U) ! ANTIGEN A1(N) = 0.25*A1(N-1); IMMUNE RESPONSE IS INACTIVE, V=0
! DX= .02  DY= .02  DZ= .02  DT= .02  NX= 101  NY= 101  NZ = 101  NT= 220
! A1(N)= 4.5  A2= 0  B1= 0  B2= 0    NU1= .0000001  NU2= .0000001
! VALUE OF U0 OUTSIDE THE TUMORS, OUTSIDE THE LYMPH NODES= 250
! THREE LYMPH NODES AT AT: ( 3 , 3 , 3 ) AND  ( 35 , 35 , 35 )
! THIRD LYMPH NODE IS AT ( 42 , 42 , 42 ) AT EACH LYMPH NODE U= 1200
! THREE TUMORS AT:  ( 13 , 13 , 13 ) WHERE U0= 10000  AT ( 26 , 26 , 26 ) WHERE U0= 5000
! AND AT ( 47 , 47 , 47 ) WHERE U0= 1000
!
*****************************************************************************************
```

There are three tumors located at (13, 13, 13) to (16, 16, 16), (26, 26, 26) to (29, 29, 29), and (47, 47, 47) to (50, 50, 50) having 10000, 5000, and 1000 cancer cells at each grid point and three affected lymph nodes at (3, 3, 3), (35, 35, 35), and (42, 42,

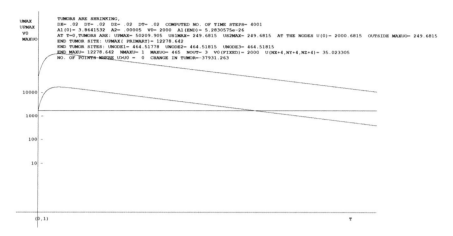

Fig. 10.4 Cancer subsided after a long time. Patient did not survive. Mathematical success is a failure in Reality

42), respectively, having 1200 cancer cells at each of them. The mass of cancer cells outside them is $u = 250$ at each point in the computational field. We will now look into the actual computations. Here, the immune system is non-functional. Drugs are used to reduce growths of cancer all over the body (Table 10.1).

Table 10.1 Figure 10.3: growth factor of cancer is getting smaller, yet the tumor is growing!!! (Please disregard Fig. 5.1H)

```
*******************************************************************************************
SYSTEM TIME =12:47:44 INTRAVENOUS AT TIME = .04   TIME STEP =  2  NT= 220  A1(N)= 1.125  FIG.5.1H
AT  42 , 42 , 42  MAXUO(N)= 1214 LARGEST USTAR= 10114
AT  15 , 15 , 15  UMAX(N)= 10113.557   CHANGE OF: UMAX= 113.55681
TOTAL NO.OF POINTS WHERE: U IS MAX=  8
AT THE PRIMARY SITE: UMAX= 10113.557
AT THE SECONDARY SITE: US1MAX= 5056.7784
AT THE SECOND SECONDARY SITE: US2MAX= 1011.3557
U AT THREE TUMORS:  USTAR1= 10113  USTAR2= 5057  USTAR3 =  1011
UNODE1= 1213.6123  UNODE1-1= 252.83892  UNODE2= 1213.6123  UNODE2+1= 252.83892
UNODE3= 1213.6123  UNODE3-1= 252.83892  U(NX+3,NY+3,NZ+3)= 252.83892  U(NX+4,NY+4,NZ+4)= 252.83892
MAX U OUTSIDE THE THREE TUMORS = 1214  AT NUMBER OF POINTS= 3
NO.OF POINTS WHERE U(N)>U(N-1)=  1157625  MAX.CHANGE FROM THE START = UOMAX - UMAX=-113.55681
*******************************************************************************************
SYSTEM TIME =12:47:59 INTRAVENOUS AT TIME = .06   TIME STEP =  3  NT= 220  A1(N)= .28125  FIG.5.1H
AT  42 , 42 , 42  MAXUO(N)= 1217 LARGEST USTAR= 10143
AT  15 , 15 , 15  UMAX(N)= 10142.512   CHANGE OF: UMAX= 28.955516
TOTAL NO.OF POINTS WHERE: U IS MAX=  8
AT THE PRIMARY SITE: UMAX= 10142.512
AT THE SECONDARY SITE: US1MAX= 5071.2562
AT THE SECOND SECONDARY SITE: US2MAX= 1013.8501
U AT THREE TUMORS:  USTAR1= 10141  USTAR2= 5071  USTAR3 =  1014
UNODE1= 1217.0191  UNODE1-1= 253.5517  UNODE2= 1217.0191  UNODE2+1= 253.5517
UNODE3= 1217.0191  UNODE3-1= 253.5517  U(NX+3,NY+3,NZ+3)= 253.5517  U(NX+4,NY+4,NZ+4)= 253.5517
MAX U OUTSIDE THE THREE TUMORS = 1217  AT NUMBER OF POINTS= 3
NO.OF POINTS WHERE U(N)>U(N-1)=  1157625  MAX.CHANGE FROM THE START = UOMAX - UMAX=-142.51233
*******************************************************************************************
SYSTEM TIME =12:48:14 INTRAVENOUS AT TIME = .08   TIME STEP =  4  NT= 220  A1(N)= .0703125  FIG.5.1H
AT  42 , 42 , 42  MAXUO(N)= 1218 LARGEST USTAR= 10150
AT  15 , 15 , 15  UMAX(N)= 10150.136   CHANGE OF: UMAX= 7.6236407
TOTAL NO.OF POINTS WHERE: U IS MAX=  8
AT THE PRIMARY SITE: UMAX= 10150.136
AT THE SECONDARY SITE: US1MAX= 5074.5676
AT THE SECOND SECONDARY SITE: US2MAX= 1014.7134
U AT THREE TUMORS:  USTAR1= 10148  USTAR2= 5075  USTAR3 =  1015
UNODE1= 1217.8609  UNODE1-1= 253.73008  UNODE2= 1217.8609  UNODE2+1= 253.73008
UNODE3= 1217.8609  UNODE3-1= 253.73008  U(NX+3,NY+3,NZ+3)= 253.73008  U(NX+4,NY+4,NZ+4)= 253.73008
MAX U OUTSIDE THE THREE TUMORS = 1218  AT NUMBER OF POINTS= 3
NO.OF POINTS WHERE U(N)>U(N-1)=  1157625  MAX.CHANGE FROM THE START = UOMAX - UMAX=-150.13597
```

At the very outset $t_n = 2$, the max value of U ($UMAX$ in the field) has jumped up by 113.56, and cancer cells started spreading faster to metastasize. At 1157625 points, all over the field they started increasing.

That is the total number of points includes the boundary points.

$$(NX + 4) \times (NY + 4) \times (NZ + 4) = 105 \times 105 \times 105 = 1157625$$

Clearly the rate of growth of cancer $a_1(t)$ is continuously decreasing. Its initial value was 4.5, at $t_n = 4$.

It has gone down to 0.0703125. Yet U increased to 10150 at the primary tumor site (15, 15, 15), meaning the tumor is still growing. Also the same is happening at the affected nodes as given by the values of $UNODE1$, $UNODE2$, $UNODE3$ ($at\ t = 0$, they were 1200 each.) Outside all the sites, U was 250. That has become 253. That means cancer keeps on growing, even though its rates of growth are decreasing.

Another noticeable aspect is as follows: At time step $t_n = 4$, $a_1(t_n) = 0.070315$ and

$$UMAX \text{ in the entire field} = U0\ MAX\,(\max\ U\ at\ t = 0) + 150.13597$$

From Table 10.2 and Fig. 10.3, let us notice what happened at $t_n = 250$: $a_1(t_n) = 5.5 \times 10^{-150}$ and

$$UMAX \text{ in the entire field} = U0\ MAX\,(\max\ U\ at\ t = 0) + 152$$

Table 10.2 Figure 10.3: rate of growth of cancer is ZERO. Still cancer is growing!!! (Fig. 5.1H in the computer output is an error)

```
*********************************************************************************************
 SYSTEM TIME -12:46:01 INTRAVENOUS AT TIME - 4.96  TIME STEP - 248  NT- 500  A1(N)- 8.7982167e-149  FIG.5.1H
 AT  42 , 42 , 42  MAXUO(N)- 1215 LARGEST USTAR- 10152
 AT  15 , 15 , 15  UMAX(N)- 10152.  CHANGE OF: UMAX- 0
 TOTAL NO.OF POINTS WHERE: U IS MAX-  8
 AT THE PRIMARY SITE: UMAX- 10152.
 AT THE SECONDARY SITE: US1MAX- 5076.
 AT THE SECOND SECONDARY SITE: US2MAX- 1015.
 U AT THREE TUMORS:  USTAR1- 10150  USTAR2- 5076  USTAR3 -  1015
 UNODE1- 1214.6246  UNODE1-1- 253.78956  UNODE2- 1214.6244  UNODE2+1- 253.78956
 UNODE3- 1214.6244  UNODE3-1- 253.78956  U(NX+3,NY+3,NZ+3)- 253.78956  U(NX+4,NY+4,NZ+4)- 253.78956
 MAX U OUTSIDE THE THREE TUMORS - 1215  AT NUMBER OF POINTS- 3
 NO.OF POINTS WHERE U(N)>U(N-1)-  2506  MAX.CHANGE FROM THE START - UOMAX - UMAX--151.99999
*********************************************************************************************
 SYSTEM TIME -12:46:16 INTRAVENOUS AT TIME - 4.98  TIME STEP - 249  NT- 500  A1(N)- 2.1995542e-149  FIG.5.1H
 AT  42 , 42 , 42  MAXUO(N)- 1215 LARGEST USTAR- 10152
 AT  15 , 15 , 15  UMAX(N)- 10152.  CHANGE OF: UMAX- 0
 TOTAL NO.OF POINTS WHERE: U IS MAX-  8
 AT THE PRIMARY SITE: UMAX- 10152.
 AT THE SECONDARY SITE: US1MAX- 5076.
 AT THE SECOND SECONDARY SITE: US2MAX- 1015.
 U AT THREE TUMORS:  USTAR1- 10150  USTAR2- 5076  USTAR3 -  1015
 UNODE1- 1214.6102  UNODE1-1- 253.78956  UNODE2- 1214.61  UNODE2+1- 253.78956
 UNODE3- 1214.61  UNODE3-1- 253.78956  U(NX+3,NY+3,NZ+3)- 253.78956  U(NX+4,NY+4,NZ+4)- 253.78956
 MAX U OUTSIDE THE THREE TUMORS - 1215  AT NUMBER OF POINTS- 3
 NO.OF POINTS WHERE U(N)>U(N-1)-  2506  MAX.CHANGE FROM THE START - UOMAX - UMAX--151.99999
*********************************************************************************************
 SYSTEM TIME -12:46:31 INTRAVENOUS AT TIME - 5  TIME STEP - 250  NT- 500  A1(N)- 5.4988855e-150  FIG.5.1H
 AT  42 , 42 , 42  MAXUO(N)- 1215 LARGEST USTAR- 10152
 AT  15 , 15 , 15  UMAX(N)- 10152.  CHANGE OF: UMAX- 0
 TOTAL NO.OF POINTS WHERE: U IS MAX-  8
 AT THE PRIMARY SITE: UMAX- 10152.
 AT THE SECONDARY SITE: US1MAX- 5076.
 AT THE SECOND SECONDARY SITE: US2MAX- 1015.
 U AT THREE TUMORS:  USTAR1- 10150  USTAR2- 5076  USTAR3 -  1015
 UNODE1- 1214.5958  UNODE1-1- 253.78956  UNODE2- 1214.5956  UNODE2+1- 253.78956
 UNODE3- 1214.5956  UNODE3-1- 253.78956  U(NX+3,NY+3,NZ+3)- 253.78956  U(NX+4,NY+4,NZ+4)- 253.78956
 MAX U OUTSIDE THE THREE TUMORS - 1215  AT NUMBER OF POINTS- 3
 NO.OF POINTS WHERE U(N)>U(N-1)-  2506  MAX.CHANGE FROM THE START - UOMAX - UMAX--151.99999
```

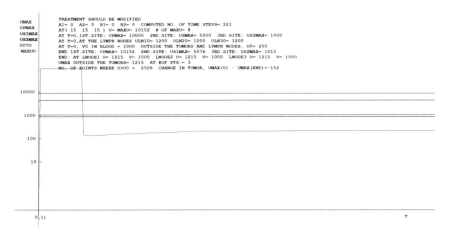

Fig. 10.5 Monoclonal drug reduces only the rate of growth of cancer and defense is inactive

There is absolutely no proliferation of cancer cells nowhere in the computational field representing the body, yet cancer cells are fighting on with full strength. However, the number of points where $U_{ijk}^n > U_{ijk}^{n-1}$ has been reduced from 1157625 to only 2506 at $t_n = 250$.

Out of sheer curiosity, we decided to go further and to 705 time steps. Results revealed an essential characteristic of cancer cells. In literature and from oncologists, we know that in many cases cancer cells could tolerate poisoning and burning. Many of them just run away from and take shelters at different organs and transfer normal cells into malignant cells.

Table 10.3 of Fig. 10.3 revealed that scenario. The number of points at which $U_{ijk}^n > U_{ijk}^{n-1}$ increased from 2506 has gone up to 3136. It means cancer cells may not replicate at every point but nevertheless they have the ability to replicate during the process of dispersion, because at the point of dispersion there is no medication to inhibit that process.

Out of sheer curiosity, we wanted to see the pattern of increase of the number of points where cancer cells are on the rise.

Figure 10.3 (mentioned here as Fig. 10.5) shows a part of this result. The variable NPTS (in green) represents the number of points in the field where $U_{ijk}^n > U_{ijk}^{n-1}$.

The graph shows that the moment cancer is attacked, the number sharply increased because many cancer cells run away from the sites where medications are strong for their own safety.

Then as time goes, the number settles down. We took a special run for 221 time steps. Results are in the Fig. 10.5 (Please note: Fig. 10.5 is the same as Fig. 10.2). These results show that by reducing the growth factor to zero, cancer may not be growing, but it could maintain its ability to remain a threat to life.

As per request of a reviewer, we took a more elaborate run with the same inputs for 1200 time steps.

Table 10.3 Figure 10.3: rate of growth is zero, yet cancer is growing. It is believed! Yet it is valid mathematically (Fig. 10.5 in the computer output is not correct)

```
****************************************************************************************************
SYSTEM TIME -10:02:29 INTRAVENOUS AT TIME - 14.06  TIME STEP -    703  NT- 1000  A1(N)- 0  FIG.5.1HA
AT 42 , 42 , 42  MAXUO(N)- 1208 LARGEST USTAR- 10152
AT 15 , 15 , 15  UMAX(N)- 10152.  CHANGE OF: UMAX- 0
TOTAL NO.OF POINTS WHERE: U IS MAX-  8
AT THE PRIMARY SITE: UMAX- 10152.
AT THE SECONDARY SITE: US1MAX- 5076.
AT THE SECOND SECONDARY SITE: US2MAX- 1015.
U AT THREE TUMORS:  USTAR1- 10150  USTAR2- 5076  USTAR3 -  1015
UNODE1- 1208.0985  UNODE1-1- 253.78956  UNODE2- 1208.0968  UNODE2+1- 253.78956
UNODE3- 1208.0968  UNODE3-1- 253.78956  U(NX+3,NY+3,NZ+3)- 253.78956  U(NX+4,NY+4,NZ+4)- 253.78956
MAX U OUTSIDE THE THREE TUMORS = 1208  AT NUMBER OF POINTS= 3
NO.OF POINTS WHERE U(N)>U(N-1)-  3136  MAX.CHANGE FROM THE START - U0MAX - UMAX--151.99999
****************************************************************************************************
SYSTEM TIME -10:02:38 INTRAVENOUS AT TIME - 14.08  TIME STEP -    704  NT- 1000  A1(N)- 0  FIG.5.1HA
AT 42 , 42 , 42  MAXUO(N)- 1208 LARGEST USTAR- 10152
AT 15 , 15 , 15  UMAX(N)- 10152.  CHANGE OF: UMAX- 0
TOTAL NO.OF POINTS WHERE: U IS MAX-  8
AT THE PRIMARY SITE: UMAX- 10152.
AT THE SECONDARY SITE: US1MAX- 5076.
AT THE SECOND SECONDARY SITE: US2MAX- 1015.
U AT THREE TUMORS:  USTAR1- 10150  USTAR2- 5076  USTAR3 -  1015
UNODE1- 1208.0842  UNODE1-1- 253.78956  UNODE2- 1208.0825  UNODE2+1- 253.78956
UNODE3- 1208.0825  UNODE3-1- 253.78956  U(NX+3,NY+3,NZ+3)- 253.78956  U(NX+4,NY+4,NZ+4)- 253.78956
MAX U OUTSIDE THE THREE TUMORS = 1208  AT NUMBER OF POINTS= 3
NO.OF POINTS WHERE U(N)>U(N-1)-  3136  MAX.CHANGE FROM THE START - U0MAX - UMAX--151.99999
****************************************************************************************************
SYSTEM TIME -10:02:47 INTRAVENOUS AT TIME - 14.1  TIME STEP -    705  NT- 1000  A1(N)- 0  FIG.5.1HA
AT 42 , 42 , 42  MAXUO(N)- 1208 LARGEST USTAR- 10152
AT 15 , 15 , 15  UMAX(N)- 10152.  CHANGE OF: UMAX- 1.8189894e-12
TOTAL NO.OF POINTS WHERE: U IS MAX-  8
AT THE PRIMARY SITE: UMAX- 10152.
AT THE SECONDARY SITE: US1MAX- 5076.
AT THE SECOND SECONDARY SITE: US2MAX- 1015.
U AT THREE TUMORS:  USTAR1- 10150  USTAR2- 5076  USTAR3 =  1015
UNODE1- 1208.0699  UNODE1-1- 253.78956  UNODE2- 1208.0682  UNODE2+1- 253.78956
UNODE3- 1208.0682  UNODE3-1- 253.78956  U(NX+3,NY+3,NZ+3)- 253.78956  U(NX+4,NY+4,NZ+4)- 253.78956
MAX U OUTSIDE THE THREE TUMORS = 1208  AT NUMBER OF POINTS= 3
NO.OF POINTS WHERE U(N)>U(N-1)=  3136  MAX.CHANGE FROM THE START - U0MAX - UMAX--151.99999
```

Inputs are the same as in the inputs for Fig. 10.3, except the number of time steps, which is 1200, here.

From Fig. 10.3 (same as Fig. 10.5), we notice that at t_{221}, the number of points where $U_{ijk}^{221} > U_{ijk}^{220}$ is 2506. However, at t_{358} this number has increased to 2530 and kept on increasing as computations moved forward. At the time step 591, it moved up to 3400, and at that time the rate of growth $a_1(n) = 0$ (Table 10.4).

So a zero growth does not mean that cancer is contained (Table 10.5).

What could be the explanation is that it must be happening in a microenvironment at the points where medications did not reach, like the boundary points. Numerical outputs are given only up to four digits after the decimal point. Please note that: MAXUO(N) = the largest value of U outside the tumor at t_n (Table 10.6).

Table 10.4 Figure 10.6 growth is zero, yet at 2530 points it is growing. At eight points, it has a max value

```
---------------------------------------------------------------------------------------------------
SYSTEM TIME -15:23:06 INTRAVENOUS AT TIME - 7.16  TIME STEP -    358  NT- 1200  A1(N)- 5.2215039e-215  FIG.10.2A
AT 42 , 42 , 42  MAXUO(N)- 1213 LARGEST USTAR- 10152
AT 15 , 15 , 15  UMAX(N)- 10152.  CHANGE OF: UMAX- 0
TOTAL NO.OF POINTS WHERE: U IS MAX-  8
AT THE PRIMARY SITE: UMAX- 10152.
AT THE SECONDARY SITE: US1MAX- 5076.
AT THE SECOND SECONDARY SITE: US2MAX- 1015.
U AT THREE TUMORS:  USTAR1- 10150  USTAR2- 5076  USTAR3 -  1015
UNODE1- 1213.0419  UNODE1-1- 253.78956  UNODE2- 1213.0415  UNODE2+1- 253.78956
UNODE3- 1213.0415  UNODE3-1- 253.78956  U(NX+3,NY+3,NZ+3)- 253.78956  U(NX+4,NY+4,NZ+4)- 253.78956
MAX U OUTSIDE THE THREE TUMORS = 1213  AT NUMBER OF POINTS= 3
NO.OF POINTS WHERE U(N)>U(N-1) - 2530  MAX.CHANGE FROM THE START - U0MAX - UMAX--151.99999
```

Table 10.5 Figure 10.6: cancer is getting stronger with a zero rate of growth (Fig. 10.4, it is Fig. 10.6)

```
*******************************************************************************
 SYSTEM TIME =16:25:37 INTRAVENOUS AT TIME = 11.98  TIME STEP =   599  NT= 1200  A1(N)= 0  FIG.10.2A
 AT   42 , 42 ,  42  MAXUO(N)= 1210 LARGEST USTAR= 10152
 AT   15 , 15 ,  15  UMAX(N)= 10152.   CHANGE OF: UMAX= 0
 TOTAL NO.OF POINTS WHERE: U IS MAX=  8
 AT THE PRIMARY SITE: UMAX= 10152.
 AT THE SECONDARY SITE: US1MAX= 5076.
 AT THE SECOND SECONDARY SITE: US2MAX= 1015.
 U AT THREE TUMORS:   USTAR1= 10150  USTAR2= 5076   USTAR3 =   1015
 UNODE1= 1209.5855   UNODE1-1= 253.78956   UNODE2= 1209.5842   UNODE2+1= 253.78956
 UNODE3= 1209.5842   UNODE3-1= 253.78956   U(NX+3,NY+3,NZ+3)= 253.78956   U(NX+4,NY+4,NZ+4)= 253.78956
 MAX U OUTSIDE THE THREE TUMORS = 1210   AT NUMBER OF POINTS= 3
 NO.OF POINTS WHERE U(N)>U(N-1)=  2941  MAX.CHANGE FROM THE START = UOMAX - UMAX=-151.99999
*******************************************************************************
```

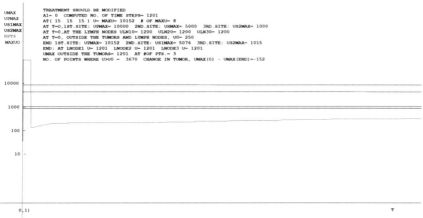

Fig. 10.6 Monoclonal drug reduces only the rate of growth of cancer and defence is inactive

Table 10.6 Figure 10.6: cancer growing and spreading with growth rate $= 0$

```
*******************************************************************************
 SYSTEM TIME =17:48:02 INTRAVENOUS AT TIME = 18.32  TIME STEP =   916  NT= 1200  A1(N)= 0  FIG.10.2A
 AT   42 , 42 ,  42  MAXUO(N)= 1205 LARGEST USTAR= 10152
 AT   15 , 15 ,  15  UMAX(N)= 10152.   CHANGE OF: UMAX= 0
 TOTAL NO.OF POINTS WHERE: U IS MAX=  8
 AT THE PRIMARY SITE: UMAX= 10152.
 AT THE SECONDARY SITE: US1MAX= 5076.
 AT THE SECOND SECONDARY SITE: US2MAX= 1015.
 U AT THREE TUMORS:   USTAR1= 10150  USTAR2= 5076   USTAR3 =   1015
 UNODE1= 1205.0617   UNODE1-1= 253.78956   UNODE2= 1205.0588   UNODE2+1= 253.78957
 UNODE3= 1205.0588   UNODE3-1= 253.78956   U(NX+3,NY+3,NZ+3)= 253.78956   U(NX+4,NY+4,NZ+4)= 253.78956
 MAX U OUTSIDE THE THREE TUMORS = 1205   AT NUMBER OF POINTS= 3
 NO.OF POINTS WHERE U(N)>U(N-1)=  3400  MAX.CHANGE FROM THE START = UOMAX - UMAX=-151.99999
*******************************************************************************
```

Initially at the boundary, the number of cancer cells was 250. Now at the boundaries (at t_{916}), this number has gone up slightly to 253.78956. These cells must have created these cancer cells entering the bloodstream.

The mathematical model: $\partial u/\partial t = a_1(n)u - a_2 uv + \kappa_1 \nabla^2 u$, since the rate constants a_1 and a_2 are certainly not promoting the replications of u, so extra cancer cells must be coming from the boundaries where at each point the number of cancer cells was 250 at the very beginning.

Analyzing all the tables, what we may notice is that when the growth of cancer cells is eliminated, then, of course, there are some noticeable positive aspects of treatment that happened:

1. Tumors are not growing anymore, even lymph nodes remain the same. The max size of the tumor in the entire field is still 10152 at the point (15, 15, 15), the primary site.
2. The number of points where $u = UMAX$ remained the same is equal to 8.
3. Number of $UMAX$ outside the tumors was initially 8 that was reduced to 3.
4. At the remotest end of the field, $x = NX + 4, y = NY + 4, z = NZ + 4$, the number of cancer cells (runaways) increased from 250 to 253.78 and remained the same.

However, the number of points where $U_{ijk}^n > U_{ijk}^{n-1}$ is increasing and neither the tumors, nor the lymph nodes shrunk a bit, on the contrary they increased slightly.

Figure 10.6 shows how slowly cancer is gaining more mass at various points in the field as time progresses.

10.5   A Deviation from the Previous Models

In all the previous chapters, we stated in accordance with the properties of our mathematical models that if in a general cancer treatment model like.

$$\partial q_i / \partial t = r_i q_1 - r_{i1} q_i + \sum_{j=1, j \neq i}^{N} r_{ij} q_i q_j + \kappa_i \nabla^2 q_i$$

$(i = 1, 2, \ldots, N, \ where \ N \ is \ the \ \text{total number of equations}).$ (10.4)

where q_i 's$(i \neq 1)$ are the responses of the immune system and drugs that fight cancer and $q_1 = $ the population of the cancer cells and r_i 's and r_{ij} 's are rate constants. The third term is negative if for some j, q_j 's are destroyers of cancer.

If $\sigma = $ the coefficient of $q_1 = (r_1 - r_{11} - \sum_{j=2}^{N} r_{1j} q_j)$, then a necessary condition that $q_1 \to 0$ as $t \to \infty$ is $\sigma < 0$. Because the conditions on the boundaries determine what will happen at the end, which is a basic property of all parabolic models like (10.1). Under certain boundary conditions, $\sigma < 0$ could be both necessary and sufficient.

The basis of this model is the laws of physical chemistry dealing with inanimate objects. In the world of bioscience, this law may be partially true which we have been observing through computational studies. This model showed a very deep fundamental property of all cancer that when globally the growth of cancer is reduced to zero, the existing cancer cells in the body can still be life threatening. So a model is developed here to inhibit all activities of cancer cells in the entire body.

10.6 Mathematical Validations of All Models Through Microstatic Analysis

Cancer moves in the body in groups. When only one or two cells move, the body's defense destroys them. "However, emerging studies have found a new migration mechanism called collective cell migration in many cancers. The collective cell migration could move as clusters with the tight cell–cell junction in the tumor microenvironments, toward the traction established by the leader cells. In addition, the collective cell migration has been shown to have higher invasive capacity and higher resistance to the clinical treatments than the single tumor cell migration. Interestingly, the collective clusters of tumor cells have been detected in the early stage of the cancer patient, which has led to the understanding of the significance of early cancer screenings" [8]. Researchers from John Hopkins University suggest that "cancer cells rarely form metastatic tumors on their own, preferring to travel in groups to increase their collective chance of survival" [9, 10].

All studies on mathematical modeling that we have done so far are aimed toward finding a suitable form of treatment of cancer. Now is the time to condense all the previous discussions and set up a compact compendium at the concluding models. Here we will validate our previous findings through mathematical modeling of proliferation, progression, propagation, and finally panacea of cancer. To do that we have adopted a rather simple mathematical model and attempted to analyze what is happening at a microlevel. Cancer just began and is totally unnoticeable. Even its activities are beyond human comprehension. To do that, we have set up linear models which could be solved analytically.

Three primary properties of all cancer cells are as follows: (i) At the outset, they start proliferating where they are, violating all the laws of homeostasis which all normal cells must obediently follow, (ii) at the same time, they start progressing annihilating neighboring normal cells in the tissues and begin progression, and (iii) while some start the process of metastasization.

All could happen at the same time, even in the microstates when neither the patients nor the doctors could comprehend that such a life-threatening state is on its way to destroy the body.

"According to Professor Trevor Graham, a Cancer Research UK-funded cancer evolution expert, the best evidence for the fact that most cancers grow slowly comes from screening. The screening programmes we have in the UK work "because there's a long period of time when a tumour may have started to develop but it's not become dangerous yet," he says."

Take bowel screening for example. For every cancer that has picked up during bowel cancer screening, four early tumors called adenomas were detected. Although most will never go on to become cancer, some will.

"It seems that tumour growth is started in lots of people, but it never quite makes it to the cancer stage," says Graham. "That really points to growth often being slow."

And this doesn't just apply to bowel cancer. Similar trends are seen in breast cancer, where 1 in 4 breast cancers picked up during screening would never have caused any problems, which points to them being slow growing [15].

For example, let us consider a simple model of the growth of cancerous cells. Let the growth of cancer be described by

$$f(u) = u_0 \exp(\alpha t), \; \alpha = 0.001 \; if \; 0 \le t \le 10$$
$$= \varepsilon, \; if \; 10 < t \le 20, \; \varepsilon \text{ is arbitrarily small}$$
$$= u_0 \exp(\alpha t), \alpha = 100 \; if \; t > 20$$

Here time t is measured in years. This model requires no explanation. Cancer could be that dangerous!!!

So we will develop mathematical models accordingly and check our previous findings and validate our procedures of treatments.

Let us consider a control volume V in which all phenomena in the microstate are happening.

Stage#1: Proliferation. Growth at the beginning. Let

$$\Phi \left(\int_V u \, dv \right) = \int_V f(u) dv \qquad (10.5)$$

where $\Phi = \partial^4/(\partial t \partial x \partial y \partial z)$.

Let us look into the microenvironment of the beginning of the invasion of cancer. Every now and then, a gene mutates. But most of the time, fortunately, the body blocks its progress, and the corresponding cell goes to cellular senescence and into apoptosis. However, sometimes a gene mutates several times, and the cell starts random proliferation. It does not work alone. A desperate group effort resumes happening in the microenvironment for an uncontrollable growth. Certainly as the body releases its immune dynamics to halt any such proliferation, the war starts on a microscale. Patients very often do not feel anything wrong. Even in the blood report, this phenomenon may not be caught.

Stage#2: Progression caused by the initial stage of metastasis.

Now we will look into basically the same phenomenon from another side which is happening at the stage #2. Here we are looking not into the core of the tumor in progression but all around it where accumulation of mass is going on due to progression of tumor developing process.

$$\partial/\partial t \left(\int_V u \, dv \right) = \int_V g(u) dv + \partial^3/\partial x \, \partial y \, \partial z \left(\int_V p(u) dv \right). \qquad (10.6)$$

Change in the control volume = net change of u inside the tumor at the core given by the first term on the right side + change of u all around the tumor along x, y, and z axes as it extends its proliferation due to progression given by the second term on the right side.

Stage #3: Propagation: Metastasis.

A cancer cell first starts evading the neighborhood in a tissue, being altered from being a normal cell because of some subtle genetic changes in it. While carrying the information of that tissue, it forms a group of its own which is the beginning of formation of a tumor. At the same time, some of these cells make up their minds to go elsewhere in the body to have new settlements. Blood and lymph carry them. They do not just float alone on blood or lymph, and they move mostly in groups attached to one another like a velcro attached to some cotton, with predetermined plans. If they can find shelter in a tissue, they assemble there, carrying the information samples of the tissues where they originated and at random start expanding by converting the residents all around them into malignant cells and move and grow bigger and bigger colonies forming tumors. This grim picture becomes very prominent when one observes MRIs of cancer patients.

Due to the change of momentum, mass of cancer cells u changes. These are changes during metastasis. Mathematically, this is given by

$$(\partial/\partial t + c \cdot \text{grad})\left(\int_V u\, dv\right) = \int_V h(u)dv \qquad (10.7)$$

where $c = ic_1 + jc_2 + kc_3$. c_1, c_2, and c_3 are velocities along the x, y, z axes and i, j, k are unit vectors along the x, y, z axes.

$f(u)$, $g(u)$, $p(u)$, and $h(u)$ denote four different states of the sizes of the cancerous masses. For the sake of simplicity, we have assumed these as follows:

$$f(u) = \alpha_1 u ,\ g(u) = \alpha_2 u,\ p(u) = \alpha_3 u \text{ and } h(u) = \alpha_4 u, \text{respectively.} \qquad (10.8)$$

All are linear.

The microstate is now taking a shape in a macro environment. A 1-cm cancer has about 100 million cells, a 0.5-cm cancer has about 10 million cells, and a 1-mm cancer has about 100 thousand cells. [www.ncbi.nlm.nih.gov›pmc›articles›PMC3320224].

10.6.1 Three Models Representing the Three States of Cancer

With (10.8), Eq. (10.5) becomes

$$\partial^4 u/(\partial t\partial x\partial y\partial z) = \alpha_1 u \quad \text{(Proliferation, Growth of a Tumor)} \qquad (10.9)$$

(To inhibit it, treatment must be conducted such that $\alpha_1 < 0$).
Equation (10.6) becomes

$$\partial u/\partial t = \alpha_2 u + \alpha_3 \partial^3 u/(\partial x \partial y \partial z) \quad \text{(Progression by penetrating tissues).} \quad (10.10)$$

$\alpha_2 = $ rate of growth of u, and $\alpha_3 = $ the rate of change of the volume ($\Delta x \times \Delta y \times \Delta z$) per unit of time. Again by appropriate treatments, $\alpha_2 < 0$ and $\alpha_3 < 0$ should be done.
Equation (10.7) becomes

$$\partial u/\partial t + c_1 \partial u/\partial x + c_2 \partial u/\partial y + c_3 \partial u/\partial z = \alpha_4 u$$
$$\text{(Propagation, Advection, Metastasis)} \quad (10.11)$$

c_1, c_2, c_3 are the velocities of u along the three axes respectively. This is advection of the mass of cancer. Equaling the accumulation of mass (no diffusion or dispersion in this part of the model).

These are what we have stated throughout this monograph. Cancer grows at any point in the body that it can occupy. In time and as they grow, they penetrate neighboring tissues spatially. At the same time, they start to metastasize, thereby they start growing like tumors.

So, (10.11) becomes

$$\partial u/\partial t = \alpha_4 u - c_1 \partial u/\partial x - c_2 \partial u/\partial y - c_3 \partial u/\partial z \quad (10.12)$$

We assume only one initial condition at $= 0$, $u = u_0 = u(0, 0, 0, 0)$. Boundary conditions are totally indeterminate. All three equations could be valid at any point/points in the body simultaneously. So there is no fixed boundary condition. In that sense, the model is incomplete. Yet it is carrying some average information of the micro and macroenvironments related to the growth of cancer and its metastasis.

10.6.2   Dimensional Analysis

So we have considered a three-dimensional mass represented by $u(x, y, z, t)$ which is changing in time.

The first equation shows the change of a three-dimensional mass M changing in time. Let L be the length and T be the time. On the right side of (10.9), $\alpha_1 = $ rate of change of a mass M *per unit volume* L^3 in time. So on each side of the Eq. (10.9), dimension is $M/(L^3 T)$.

For the second Eq. (10.10), the left side has a dimension M/T. The first term of the right side has a dimension M/T, and the second term's dimension is $(L^3/T) \times (M/L^3) = M/T$. In the third Eq. (10.11), the term on the right side has a dimension

M/T, and the dimension of the right side is $M/T + (L/T)(M/L) + (L/T)(M/L) + (L/T)(M/L)$. So all the equations are dimensionally well balanced.

We will attempt now to solve this model analytically. No computations are necessary.

10.6.3 Analytical Solution

Of course, this is a simplified model. Nevertheless, it may point out how treatments could proceed.

It will also be interesting to see how one solution could fit all the models. Let us first consider the Eq. (10.9). We will solve analytically by applying the separation of variables. To solve this, we substitute into the equation (10.9):

$$u(x, y, z, t) = G(t)F(x, y, z). \qquad (10.13)$$

And get:

$$G'(t)F_{xyz} = \alpha_1 G(t)F(x, y, z) \qquad (10.14)$$

where $F_{xyz} = \partial^3 F/(\partial x \partial y \partial z)$.

We may assume that $F(x, y, z)$ and F_{xyz} must be bounded because of the physical limitations of the body.

From (10.10), we get:

$$G'(t)/(\alpha_1 G(t)) = F(x, y, z)/F_{xyz} = \lambda \text{(which is a constant)} \qquad (10.15)$$

If we can choose medications to make $\alpha_1 < 0$, meaning that the rate of growth is negative and $\lambda > 0$, then $\lambda \alpha_1 < 0$.

$$G'(t)/G(t) = \xi \text{ where } \xi = \alpha_1 \lambda. \text{ So, } \xi < 0 \qquad (10.16)$$

Let $\xi = -\eta$, $\eta > 0$.

Since $F(x, y, z) = $ the mass of a tumor, $\forall x, y, z$, $F(x, y, z) > 0$. Also, since this mass tends to increase in all directions, F_x, F_y, F_z, F_{xyz} are all positive. So, $\lambda > 0$ (choice of $\alpha_1 < 0$ indicates that medications have been applied to reduce the rate of growth of cancer cells which tend to grow). Thus, our assumption that $\lambda > 0$ and $\alpha_1 < 0$ are justifiable, or in other words, they do not contradict each other. Or in other words, while tumor tries to grow, medicines try to bring that growth down.

Integrating (10.15), we get:

$$G(t) = G(t_0) \exp(-\eta t), \qquad \eta > 0 \qquad (10.17)$$

So as $t \to \infty$, $G(t) \to 0$ (Assuming $G(t_0)$ is bounded. For simplicity, we set $t_0 = 0$).

From (10.15), we get:

$$F_{xyz} = (1/\lambda)F(x, y, z) \quad \lambda > 0 \tag{10.18}$$

Now, at this point, let us consider Eq. (10.10). Using the same substitution (10.13), we get:

$$G'(t) \cdot F = \alpha_2 G(t)F + \alpha_3 G(t)F_{xyz}$$
$$G'(t)/G(t) = \alpha_2 + \alpha_3 F_{xyz}/F$$

Using (10.18), we get:

$$G'(t)/G(t) = \alpha_2 + \alpha_3/\lambda \quad \text{where } \lambda > 0 \tag{10.19}$$

Since applying proper medications, we can make $\alpha_2 < 0$ and $\alpha_3 < 0$ (10.19) is compatible with the previous assumption making $G'(t)/G(t) = -\eta$.

Using (10.13) in the third assumption, from (10.12) we get $G'(t)/G(t) = \alpha_4 - (c_1 F_x + c_2 F_y + c_3 F_z)/F = -\eta$.

Then,

$$F = 1/(\alpha_4 + \eta)(c_1 F_x + c_2 F_y + c_3 F_z) = \beta_1 F_x + \beta_2 F_y + \beta_3 F_z \tag{10.20}$$

where $\beta_i/3 = c_i/(\eta + \alpha_4) \ i = 1, 2, 3$.

Let us first consider a one-dimensional model $\gamma \, df/dx = f$. Then a solution is $f = f(0) \exp(x/\gamma), f(0) = value \ of \ f \ at \ x = 0$. Following this, the general solution of (10.20) is as follows:

$$F(x, y, z) = F(0, 0, 0) \exp(x/\beta_1 + y/\beta_2 + z/\beta_3) \tag{10.21}$$

So, the final solution is, combining (10.17) and (10.22), of the following form:

$$u(x, y, z, t) = u(0, 0, 0, 0) \exp(-\eta t) \exp(x/\beta_1 + y/\beta_2 + z/\beta_3) \tag{10.22}$$

Regardless of the values of β_i's, $u(x, y, z, t) \to 0$, as $t \to \infty$, because $\eta > 0$.

All three stages related to proliferation, progression, and propagation could be inhibited applying a specific treatment that makes $\alpha_i, i = 1, 2, 3$ all negative.

These analytical studies consider the attack of cancer not just on one organ, and it is an attack all over the body where cancer may have metastasized. So the war zone may consist of 10^{14} points. Because every cell could be a victim of this most brutal enemy. And that is how this disease should be handled.

Not only applying monoclonal drugs, but cryoablation (killing cancer cells by freezing them and then thawing them) or with radiofrequency ablation where cancer cells are heated with high frequency electrical energy causing their rapid death and death of some neighboring cells may also be applied. This is not the standard radiation treatment.

Validation

It may be wise to validate that the analytical solution (10.22) that we have satisfies all Eqs. (10.9), (10.10), and (10.11).

If we substitute (10.22) into (10.9), we get: $u_{txyz} = -\eta/(\beta_1\beta_2\beta_3)$. $u = \alpha_1 u$. So $\alpha_1 < 0$. This is compatible with our assumption that $\alpha_1 < 0$. $\alpha_1 = \eta/(\beta_1\beta_2\beta_3)$. This is how α_1 should be chosen.

If we substitute into (10.10), we get: $-\eta u = \alpha_2 u + \alpha_3/(\beta_1\beta_2\beta_3)u$. So if we assume that α_2 and α_3 are both negative, then on both sides coefficients of u are negative, so two sides are compatible. And certainly (10.22) satisfies (10.11). So $\alpha_2 = -\eta/2$ and $\alpha_3 = -\eta(\beta_1\beta_2\beta_3)/2$.

These results clearly show all the three models (10.9), (10.10) and (10.11) work together, in general from the very micro static states when cancer strikes and that the faster all the three models, (10.9), (10.10) and (10.11) will be successfully resolved, the better for the patient. In this regard, the drug the Skilled Killer Drug (SKD) may work best.

Substituting (10.22) in (10.11), we get: $-\eta u = \alpha_4 u - (c_1/\beta_1 + c_2/\beta_2 + c_3/\beta_3)u$. That gives

$$\eta + \alpha_4 = (c_1/\beta_1 + c_2/\beta_2 + c_3/\beta_3). \tag{10.23}$$

We have

$$\beta_i = 3c_i/(\eta + \alpha_4) \ i = 1, 2, 3. \tag{10.24}$$

Substituting (10.24) in (10.23), we see that both sides are equal.

10.6.4   The Treatment

To obtain the solution (10.17) what we had to assume for the parameters α_i, $i = 1, 2, 3$ are to be fulfilled by appropriate treatment plans as discussed before.

Numerical solutions done in the previous chapters give some ideas on quantifications of these parameters. Here that cannot be done directly.

This relatively simple linear model reveals that at least two drugs are needed to treat cancer:

(1) An anti-proliferation drug which will reduce the growth of a mass of cancer and annihilate it regardless of how insignificantly small it is. That will attack every microenvironment where cancer is striking, all over the body.

(2) An anti-progression drug to destroy all pathways to form tumors.

If these two have been achieved, cancer will be paralyzed and its progress from a microstate to a macrostate will be terminated. These two types of drugs may be SCD1 and SCD2, or with SKD.

In these regards, as we have discussed in rather details in Chap. 5, applications of immunotherapy with monoclonal drugs could be used and Chap. 7. All these drugs should be done intravenously. The decision is certainly up to the experts in oncology.

By all means, every microenvironment of cancer must be thoroughly eradicated, and these models could be helpful in this regard. Our final model consists of all the forms of treatments referred in the Mayo Clinic report [14].

10.7   Conclusion

Treatments of cancer are far more different than treatments of other diseases. Each cancer is unique, so is each patient. Cancer is the generic name of a class of diseases with one common property that they disregard the fundamental properties of all normal cells. They are mostly deathless. So dynamics of treatments vary from person to person. To treat just one patient, very often a team of experts starts working together. Such a team consists of not only just oncologists, but cardiologists, pulmonologists, endocrinologists, hematologists, radiologists, neurologists, psychiatrists, urologists, internists, surgeons, ophthalmologists, dietetics, physiotherapists, nutritionists, and constant assistance of expert nurses. Possibly, there is no other disease which requires the cooperation of so many experts.

Now mathematicians and computer specialists are going to participate actively in this group. With just one aim: cancer must be crushed and the war must be won.

References

1. Pinchover, Y., Rubinstein, J.: An Introduction to Partial Differential Equations. Cambridge, New York (2005)
2. Farlow, S.J.: Partial Differential Equations for Scientists and Engineers. Dover (1982)
3. Seyfried, T.N., Huysentruyt, L.C.: On the Origin of Cancer Metastasis. Crit. Rev. Oncog. **18**(1–2), 43–73 (2013)
4. Anand, P., Kunnumakara, A.B., Sundaram, C., Harikumar, K.B., Tharakan, S.T., Lai, O.S., Sung, B., Aggarwal, B.B.: Cancer is a preventable disease that requires major lifestyle changes. Pharm. Res. **25**(9) (2008)
5. Steinmetz, K.A., Potter, J.D.: Vegetables, fruit, and cancer prevention: a review. J. Am. Diet Assoc. **96**, 1027–1039 (1996)
6. Greenwald, P.: Lifestyle and medical approaches to cancer prevention. Recent Results Cancer Res. **166**, 1–15 (2005)
7. Vainio, H., Weiderpass, E.: Fruit and vegetables in cancer prevention. Nutr. Cancer. **54**, 111–142 (2006)
8. Yang, Y., Zheng, H., Zhan, Y., Fan, S.: An emerging tumor invasion mechanism about the collective cell migration. Am. J. Transl. Res. **11**(9) (2019)

9. Theveneau, E., Mayor, R.: Collective cell migration of epithelial and mesenchymal cells. Cell Mol. Life Sci. **70** (2013)
10. John Hopkins study: Cancer cells travel in groups to forge metastases with greater chance of survival, study suggests. Report, 2 Feb 2016
11. scienceblog.cancerresearchuk.org/2018/10/18/
12. Chis, O., Opris, D.: Mathematical analysis of stochastic models for tumor-immune systems (2009)
13. Chignola, R., et al.: Forecasting the growth of multicellular tumour spheroids: implications for the dynamic growth of solid tumours. Cell Prolif. **33**, 219–229 (2000). https://doi.org/10.1046/j.1365-2184.2000.00174.x
14. https://www.mayoclinic.org/tests-procedures/cancer-treatment/

Chapter 11
Conclusion

Abstract We have briefly discussed how in some cases cancer does pose a threat to life which is irreversible. We also stated why necrotic tumors often pose severe metastasis. On a positive side, we discussed how a change in lifestyle could help many cancer patients.

11.1 Rationale

Cancer is that creation of nature which can defy practically all laws of nature. They could be unbounded and deathless. They can survive burning and poisoning. They can survive lack of oxygen. They can even replicate after being murdered. Yet they are essentially not foreign. Once they were just normal cells obeying all laws of homeostasis. Once they helped the body to grow, to fight disease, to build muscles, to strengthen nerves, to maintain the health of the entire endocrine system and what not. Now they turned against the basic safety measures of the body. They are antilife. They are doing all these because of genetic mutations. In fact, several mutations take place inside the body that the body failed to correct.

> "It is important to remember that statistics on the survival rates for people with breast cancer are an estimate. The estimate comes from annual data based on the number of people with this cancer in the USA. Also, experts measure the survival statistics every 5 years. So, the estimate may not show the results of better diagnosis or treatment available for less than 5 years" [1].

So let us understand the rationale for this conclusion.

We are backtracking now. In 2020, there were 2.3 million women diagnosed with breast cancer and 685 000 deaths globally according to WHO. As we write in 2021, an estimated 281,550 new cases of invasive breast cancer are expected to be diagnosed in women in the USA, along with 49,290 new cases of noninvasive (in situ) breast cancer. Still the grim picture of breast cancer remains almost as grim as before. Still 1 in 8 US women (about 12%) will develop invasive breast cancer over the course of her lifetime. It is somewhat difficult for most of us to think about the statistical truth that "About 85% of breast cancers occur in women who have no family history

© Springer Nature Singapore Pte Ltd. 2021   323
S. Dey and C. Dey, *Mathematical and Computational Studies on Progress, Prognosis,
Prevention & Panacea of Breast Cancer*, Forum for Interdisciplinary Mathematics,
https://doi.org/10.1007/978-981-16-6077-1_11

of breast cancer. These occur due to genetic mutations that happen as a result of the aging process and life in general, rather than inherited mutations" [2].

It is true that with better and better diagnostic studies, prognosis is improving, but we have to remember that the population is increasing too and with that urbanization, air pollution, water pollution, food pollution, level of stress all are on the rise making cancer more prevalent and more aggressive.

We believe in such cases lifestyle is also a major factor. In this regard, we have developed several models to point out that nutritional treatment could be very helpful as preventive treatments of this disease. That is what mathematicians may suggest for all to do to inhibit this deadly disease.

The simplest model to inhibit practically any ailment, especially cancer, is to think of the model:

$$du/dt = (a - bV)u, \ at \ t = 0, u = u_0, \text{with a solution } u = u_0 \exp((a - bV)t)$$

$u = a \ mass \ of \ cancer \ sticking \ together$ (cancer cells, which are mutated normal cells, move in a group. If they move alone or in a very small group, leucocytes generally find them and destroy them). $u_0 = the \ value \ of \ u \ at \ t = 0, a = the \ rate \ of \ growth \ of \ u, \ and \ b = the \ rate \ at \ which \ each \ element \ of \ u \ is \ destroyed \ by \ each \ element \ of \ V$.

So in the plain simple language $a =$ the rate of growth which is the strength of the attacker and $bV =$ the strength of the defender. Attackers get stronger when the defense gets weaker and vice versa. Our objective is how to strengthen the defender. So both b and V must be increased.

The strength of the defenders depends to some extent on our choice of lifestyles, our choice of food, drinks, keeping the environment pollution free, and even socialization which is often a primary source of stress. Statistics have now backed it up.

To kill this disease completely is next to impossibility. But they could be safely kept under control. And this cannot be done just with medications. When it attacks the body, it needs a team effort of doctors, nurses, dieticians, physiotherapists, pharmaceutical companies, psychologists, and even social workers to fight back. In this regard, the principles and practices that a person used to do as a lifestyle may require a thorough change. Truly, the entire lifestyle of a patient needs to be changed together with changing modes of treatments from time to time.

We have focused on the treatment of breast cancer because the first author had the opportunity to work with several oncologists dealing with this disease where cancer was strongly metastatic. And furthermore, both authors received funding from NASA Ames on this topic of study.

Mathematical modelings [3–5] that have been done so far are far from looking into all the facets of cancer. Most researchers treated cancer cells as ordinary chemicals or biochemicals dividing, rushing, and running inside the body. All these are true. But they do a lot more. They play hide and seek. They may go into a state of remission and stay inactive for years. They may then attack at random which may be completely unpredictable. They disregard all the laws of natural science and pose some unparalleled challenges in the study of bioscience. We have discussed many of these aspects of cancer throughout the monograph.

11.2 An Observation on Mathematical Modeling

Mathematical modeling is a huge challenge because of the diversities of the nature of cancer. Two patients with exactly the same physiological conditions may need two entirely different procedures for cancer treatment. The reaction–diffusion (inward) that we have used and named reaction–dispersion equation depends on considering growth of cancer cells inside a control volume (CV):

Rate of growth of cancer = rate of growth inside the CV − rate of death inside the CV + rate of arrival of cancer cells inside the CV from outside.

This obviously follows the principle of balance of mass. However, if we look into the microstatic environment at the time of growth of cancer, we find that when cancer cells are just growing in a tissue and forcibly making room for their growth destroying functions of normal cells around them, nothing is coming in from outside at that very moment. Truly, at the very beginning, cancer cells do not come in, they go out to metastasize, and the cells inside as well as outside just grow meaning they replicate. This we have discussed in Chap. 10. However, they may go into a location where there is already a growing tumor or a tumor could be just started. So the principle of conservation of mass, the way it has been applied in Chap. 4, in a microenvironment near the close proximity of cancer cells, does not always necessarily depict the real picture which is changing continuously.

Another modeling challenge was the fact that the nonlinear partial differential equations defining the models derived in the Chap. 4 are not well posed mathematically because the boundary conditions are indeterminate. Even the existence of the solutions cannot be ascertained. What we have proved is that computationally all the models are well posed. We ran many cases with virtual changes of initial/boundary conditions and noticed no significant changes in the computational results.

11.3 The Graphs

In this monograph, there are lots of colorful graphs. They have been carefully chosen out of hundreds of similar graphs. These graphs relate to unique story of some unique patient. A story of triumph or a story of tears. We selected mostly those where a success of treatment has been recorded. However, we have tried to look into some cases where mathematically treatments were successful at the beginning, yet results were dismal in the end. There are cases where mathematicians see success, but doctors see just the opposite. Because theoretically treatments did work, but in reality they failed. The primary reason was that the patient could not tolerate the side effects of medications for that length of time when she was expected to be free from cancer. These are mostly breast cancer patients. Nowadays, because of early detection and better medications, the number of breast cancer cases is on the decline, and in general, cancer survivors are increasing. Tests are more precise. Radiology, mammograms, MRIs are all very advanced and accurate.

11.4   Numerical Challenges in the Models

The graphs that we have drawn are the max values of the variables at each time step. That means $u_{\max_{ijk}}$, $v_{\max_{ijk}}$, etc., over the entire computational field, at a given time t_n have been graphed. So at each time step t_n, when we get computationally solutions of $u(t, x_i, y_j, z_k)$, $v(t, x_i, y_j, z_k)$, etc., we find $u_{\max}(t_n)$, $v_{\max}(t_n)$, etc., at each time step and we plot them. Those are the graphs given in this text. These graphs reveal success/failure of the methods of treatment as time goes on.

The next challenge was solving stiff models dealing with biochemical kinetics. The equations that we have solved are of the form:

$du/dt = f_1(u, v, ...)$; $dv/dt = f_2(u, v, ...)$, etc. In the steady state, they took the form $f_1(u, v, ...) = 0$, $f_2(u, v, ...) = 0$. These are stiff algebraic nonlinear equations and numerical solutions applying a marching forward finite difference method is very challenging. For example use of Charlie's Predictor/Corrector Method is challenging [6].

$x_1 + x_2 = 2$, $(1 + \varepsilon)x_1 + x_2 = 2 + \varepsilon$, ε is arbitrarily small. The solutions are: $x_1 = 1$ and $x_2 = 1$.

If we choose $\varepsilon = 10^{-5}$ and rewrite the equations as: $x_1 + x_2 = 2$, $(1 + \varepsilon)x_1 + x_2 = 2$, (omitting ε on the right side), the solutions become: $x_1 = 0$ and $x_2 = 2$. So a small change in data caused a huge change in the solutions. This is the essential nature of stiff models. If we represent this system in a matrix form, it will be $Ax = b$, where $x = (x_1, x_2)^T$. The eigenvalues of A are: $\lambda_1 \approx 2$, $\lambda_2 \approx -4.9999e - 6$. So stiffness is of the order $|\lambda_{\max}|/|\lambda_{\min}| \approx 4.0 \times 10^5$. The system is very stiff.

Such challenges appeared in our models constantly which in a very simplistic way we may present as follows.

11.4.1   An Extended Charlie Model

Let us consider a linear model

$$dX_1/dt = (1 + \varepsilon)X_1 + X_2 + X_3$$

$$dX_2/dt = X_1 + (1 + \varepsilon)X_2 + X_3$$

$$dX_3/dt = X_1 + X_2 + (1 + \varepsilon)X_3$$

If we choose $\varepsilon = 0.000001$ and write the system in a matrix form as: $dX/dt = \mathbf{AX}$, the eigenvectors are: $V_1 = (1, 1, 1)^T$, $V_2 = (-1, 0, 1)^T$, $V_3 = (-1, 1, 0)^T$, and the eigenvalues are:

$\lambda_1 \approx 3$, $\lambda_2 = 10^{-6}$, $\lambda_3 = 10^{-6}$. So stiffness $\approx 3 \times 10^6$, which means that the system is stiff.

We had to overcome such challenges in almost each model that we have solved.

In the Appendix B, we have shown that if a one-dimensional parabolic PDE is approximated by finite differences to be solved in a field of the size I, we get a system of equations, whose stiffness is of the order I^2. If the field is three-dimensional for a three-dimensional parabolic model, then the stiffness will be of the order I^6. So larger the system means the higher the stiffness. These discussions are not directly applicable to nonlinear systems. Here, we linearize the system at a given time and compute the eigenvalues and stiffness. However, just looking at the graphs one can understand how stiff the system is. For example, we consider the following graphs.

11.5 A Stark Reminder from Previous Studies

To validate what we have discussed in Sect. 11.3, we have imported here three graphs on a tumor growth, untreated at the beginning, mentioned in the previous chapters (Fig. 11.1).

This shows that a slow growing tumor may suddenly take a deadly turn. This could baffle even some expert doctors. This indicates a situation when a patient had no complaints about any discomfort in the body for years, suddenly very strong

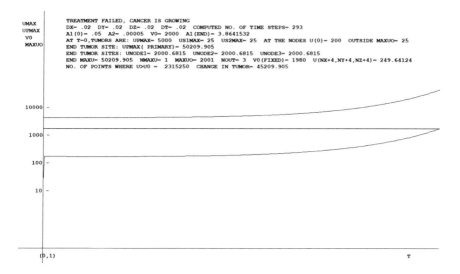

Fig. 11.1 Untreated tumor growth with a sad end

discomforts were felt, and a large tumor was identified. By that time, cancer have already spread to other organs.

Stiffness is very low at the outset and becomes very strong later because linearization at an instant means observing slopes of tangent lines to the curves. As time goes they change significantly. One common property of all stiff equations is: For numerical solutions, they require very small step sizes.

A very simple example is:

$$du/dt = -\lambda(t)(u - f(t)) + f'(t), \text{ with a solution } u = f(t).$$

At $0 \le t \le 100$, $\lambda(t) = 1$ and at $t > 100$, $\lambda(t) = 1000$.

This equation is non-stiff for $0 \le t \le 100$. Right after that, it suddenly became stiff, although the solution $u = f(t)$ did not change.

Such instances happen quite often in mathematical modeling for the treatment of cancer. But in the codes we mostly chose $\Delta t = 0.02$. That is a strong property of the CDey-Simpson numerical method as discussed in the text.

An excellent reference on stiff equation is [7, which may be downloaded].

Next we consider Fig. 11.2.

$(ag)_{max}$, the growth of cancer, is given by the red curve. It first started decreasing sharply, which could be noticed in MRIs. Patients and their doctors will feel happy about that. But as time goes by $(ag)_{max}$ increases. Not only that, outside the primary site $(MAXAGO)$ also increases which indicates that cancer is spreading outside the tumor. So the long-term results are pathetic.

Eigenvalues of a linear system indicate how the graphs for the variables should look like. For linearized nonlinear systems, the slopes of the graphs show what stiffness could be. If we compare the slopes of the graphs in Fig. 11.2, it is evident

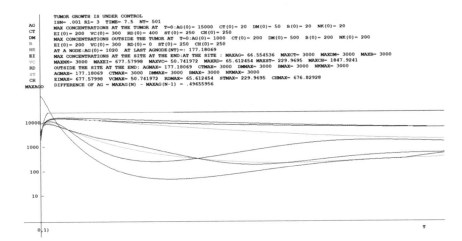

Fig. 11.2 Drugs are self adjusted as AG changes

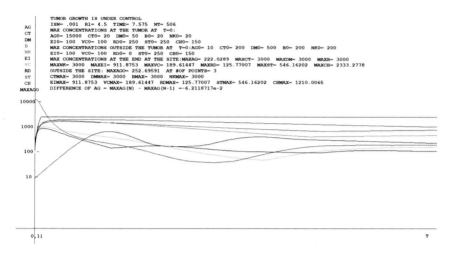

Fig. 11.3   No modification of conditions for Fig. 11.2

that whereas some slopes are very small, some are very high, which means the stiffness is very large.

Let us consider another such case (Fig. 11.3). Mathematically, this graph reveals how a patient could survive after having a long-term treatment.

In Fig. 11.3, the slopes of some of the graphs up to about 300 time steps are changing drastically even on a log scale. The system is very stiff.

These graphs are not necessarily telling the stories of any happy endings, although they seem to be all going into a steady state meaning that cancer is under control. The reason is it is taking a large amount of time for this steady state to happen. By that time a patient may not survive.

The first author did experience at least one such state. With regard to one of his near and dear patients, two doctors said there is no trace of cancer in the body. The patient is breathing fine. "LUNGS sound perfect... All medical reports and vitals are quite good." But one senior female doctor said that the body is on the verge of death. There is too much poison in the body. The patient breathed her last within a few days. And the story ended there.

So mathematically curing cancer may not mean saving life!!! This is the grim truth. The grim reality. The grim end. It must be done within a reasonable amount of real time. And there lies the challenge.

Even when a patient becomes cancer-free, at that point life should be supported with a more life sustaining lifestyle [8]. We have discussed all these, but now we must reiterate in our final comments.

11.6   Breast Tumor Detection

We have mentioned several times that the earlier the cancer is detected, the better it is to make a person cancer-free for a long time because the growth of cancer may be given by:

$du/dt = \lambda u$ with $u(0) = u_0$, where $u_0 = a$ *small amount of malignant mass*, and $\lambda = $ *the rate of growth*. Solution is $u(t) = u_0 \exp(\lambda t)$.

With some medication, the equation could become: $du/dt = (\lambda - b)u$, giving a solution $u(t) = u_0 \exp((\lambda - b)t)$. So if $b > \lambda$, in due time $u(t)$ will be damped out. This is indeed mathematically correct.

In some cases, in situ breast cancer treatment may not be that simple. Many experts noticed that the larger the size of the tumor, the worse it is for treatment. The simple reason is it could spread faster and could increase its rate of growth. So in that case, assuming that it may still be in situ, the model will be: $du/dt = \lambda(t)u$, where $\lambda'(t) > 0$. So if $\lambda(t) = at^2 + b$, where $a > 0$ and $b > 0$, the solution will be $u(t) = u_0 \exp(at^3/3 + bt)$. The tumor is growing very fast, and very likely it will spread very fast. "It is thought that the risk of developing metastases increases monotonically with tumor size, because the larger the cancer at diagnosis, the more cells are available to metastasize" [9]. That is how breast cancer works. "Tumor size is an important factor in breast cancer staging, and it can affect a person's treatment options and outlook. Tumors are likely to be smaller when doctors detect them early, which can make them easier to treat." [https://www.medicalnewstoday.com/articles/325669].

So it is very important that all adult women must do as often as possible a clinical breast examination (CBE). In this regard, we would like to mention the work of Barton et al. [10]. The authors described in detail how this examination will be done. We recommend all adult women to do this.

11.7   Tumor Necrosis

After performing a surgery on a patient, when the surgeon told the first author that the tumor was big but very necrotic and explained what necrotic means, the first author was happy. But when the surgeon explained what it could mean, soon that happiness got dried up just like a drop of water on a red hot surface.

Apparently, a necrotic tumor implies that the cancer cells at the core of the tumor are dead. Naturally it sounds like cancer is dying. In fact, it is the hallmark that with full speed cancer has become aggressive and metastasizing [11, 12]. A necrotic core indicates high-grade, aggressive tumor progression associated with very poor prognosis [13].

One reason could be that cancer cells could fight among themselves for glucose and oxygen and many decide to look for a different organ to infect aggressively to survive. That exigence could be another cause for faster metastasis.

A simple model is: $du/dt = au - bu^2, a > 0, b > 0$ at $t = 0, u = u_0 = a/(1+b)$. Integration is elementary. The solution is: $u(t) = a/(b + exp(-at))$. Tumor is slowly growing. But it will not increase beyond a/b. Tumor will get larger so long $u < a/b$, regardless what a and b are.

This model is a logistic model. In general, it may be applicable only to a benign tumor.

11.8 Stress Reduction

Stress hormones have very serious negative effects on women's health. No wonder that from ancient times stress was considered to be a main culprit causing breast cancer.

There are many Web sites relating to this information. One of them is: [https://drannacabeca.com/blogs/keto-alkaline-diet/hormones-health-what-many-women-don-t-know]. Here we found: "What's even more startling is that, according to new research, forty-three percent of women say hormones have negatively affected their overall well-being. The same survey found that, of 2000 American women aged 30–60, nearly half have experienced the symptoms of a hormonal imbalance."

Controlling stress is a major requirement to fight against breast cancer and keeping it under control. The authors Niazi and Niazi [14] stated that: "Mindfulness Based Stress Reduction (MBSR) therapy is a meditation therapy, though originally designed for stress management, it is being used for treating a variety of illnesses such as depression, anxiety, chronic pain, cancer, diabetes mellitus, hypertension, skin and immune disorders."

There are a host of studies pointing out that meditation reduces stress level and improves health. It is an essential tool to keep stress under control for breast cancer patients. "Studies show many benefits of meditation for breast cancer patients, such as decreased stress and anxiety, improvement of psychosocial factors, sleep quality and life perspective and feelings of empowerment, competence, personal strength, sense of calm, serenity, and balance" [15].

This may be substantially backed up by mathematics. We may think of all information as elements of a set to begin with. That means each individual information identifies a unique event which is an element of the set. As time goes and more and more events happen, the set gets bigger and bigger. This is done by our mind. Our mind started processing information since we were babies. As we grow up, our mind becomes more and more eager to process information. In fact, the worse the information is, which hurts us most, we are more eager to process that information. That causes stress. So stress may be considered as a measure of information. Through stress management, we may reduce that measure and dismantle some of its machineries that dangerously destroy our peace of mind.

Let $S(t_n)$ be an information set at a time t_n. Let μ_n be its measure. Since mind collects information continuously, μ_n is an increasing function of time. During mindful meditation, processing of information slows down. Mind becomes trained to ignore many incoming thoughts. So instead of being an increasing function of time, μ_n becomes a decreasing function. This is a stress reduction process [16]. There is a flood of scientific articles on the Internet.

For many, this is a change of the day-to-day lifestyle.

11.9 Some Final Comments

We have to admit the fact that:

Starting from the third chapter, we noticed that fighting cancer is indeed a real war. All bacterial, viral diseases, all infections are confrontations between pathogens carrying foreign antigens, and the body's immune system. So these are battles that the body must fight every second, often without us being aware of it. Most of the time these fights cause awareness because they cause pain, fever, inflammations, nausea, and many other symptoms. But at the beginning, even for months and for long years, the body may not show any sign of sickness when cancer attacks, because it may cause absolutely no noticeable symptoms, yet it could progress steadily and slowly, till it becomes deadly.

Cancer is the byproduct of four primary factors: (1) stress, (2) carcinogenic environment, (3) unhealthy lifestyle, and finally (4) inherited genes. If the first three could be kept under control, cancer may be prevented for many people.

Most models were developed following these discussions with the specialists. Yet we always keep in our mind that in 2021, regardless of better treatments, the grim aspect is that about 1 in 8 US women will develop invasive breast cancer, and the death rate is still about 2.6%. That is too much.

> The fact that only 5–10% of all cancer cases are due to genetic defects and that the remaining 90–95% are due to environment and lifestyle provides major opportunities for preventing cancer. Because tobacco, diet, infection, obesity, and other factors contribute approximately 25–30%, 30–35%, 15–20%, 10–20%, and 10–15%, respectively, to the incidence of all cancer deaths in the USA, it is clear we can prevent cancer [8].

We believe Preetha et al. [8] and her team have done a very excellent study showing that following a healthy lifestyle cancer may be avoidable. Most elaborately they described their findings in [8]. The 10 commandments of cancer prevention are below.

"Scientists at the Harvard School of Public Health estimate that up to 75% of American cancer deaths can be prevented. If we do the following:

1. *Avoid tobacco* in all its forms, including exposure to secondhand smoke. You do not have to be an international scientist to understand how you can try to protect yourself and your family.

2. *Eat properly.* Reduce your consumption of saturated fat and red meat, which may increase the risk of colon cancer and a more aggressive form of prostate cancer. Increase your consumption of fruits, vegetables, and whole grains.

3. *Exercise regularly.* Physical activity has been linked to a reduced risk of colon cancer. Exercise also appears to reduce a woman's risk of breast and possibly reproductive cancers. Exercise will help protect you even if you do not lose weight.

4. *Stay lean.* Obesity increases the risk of many forms of cancer. Calories count; if you need to slim down, take in fewer calories and burn more with exercise.

5. *If you choose to drink, limit yourself to an average of one drink a day.* Excess alcohol increases the risk of cancers of the mouth, larynx (voice box), esophagus (food pipe), liver, and colon; it also increases a woman's risk of breast cancer. Smoking further increases the risk of many alcohol-induced malignancies.

6. *Avoid unnecessary exposure to radiation.* Get medical imaging studies only when you need them. Check your home for residential radon, which increases the risk of lung cancer. Protect yourself from ultraviolet radiation in sunlight, which increases the risk of melanomas and other skin cancers. But do not worry about electromagnetic radiation from high-voltage power lines or radiofrequency radiation from microwaves and cell phones. They do not cause cancer.

7. *Avoid exposure to industrial and environmental toxins* such as asbestos fibers, benzene, aromatic amines, and polychlorinated biphenyls (PCBs).

8. *Avoid infections that contribute to cancer*, including hepatitis viruses, HIV, and the human papillomavirus. Many are transmitted sexually or through contaminated needles.

9. *Make quality sleep a priority.* Admittedly, the evidence linking sleep to cancer is not strong. But poor and insufficient sleep is associated with weight gain, which is a cancer risk factor.

10. *Get enough vitamin D.* Many experts now recommend 800–1000 IU a day, a goal that is nearly impossible to attain without taking a supplement. Although protection is far from proven, evidence suggests that vitamin D may help reduce the risk of prostate cancer, colon cancer, and other malignancies. But do not count on other supplements" [17].

We hope and pray that all of us must make every attempt to follow most of these guidelines.

References

1. https://www.cancer.net/cancer-types/breast-cancer/statistics
2. https://www.breastcancer.org/symptoms/understand_bc/statistics
3. Dey, S.K.: Computational modeling of the breast cancer treatment by immunotherapy, radiation and estrogen inhibition. Sci. Math. Jpn **58**(2) (2003)
4. Tabassum, S., et al.: Mathematical modeling of cancer growth process: a review. J. Phys. Conf. Ser. **1366**, 012018 (2019)
5. Quaranta, V., et al.: Mathematical modeling of cancer: the future of prognosis and treatment. Clin. Chim. Acta (2005)

6. Dey, S.K., Dey, C.: An explicit finite difference solver by parameter estimation. In: Adey (ed.) Proceedings of the 4th International Conference on Engineering Software IV, Kensington Exhibition Center, London, England, June 1985. Springer

7. Numerical Methods for Evolutionary Systems, Lecture 2 C. W. Gear Celaya, Mexico, January 2007 Stiff Equations

8. Preetha, A., et al.: Cancer is a preventable disease that requires major lifestyle changes. Pharm. Res. 25 (2008)

9. Sopik, V., Narod, S.A.: The relationship between tumour size, nodal status and distant metastases: on the origins of breast cancer. Breast Cancer Res. Treat. 170(3), 647–656 (2018)

10. Barton, M.B. et al.: Does this patient have breast cancer? The screening clinical breast examination: should it be done? How? JAMA 262(13) (1999)

11. Bredholt, G. et al.: Tumor necrosis is an important hallmark of aggressive endometrial cancer and associates with hypoxia, angiogenesis and inflammation responses. Oncotarget 6(37) (2015)

12. Liu, Z.-G., Jiao, D.: Necroptosis, tumor necrosis and tumorigenesis. Cell Stress 4(1) (2020)

13. Tomes, L.E. et al. Necrosis and hypoxia in invasive breast carcinoma. Breast Cancer Res. Treat. 81(1), 61–69 (2003)

14. Niazi, A.K., Niazi, S.K.: Mindfulness-based stress reduction: a non-pharmacological approach for chronic illnesses. N. Am. J. Med. Sci. 3(1), 20–23 (2011)

15. Castanhel, F.D., Liberali, R.: Mindfulness-based stress reduction on breast cancer symptoms: systematic review and meta-analysis. Einstein (Sao Paulo) 16(4) (2018)

16. Dey, S.K.: Mathematical studies of convergence of information, informative and informativism. Cybernetica XLI(2/3/4) (1998)

17. https://www.health.harvard.edu/newsletter_article/the-10-commandments-of-cancer-prevention

18. Bendahmane, M.: Existence of solutions for reaction-diffusion systems with L1 data. Adv. Differ. Equ. 7(6), 743–768

19. Shampine, L.F., Gear, C.W.: A user's view of solving stiff ordinary differential equations. SIAM J. Numer. Anal. 21(1), 1–17 (1979)

Appendix A

A.1. The Algorithm of CDey-Simpson Method

Let us consider a partial differential equation:

$u_t = F(u, u_x, u_y, u_z, u_{xx}, u_{yy}, u_{zz})$ subject to certain initial/boundary conditions.

If we approximate this equation by finite differences, we get a system of ordinary differential equations of the form

$$dU_{ijk}/dt = f_{ijk}(U_{ijk}, U_{i-1,j,k}, U_{i,j-1,k}, U_{i,j,k-1}, U_{i+1,j,k}, U_{i,j+1,k}, U_{i,j,k+1}).$$

U_j = the net function corresponding to the value of u_j.

$$i = 1, 2, \ldots, I; \ j = 1, 2, \ldots, J; k = 1, 2, \ldots, K$$

To study CDey-Simpson method, let us consider a general system of ordinary differential equation like:

$$du_j/dt = f_j(u_1, u_2, \ldots, u_J), j = 1, 2, \ldots, J \tag{A.1}$$

subject to a set of initial conditions: $u(t = 0) = u_j^0$.

If we approximate (A.1) by Euler forward, then we get:

$$U_j^{n+1} = U_j^n + hf_j(U_1^n, U_2^n, \ldots, U_J^n), \quad h = \Delta t. \tag{A.2}$$

$$j = 1, 2, \ldots, J$$

U_j^n = the net function corresponding to u_j^n.

Let us write (A.2) as: $\hat{U}_j = U_j^n + hf_j(U_1^n, U_2^n, \ldots, U_J^n)$.

Charlie's convex corrector is:

© Springer Nature Singapore Pte Ltd. 2021
S. Dey and C. Dey, *Mathematical and Computational Studies on Progress, Prognosis, Prevention & Panacea of Breast Cancer*, Forum for Interdisciplinary Mathematics, https://doi.org/10.1007/978-981-16-6077-1

$$U_j^{n+1} = (1 - \gamma)\hat{U}_j + \gamma(U_j^n + hf_j(\hat{U})), \tag{A.3}$$

where $\hat{U} = (\hat{U}_1, \hat{U}_2, \ldots, \hat{U}_J)^T, 0 < \gamma < 1.$

The Algorithm

The algorithm of CDey-Simpson is:

(1) Apply 1/2 step Euler forward and compute: $\hat{U}_j = U_j^n + (h/2)f_j(U_1^n, U_2^n, \ldots, U_J^n), j = 1, 2, \ldots J$

(2) Correct it by Charlie's convex corrector:

$$U_j^{n+1/2} = (1 - \gamma_j)\hat{U}_j + \gamma_j(U_j^n + (h/2)f_j(\hat{U}_1, \hat{U}_2, \ldots, \hat{U}_J))$$

$$j = 1, 2, \ldots J$$

(3) Apply 1/2 step Euler forward again and compute:

$$\underline{U}_j^{n+1} = U_j^{n+1/2} + (h/2)f_j\left(U_1^{n+1/2}, U_2^{n+1/2}, \ldots, U_J^{n+1/2}\right)$$

$$j = 1, 2, \ldots J$$

(4) Correct it by Charlie's convex corrector:

$$UC_j^{n+1} = (1 - \gamma_j)\underline{U}_j^{n+1} + \gamma_j(U_j^{n+1/2} + (h/2)f_j(\underline{U}_1^{n+1}, \underline{U}_2^{n+1}, \ldots, \underline{U}_J^{n+1}))$$

$$j = 1, 2, \ldots J$$

(5) Having three points at each $j = 1, 2, \ldots J$, at $t = t_n, t_{n+1/2}$, and t_{n+1} compute: using Simpson's (1/2) rule:

$$US_j^{n+1} = U_j^n + (h/6)\{f_j(U_1^n, U_2^n, \ldots, U_J^n) + 4f_j\left(\underline{U}_1^{n+1}, \underline{U}_2^{n+1}, \ldots, \underline{U}_J^{n+1}\right)$$
$$+ f_j(UC_1^{n+1}, UC_2^{n+1}, \ldots, UC_J^{n+1}))\}$$

$$j = 1, 2, \ldots J$$

(6) Correct the step # (5) again by Simpson's formula and get:

$$U_{CSj}^{n+1} = U_j^n + (h/6)\{f_j(U_1^n, U_2^n, \ldots, U_J^n) + 4f_j\left(\underline{U}_1^{n+1}, \underline{U}_2^{n+1}, \ldots, \underline{U}_J^{n+1}\right)$$
$$+ f_j(US_1^{n+1}, US_2^{n+1}, \ldots, US_J^{n+1}))\}$$

$$j = 1, 2, \ldots J$$

(7) If we correct this again by Simpson's rule that is CDey-Simpson method:

$$U_j^{n+1} = U_j^n + (h/6)\{f_j(U_1^n, U_2^n, \ldots, U_J^n) + 4f_j(\underline{U}_1^{n+1}, \underline{U}_2^{n+1}, \ldots, \underline{U}_J^{n+1}) + f_j(US_{CS1}^{n+1}, US_{CS2}^{n+1}, \ldots, US_{CSJ}^{n+1}))\}$$

$$j = 1, 2, \ldots J$$

At $n = 2$, instead of Simpson's rule we have used the trapezoidal rule. In some cases, we have used Charlie's formula only.

A.2. Eigenvalue Analysis

Let us express the system (A.1) as:

$$d\mathbf{U}/dt = \mathbf{F}(\mathbf{U}), \text{ where } \mathbf{U} = (u_1, u_2, \ldots, u_J)^T \in D \qquad (A.4)$$

D is a J—*Dimensional CDey Simpson space*, as defined in Chap. 4. Let us linearize this nonlinear system at t_n. Then it becomes

$$d\mathbf{U}/dt = \mathbf{F}(\mathbf{U}(t_{n-1})) + \mathbf{F}'(\mathbf{U}(t_{n-1}))(\mathbf{U}(t_n) - \mathbf{U}(t_{n-1})). \qquad (A.5)$$

where $\mathbf{F}'(\mathbf{U}(t_{n-1})) =$ the Frechet derivative of $\mathbf{F}(\mathbf{U}(t_{n-1}))$ which is known (sometimes we call it Jacobian matrix). Since at $t = t_{n-1}$ we may assume that the solutions are known, $\mathbf{F}'(\mathbf{U}(t_{n-1}))$ is a linear operator (a $J \times J$ matrix). So, (A.5) is now a linear system of equations, linearized at $t = t_{n-1}$. We have a linear system now, may be expressed as:

$$d\mathbf{U}/dt = \mathbf{A}\,\mathbf{U} + \mathbf{B}, \text{ where } \mathbf{A} = \mathbf{F}'(\mathbf{U}(t_{n-1})) \qquad (A.6)$$

General solution is given by:

$$\mathbf{U} = \sum_{j=1}^{J} \mathbf{C}_j exp(\lambda_j t), \qquad (A.7)$$

where λ_j's are the eigenvalues of $\mathbf{A}$.
   $\mathbf{C}_j$'s are vectors which are arbitrary constants, with $j = 1, 2, \ldots, J$ components. The measure of stiffness is given by

$\xi = |real\ \lambda_{max}|/|real\ \lambda_{min}|$, provided $|\lambda_{min}| \neq 0$,

where $|real\ \lambda_{max}| =$ *the max value of the real part of* λ and $|real\ \lambda_{min}| =$ *the min value of the real part of* λ.
   Let us consider the stability properties of the CDey-Simpson rule.

To consider the stability of numerical solutions of the nonlinear system (A.1), we may apply D-Mapping analysis [1]. Here, we use linearization process to look into the stability analysis as follows:

Let us consider a linear system

$$dX/dt = AX \tag{A.8}$$

where $X = (X_1, X_2, \ldots, X_J)^T$, $A = a \; J \times J \; square \; matrix$. We assume that A has J number of linearly independent eigenvectors. Let the eigenvectors be $v_1, v_2, \ldots, v_J$. Then $M = the \, modal \, M matrix$ of A is

$$M = [v_1 \, v_2 \ldots v_J], \, the \, columns \, are: v_j = (v_{j1}, v_{j2}, \ldots, v_{jJ})^T$$

Then the spectral matrix of A is Λ, where $\Lambda = M^{-1}AM$. So $M\Lambda M^{-1} = A$. Substituting this into (A.8) we get:

$dX/dt = M\Lambda M^{-1}X$. So, $M^{-1}dX/dt = \Lambda M^{-1}X$. That is the same as $d(M^{-1}X)/dt = \Lambda(M^{-1}X)$, because this is a linear system.

Let $U = (M^{-1}X)$. Then the system is $dU/dt = \Lambda U$.

Since $\Lambda (the \, spectral \, matrix \, of \, A) = diag(\lambda_1, \lambda_2, \ldots, \lambda_J) = A \, diagonal \, matrix \, and \, \lambda_j = an \, eigenvalue \, of \, A$, in the element form we get: (for stability λ_j 's must be negative [2])

$$du/dt = -\lambda u. \tag{A.9}$$

At $t = t_0, u = U^0$

Several people, including some mathematicians, asked me why Equation (A.1) is drastically simplified into (A.9). So we have derived (A.9).

The second author explained this in the most simplistic way at a colloquial session of Applied Mathematics Department of Calcutta University, in December 1983 (he was then just 11). He said, "at each time step we need to study the local behavior of the nonlinear systems by looking into their eigen properties, namely the eigenvalues which are λ_j 's. We look into the largest value and the least value, discarding $\lambda = 0$ $if \, there \, is \, any$. That shows how fast and/or how slow the solutions are changing with regard to each other at the same time. If the ratio of changes between two solutions is high, then the system is stiff. This ratio gives a measure of the stiffness and that requires use of a numerical method which may handle that stiffness." This is the idea behind setting up Eq. (A.9).

A.3. Stability Analysis

We will consider here (A.9).

$$Let\ t_{n+1} = t_n + h,$$

$\forall n$, *his the time step. Let* U^n *be the net function corresponding to* u^n *the analytical solution att* $= t_n$

To compute U^{n+1}, knowing the value of U^n, the algorithm Charlie–Simpson consists of the following steps: (i) Compute *half − step forward* $U_E^{n+1/2}$ by Euler forward; (ii) correct it by Charlie's method and obtain $U_{CH}^{n+1/2}$; (iii) compute next *half − step*, U_{EU}^{n+1} by Euler forward; (iv) correct it and compute U_{CH}^{n+1} applying Charlie's corrector; (v) using these three values, namely U^n, $U_{CH}^{n+1/2}$, and U_{CH}^{n+1} apply Simpson's to correct it and obtain U_{SIP}^{n+1}; (vi) use U_{SIP}^{n+1} and correct it by Simpson again and obtain U^{n+1}, the final value of U at t_{n+1}. This is an *explicit Finite Difference method. It is marching forward with time.* There is no iterative process within the field of computation. So, computationally it is very economic. The algorithm is fully vectorized and can easily be parallelized for large-scale computations by a supercomputer.

For finite difference solutions, parabolic models, which our models are, are very stiff especially when the field is large. The present algorithm seems to handle this stiffness quite effectively. This is reflected through the stability analysis. We use Eq. (A.9).

Let us assume U^n is known.

(i). $U_E^{n+1/2} = U^n + (x/2)U^n = (1 + x/2)U^n$, $x = -\lambda h$.

(ii). $U_{CH}^{n+1/2} = (1 - \gamma_1)U_E^{n+1/2} + \gamma_1\left(U^n + (x/2)U_E^{n+1/2}\right) = \sigma_1 U^n$! Half-step
        Charlie, obtained by substituting (i) in (ii),

where $\sigma_1 = 1 + x/2 + \gamma_1\left(x^2/4\right)$

(iii). $U_E^{n+1} = U_{CH}^{n+1/2} + (x/2)U_{CH}^{n+1/2}$

(iv). $U_{CH}^{n+1} = (1 - \gamma_2)U_E^{n+1} + \gamma_2\left(U_{CH}^{n+1/2} + (x/2)U_E^{n+1}\right) = \sigma_2 U_{CH}^{n+1/2}$

where $\sigma_2 = 1 + x/2 + \gamma_2\left(x^2/4\right)$

Hence $U_{CH}^{n+1} = \sigma_3 U^n$, where $\sigma_3 = \sigma_1\sigma_2$ (double 1/2 Charlie).

First Corrector of Simpson: (Double 1/2 Charlie and once corrected by Simpson).

(v). $U_{CHS1}^{n+1} = U^n + (x/6)(1 + 4\sigma_1 + \ldots \sigma_3)U^n = \sigma_4 U^n, \sigma_4 = 1 + (x/6)(1 + 4\sigma_1 + \sigma_3)$.

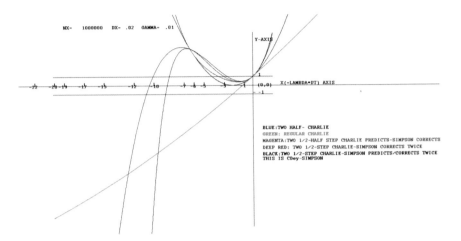

Fig. A.1 Stability contours of CDey-Simpson and other methods

Second Corrector using Simpson: (Double 1/2 Charlie and twice corrected by Simpson).

(vi). $U_{CHS2}^{n+1} = U^n + (x/6)(1 + 4\sigma_1 + \sigma_4)U^n = \sigma_5 U^n, \sigma_5 = 1 + (x/6)(1 + 4\sigma_1 + \sigma_4).$

Finally, CDey-Simpson method is:

(vi). $U_{CHSF}^{n+1} = U^n + (x/6)(1 + 4\sigma_1 + \sigma_5)U^n = \sigma_6 U^n, \sigma_6 = 1 + (x/6)(1 + 4\sigma_1 + \sigma_5).$

U_{CHSF}^{n+1} = the final computed value from CDey-Simpson.

For computational/numerical stability, $|\sigma_i| < 1$ for any given i.

Dr. Harvard Lomax (Divisional Chief, Computational Fluid Dynamics, NASA Ames) did a similar analysis of Charlie's original method [6] in 1982 [An explicit predictor–corrector solver with applications to Burgers' equation, NASA Ames, October 1983].

A.4. Stability Graphs

It is now needed to see some of the stability contours showing why CDey-Simpson worked so well.

In most cases, we have used the following values of $\gamma = 0.01, 0.05$ and 0.1. These are given by Figs. A.1, A.2, and A.3, respectively. $x = -\lambda \Delta t$ (x-axis is the $t - axis$); the two horizontal lines above and below are given by $y = 1$ and $y = -1$. These are the upper and lower values of σ (*border lines of stability*). We have considered the real part of the values of λ. For stability, the values of $(-\lambda \Delta t)$ must lie between these boundaries. This means the parts of the curve bounded between

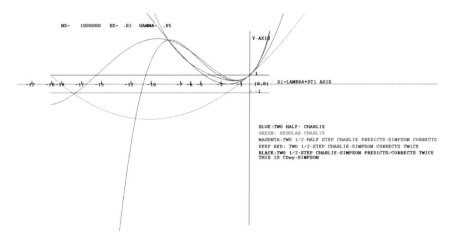

Fig. A.2 Stability contours of CDey-Simpson and other methods

$y = -1$ *and* $y = 1$. define the region/regions of numerical stability. That means we must choose the values of Δt such that we stay within these two lines, else the method will fail. The blue curve represents $y = \sigma_3$, two half-step Charlie [6].

The green curve represents $y = 1 + x + \gamma x^2$, $\gamma = a\ convex\ parameter$.

The magenta curve represents $y = \sigma_4$, two half-step Charlie with Simpson.

The deep red curve represents $y = \sigma_5$, two half-step Charlie and two Simpson.

The black curve represents $y = \sigma_6$, CDey-Simpson.

We first chose $\gamma = 0.01, 0.05$ *and* 0.1, because we mostly used these values of γ. Later we chose $\gamma = -0.1$ and $\gamma = 1.25$ for just educational purposes. For $\gamma = 0.5$, Charlie's method becomes the second-order Runge–Kutta method. We did that too.

The graph in blue color represents the stability of a numerical method where Charlie's method is used with two half-steps. The range of stability is $-4 \leq -\lambda(\Delta t) < 0$ or in other words $0 < \Delta t < 4/\lambda$. This reveals how to choose Δt so that numerically the method will remain stable; (larger value of λ requires smaller values of Δt). For CDey-Simpson, black curve is to be used. There are two regions of stability. The first one is: $0 < \Delta t < 3/\lambda$, and another is $8.75/\lambda < \Delta t < 9.75/\lambda$. This is very interesting.

For $\gamma = 0.05$, the results are more interesting. The green curve (original Charlie) has two regions of stability. One is $(0, -2)$ and the other is $(-18, -20)$. That means $for\ 0 < \Delta t \leq 2/\lambda$ as well as $18/\lambda \leq \Delta t \leq 20$ the method must remain stable. While just playing with numerical methods, the second author noticed in 1983 and Dr.Harvard Lomax, the Chief of Computational Fluid Dynamics Division of NASA Ames, found out this novel truth and called it "Charlie method" (Charlie was just 11).

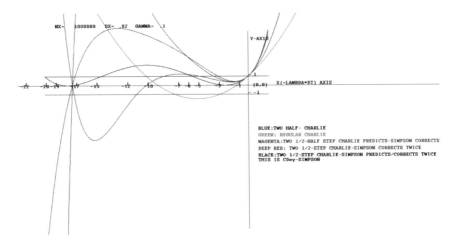

Fig. A.3 Stability contours of CDey-Simpson and other methods

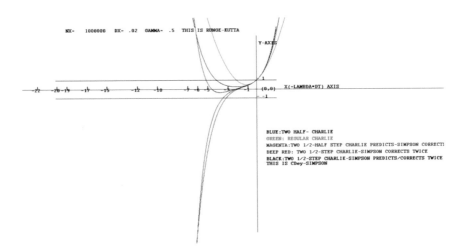

Fig. A.4 Stability contours of CDey-Simpson and other methods

Considering $\gamma = 0.1$, the stability for CDey-Simpson again has two regions as could be seen in Fig. A.3. The regions are: $0 < \Delta t \leq 3.5/\lambda$ and $17.25 \leq \Delta t \leq 17.45$. The red graph (two half-Charlie with Simpson) displays three regions of stability. These are: $0 \leq \Delta t \leq 4.75$, $8/\lambda \leq \Delta t \leq 11/\lambda$, $17.25/\lambda \leq \Delta t \leq 17.5$.

For $\gamma = 0.5$, Charlie's method becomes the same as the second-order Runge–Kutta. It has much poorer stability properties as shown in Fig. A.4.

Sometimes even for $\gamma < 0$, it worked. Fig. A.5 shows why.

Sometimes CDey-Simpson did work for $\gamma > 1$. This graph in Fig. A.6 explains how it could be possible.

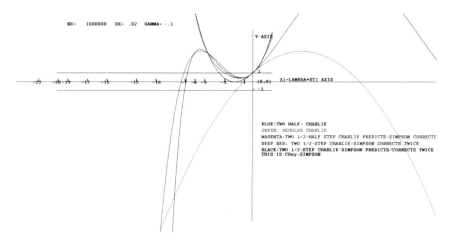

Fig. A.5 Stability contours of CDey-Simpson and other methods

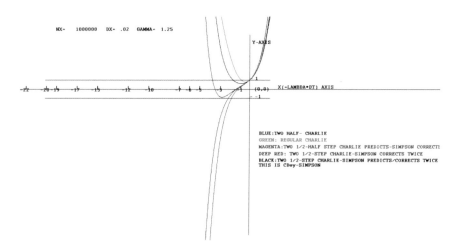

Fig. A.6 Stability contours of CDey-Simpson and other methods

References

1. Dey, S.K.: D-mapping analysis for numerical solution of nonlinear systems with applications to differential equations. Int. J. Modell. Simul. **16**(1), 6–14 (1996)
2. Jain, M.K, Iyengar, S.R.K, Jain, R.K.: Numerical Methods for Scientific and Engineering Computations. New Age International Publication (P) Ltd
3. Dey, S.K.: Nonlinear discretization errors in partial difference equations. BIT Numer. Math. (1980)
4. Dey, S.K.: Numerical Studies of a Perturbed Iterative Scheme with Applications to Coupled Nonlinear Systems of Equations. Applied Physics Division. Argonne

National Lab. ANL/FPP Technical Memo. No.96, US Dept. of Energy. Nov. 1977

5. Ames, W.F.: Numerical Methods for Partial Differential Equations, Barnes & Noble, Inc. New York (1969)

6. Dey, C., Dey, S.: Explicit finite difference predictor and convex corrector with applications to hyperbolic partial differential equations February 1983. Comput. Math. Appl. **9**(3). https://doi.org/10.1016/0898-1221(83)90,068-8S. Charlie Dey, Suhrit Dey

Appendix B

B.1. Why are Parabolic PDEs Stiff?

Let us consider the parabolic PDE in one dimension,

$$u_t = \alpha u + \kappa u_{xx} \tag{B.1}$$

subject to the conditions: $u(x, 0) = u_0$ and $u(0, t) = u(1, t) = u_B$, $\kappa =$ the heat conduction coefficient.

Approximating by Euler forward, we get:

$$U_i^{n+1} = (1 + \alpha \Delta t)U_i^n + \left(\kappa \Delta t/\Delta x^2\right)\left(U_{i-1}^n - 2U_i^n + U_{i+1}^n\right), i = 1, 2, \dots, I \tag{B.2}$$

where U_i^n is the net function corresponding to $u(x_i, t_n)$. Clearly, (B.2) is a linear system which may be expressed as:

$$U^{n+1} = AU^n \tag{B.3}$$

where $U^n = (U_1^n, U_2^n, \dots, U_I^n)^T \varepsilon D \subset R^I$, where D is the Banach space and R^I is I-dimensional real space and $A = tridiag(\beta, 1 + \alpha \Delta t - 2\beta, \beta)$, a tridiagonal matrix, and $\beta = \kappa \Delta t/\Delta x^2$. The eigenvalues are given by

$$\lambda_i = (1 + \alpha \Delta t - 2\beta) + 2\beta cos(n\pi/(I + 1)) \tag{B.4}$$

So, $\lambda_1 = 1 + \alpha \Delta t - 2\beta + 2\beta cos(\pi/(I + 1))$
And $\lambda_I = 1 + \alpha \Delta t - 2\beta + 2\beta cos(I\pi/(I + 1))$
Applying the $cos(x)$ series: $cos(x) = 1 - x^2/2! + x^4/4! - \dots$ and approximating by the first two terms, we get

© Springer Nature Singapore Pte Ltd. 2021
S. Dey and C. Dey, *Mathematical and Computational Studies on Progress, Prognosis, Prevention & Panacea of Breast Cancer*, Forum for Interdisciplinary Mathematics, https://doi.org/10.1007/978-981-16-6077-1

$$\lambda_1 = (1 + \alpha\Delta t - 2\beta) + 2\beta(1 - \pi^2/(2(I + 1)^2)) = (1 + \alpha\Delta t) - \beta\pi^2/(I + 1)^2,$$

substituting the value of β and multiplying both sides by $(I + 1)^2(\Delta x)^2$ gives

$$\lambda_1(I + 1)^2(\Delta x)^2 = (1 + \alpha\Delta t)((I + 1)^2(\Delta x)^2) - \kappa\Delta t\pi^2 = (1 + \alpha\Delta t) - \kappa\Delta t\pi^2$$

Similarly,

$$\lambda_I(I + 1)^2(\Delta x)^2 = (1 + \alpha\Delta t) - \kappa\Delta t\pi^2 I^2.$$

Evidently lambda_1 is the least and lambda_I is the largest (because I is much larger than the first two terms). These give:

$\lambda_I/\lambda_1 = (1 + \alpha\Delta t - \kappa\Delta t\pi^2 I^2)/(1 + \alpha\Delta t - \kappa\Delta t\pi^2)$, where Δt is small.

Clearly, $|\lambda_I|/|\lambda_1| = (\mu - I^2)/(\mu - 1)$, where $\mu = (1 + \alpha\Delta t)/(\kappa\Delta t\pi^2) = $ a constant.

So, if I is very large, $Stiffness = |\lambda_I|/|\lambda_1|$ becomes large (note: $(I + 1)\Delta x = 1$, the boundary).

The stiffness is of the order I^2. If we consider a three-dimensional system where $I = J = K$, the stiffness is or the order of I^6. So the system of finite difference equations is very stiff. And these models have been successfully solved by the CDey-Simpson marching forward numerical method.

In Appendix A, the numerical stability properties of CDey-Simpson have been studied.

Appendix C

C.1. Real-Time Synchronous Massively Parallel Computing with Shared Memory.

An adult human body has about one hundred trillion cells (10^{14}). We have considered a field with about one million (10^6) points, each point representing a mass of biochemicals. When we solve 10 nonlinear partial differential equations, we use ten million variables in the codes. They are all chemicals and biochemicals, and theoretically each could reach every cell in the body. As such, we have depicted only a small part of the entire picture.

Assuming that all the mappings are surjective, which many medical tests support (like blood tests), we get some idea of what is going on in the entire body, but certainly that may not be enough, especially when we deal with the treatment of a deadly disease like cancer which has the ability to befool all of its inhibitors. Certainly use of MRI's, CT scans, ultrasounds, and angiograms will be applied for verification of computational findings.

The code that doctors may need for the real-life treatment could require much bigger computational fields. So our codes may have to be rewritten to be applied on a massively large-scale basis to obtain information covering as many parts of the body as possible. This may have to be accomplished for real-time computations with applications of AI and deep learning for analysis. So codes ought to be parallelized and that will require a set of massively parallel supercomputers for processing and execution. With a given set of data, doctors will prefer to see what the results are as fast as possible, so that they should be able to make adjustments of input date, if needed, and check the results again as they feel necessary to carry out their treatment plans. So codes must be adopted for real-life interactive or synchronous computation.

C.2. Organization of the Equations for Parallel Computations

How such a code may be organized is the topic of our present discussion. We will show a flowchart for our algorithm in this regard.

© Springer Nature Singapore Pte Ltd. 2021

S. Dey and C. Dey, *Mathematical and Computational Studies on Progress, Prognosis, Prevention & Panacea of Breast Cancer*, Forum for Interdisciplinary Mathematics, https://doi.org/10.1007/978-981-16-6077-1

Let $R^M = $ Real M $dimensional$ $CDey - Simpson$ space (as defined in Chap. 4).
Let $D \subset R^M$.

Let us consider a system of nonlinear time-dependent difference equations to be solved by a marching forward explicit method at each time step. Let the system be represented by the equations:

$$\Phi^{n+1} = G(\Phi^n), \quad \Phi^n \in D \tag{C.1}$$

where $\Phi^n = (U^n, V^n, W^n)^T$ and $U^n = (U_{1jk}^n, U_{2jk}^n, \ldots, U_{Ijk}^n)^T$, and the same is true for V^n and W^n, $n = 1, 2, \ldots NT$. These are the variables to be solved by the CDey-Simpson algorithm.

There are three sets of variables whose values at each time step need to be computed. So, Φ^n is a block of block vectors.

In the element form, (C.1) is expressed as

$$u_{ijk}^{n+1} = g_1\left(u_{ijk}^n, v_{ijk}^n, w_{ijk}^n\right), \ i = 1, 2, \ldots, I; \ j = 1, 2, \ldots, J; k = 1, 2, \ldots, K \tag{C.2}$$

$$v_{ijk}^{n+1} = g_2\left(u_{ijk}^n, v_{ijk}^n, w_{ijk}^n\right)$$

$$w_{ijk}^{n+1} = g_3\left(u_{ijk}^n, v_{ijk}^n, w_{ijk}^n\right)$$

These three equations are represented in a vector formula in (C.1). All information at the time step t_n is stored in a common shared memory (CSM). In general, $I = J = K$. So total number of equations is $3I^3$.

For CDey-Simpson explicit finite difference vectorized scheme, all data at t_n will be available from the stored data at the previous time step solutions of the system. So these could be available from the CSM.

Keeping in mind that I, J, K are extremely large, we consider $N\#$ of primary processors $P^1, P^2, \ldots, P^N$ for three equations in (C.2) and each consists of NR number of secondary processors, namely $P_r^s, r = 1, 2, \ldots NR$. for each $s = 1, 2, \ldots, N$. ($N = 3$, if we use only three sets of equations).

Let the processor P_r^s solve $u_{ijk}^{n+1} = g_1\left(u_{ijk}^n, v_{ijk}^n, w_{ijk}^n\right)$ for $i = 1, 2, \ldots, I_r, j = 1, 2, \ldots, J_r, k = 1, 2, \ldots, K_r$. Thus, P_r^s will solve the subsystem of the order $I_r \times J_r \times K_r$. If $I_r = J_r = K_r$, and then the subsystem is of the order I_r^3. Thus, $\sum_{r=1}^N I_r^3 = I^3$. All inputs are from the previous time step solutions stored in the CSM.

The three processors $P^s, s = 1, 2, 3$ will work totally independent of each other grabbing all inputs from the previous time step solution stored in the CSM (that means at no time during computations they exchange the results among themselves at any time step, because all the previous time step computations, required at the beginning of computations, will be considered as known pseudo boundary conditions).

Now let us look into the subprocessors.

The boundary conditions will be used by P_1^s at the very beginning and P_N^s at the very end as given entries. However, the processor P_r^s will require the value of u at $i = I_{r-1} \forall j, k$, $j = J_{r-1} \forall i, k$, $k = K_{r-1} \forall i, j$ at t_n to begin computations at t_{n+1}. These values will come from the CSM. So each subprocessor works independently of each other.

To see a more clear picture, we represent the system in a general vector form:

Let $\Phi^n = (U^n, V^n, W^n)^T$

$U^{n+1} = G_1(U^n, V^n, W^n)$ (forward finite difference equations, solutions at the time step t_{n+1}).

Let

$(G_1)_I = $ an $I - order\ vector$, with $J - order\ vectors$ as elements.

$\left(G_{1_I}\right)_J = $ a $J - order\ vector$, with $K - order\ vectors$ as elements where

$\left(G_{1,J}\right)_K = $ a $K - order\ vector$.

Similarly, we define (G_2), and (G_3).

To define the boundary conditions for U at each time step, we consider:

$(G_1)_1 = $ a $J - order\ vector$, with $K - order\ vectors$ as elements.

$(G_{11J})_K = $ a $K - order\ vector$.

$(G_1)_{I+1} = $ a $J - order\ vector$, with $K - order\ vectors$ as elements.

$\left(G_{1\ I+1,J}\right) = $ a $K - order\ vector$.

Similarly, we can define the other boundary conditions for V and W, respectively.

This gives us a very clear picture of what we are computing and why this entire process of computation is a vector process.

C.3. Description of the Flowchart

Computation at the very first time step t_1 is indeed very crucial. There should be a short time pause, and all results should be thoroughly checked. This will be done by the set of processors $P^1, P^2, \ldots, P^N$ (note that N has nothing to do with the number of time steps). These processors are independent of each other.

All data will be sent to the data analyzer R. For several cases, after computations of a few time steps, about 10 at most, some checking ought to be done to see computations are moving toward the right direction. For instance, computationally we should check (1) the principle of chemotaxis is maintained, (2) dosages of medications are not going up or down abruptly, (3) radiation is primarily targeting the tumors and other malignant cells, etc.

We have checked all these ourselves. But when the computations are massive and many supercomputers are doing them simultaneously, these should be checked by some form of a thinking machine possibly on AI (or maybe artificial consciousness). R(AI) does not make any changes in case the inputs have errors. It passes all the information to CSM where the final AI will make its own evaluations. If modifications are needed, code ends and goes to the team of doctors with a report.

R is that machine. It collects data and does some preliminary analyses. R does not change the initial conditions and/or does data corrections. In case it detects any problem, it sends all data to the common shared memory (CSM) where the next AI will be activated, computations will end, and all data will be sent to the teams of doctors. Only doctors could make these changes.

Computations will begin using all changes done by the doctors.

If the second AI does not notice any changes, parallel computations at each time step begin. It should be noted that these time-dependent computations are not the same as we do in most engineering models. Here, artificial intelligence (AI) will play a role. AI will determine if there are any special conditions/corrections to be adopted because of medical/computational requirements.

For instance, if some special medication is needed, or if growths of T-cells, B-cells, NK-cells, etc., should be controlled more or less than what was expected or any medication should be increased or decreased, or any medication must be added, those conditions in the codes must be turned off or turned on. Or Δt, Δx, etc., should be adjusted, those must be looked into, and codes should be modified accordingly. The codes should have these options.

(PS. We have done these in our codes ourselves, evidently without the help of AI. It was possible because we have used only one processor and computations have been done on laptops in a field with only $105 \times 105 \times 105$ number of points at most).

The clusters are the parallel processors $P_1, P_2, \ldots, P_M$ (note: M has nothing to do with the $M - dimensional\ space$) at each time step. How many time steps will be required, that may be determined by the AI connected with the CSM, or the team of doctors may assign that as inputs. These processors will solve the model and send them back to AI.

If and when the steady-state solutions are found, AI [5] controlling the clusters will send all results graphically, and/or by dynamic simulation and/or by other forms to the team of doctors in charge of the patient.

These results may or may not be approved by the team. If not, they will make necessary changes, as inputs and new sets of computations will resume.

Certainly, this is a very short description of what we have done ourselves. Experts may change this flowchart as needed.

If AI with CSM does not augment data structure and proceed with the initial data during all time step computations, the processors $P_1, P_2, \ldots, P_M$ may not need any exchange of data during computations among themselves, because the $CDey - Simpson$ being a vectorized algorithm, all necessary values at the beginning and the end of the column vectors like $U_{i,j,k}^{n-1}$ will be treated like pseudo boundary conditions used in the case of when only one processor in a laptop is used.

The Flowchart

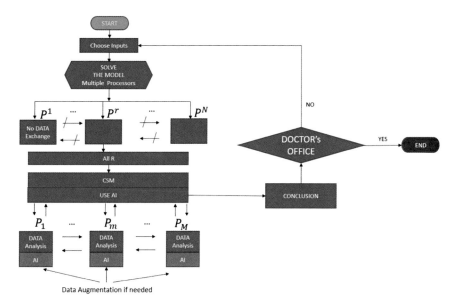

  This is a preliminary plan for applications of our models in the real world in real time.

  Our flowchart is very much simplified. In IEEE transactions, several recent works on this topic may be found.

References

1. Plattner, H: A Course in In-Memory Data Management. Springer (2014)
2. Kojima, O: Parallel Data Transmission Method and Parallel Data Transmission System (2007)
3. Woodard, L.: Introduction to Parallel Programming. (woodard@cac.cornell.edu) June 11, 2013
4. Dey, S.K.: Parallel solvers for large-scale linear and non-linear systems. Neural Parall. Sci. Comput. (1999)
5. Cohen, P.R.: Feigenbaum: The Handbook of Artificial Intelligence, Vol. III, HeurisTech Press. Stanford, California (1982)